Fuel Taxes and the Poor

Fuel Taxes and the Poor

The Distributional Effects of Gasoline Taxation and Their Implications for Climate Policy

EDITED BY

Thomas Sterner

Routledge
Taylor & Francis Group

LONDON AND NEW YORK

First published in 2012 by RFF Press

Published 2014 by Routledge
2 Park Square, Milton Park, Abingdon, Oxfordshire OX14 4RN

Simultaneously published in the USA and Canada by Routledge
711 Third Avenue, New York, NY 10017

Routledge is an imprint of the Taylor and Francis Group, an informa business

First issued in paperback 2015

British Library Cataloguing in Publication Data
A catalogue record for this book is available from the British Library

Library of Congress Cataloguing-in-Publication Data
Sterner, Thomas, 1952-
  Fuel taxes and the poor : the distributional effects of gasoline taxation and their implications for climate policy / Thomas Sterner.
     p. cm.
Includes bibliographical references.
  ISBN 978-1-61726-092-6 (hbk. : alk. paper) — ISBN 978-1-936331-92-5 (ebook) 1. Gasoline—Taxation—United States. 2. Motor fuels—Taxation—United States. 3. Poor—United States. 4. Environmental policy—United States. I. Title.
  HD9579.G5U5783 2011
  336.2'78665538270973—dc23   2011038394

ISBN 978-1-61726-092-6 (hbk)
ISBN 978-1-138-18423-7 (pbk)
ISBN 978-1-93633-192-5 (ebk)

About Resources for the Future *and* RFF Press

Resources for the Future (RFF) improves environmental and natural resource policymaking worldwide through independent social science research of the highest caliber. Founded in 1952, RFF pioneered the application of economics as a tool for developing more effective policy about the use and conservation of natural resources. Its scholars continue to employ social science methods to analyze critical issues concerning pollution control, energy policy, land and water use, hazardous waste, climate change, biodiversity, and the environmental challenges of developing countries.

RFF Press supports the mission of RFF by publishing book-length works that present a broad range of approaches to the study of natural resources and the environment. Its authors and editors include RFF staff, researchers from the larger academic and policy communities, and journalists. Audiences for publications by RFF Press include all of the participants in the policymaking process—scholars, the media, advocacy groups, NGOs, professionals in business and government, and the public.

About the EfD initiative

The **Environment for Development (EfD)** initiative (www.environment-fordevelopment.org) supports poverty alleviation and sustainable development through the increased use of environmental economics in the policymaking process. The EfD initiative is a capacity-building program focusing on research, policy advice, and teaching.

The EfD is managed by the Environmental Economics Unit of the University of Gothenburg. Financial support is provided by the Swedish International Development Cooperation Agency (Sida). The six EfD centers in Central America, China, Ethiopia, Kenya, South Africa, and Tanzania are hosted by universities or academic institutions in each respective region or country. Resources for the Future and RFF Press are partners in EfD through research collaboration, communications support, and publications, including the EfD book series.

Resources for the Future

Contents

Contributors

Wisdom Akpalu is assistant professor of economics at Farmingdale State College, State University of New York. His research has focused on natural resource management in developing countries. He has recently been elected a member of the executive committee of the International Institute of Fisheries Economics and Trade (IIFET). Some of his recent works have been published in the Journal *Ecological Economics, Journal of Agricultural and Resource Economics*, and *Environment and Development Economics*.

Francisco Alpizar is coordinator and senior research fellow at the Environment for Development Center at CATIE, a leading research institute based in Costa Rica. He is founder and Deputy Director of the Latin American and Caribbean Environmental Economics Program (LACEEP). Alpizar's current research focuses on experiments in behavioral economics, examining incentives that influence the creation and functioning of markets for ecosystem services, financing of protected areas, solid waste management policies and effectiveness of transport regulation. His publications include contributions to the *Journal of Public Economics*, the *Journal of Economic Behavior and Organization, and Energy Policy*

Rahel Deribe Bekele is junior research fellow at Environmental Economics Policy Forum for Ethiopia (EEPFE) of Ethiopian Development Research Institute (EDRI). Her recent research and publication records have focused on Natural Resource and Environmental Economics and Agricultural Economics. Also, she was the second master's thesis award winner of The African Thesis Award 2008 which was prepared by the African Studies Center (ASC) of Leiden, Netherlands.

Antonio M. Bento is an associate professor in Cornell's Charles H. Dyson School of Applied Economics and Management. Bento earned a Ph.D. in agricultural and resource economics from the University of Maryland in 2000. Most of his research lies at the boundaries of environmental, energy, urban, and public economics, and uses state-of-the-art econometric and computable general equilibrium methods, as well as geographical information (GIS) tools. Professor Bento's research has appeared in numerous journals and books, including the *American Economic Review, Review of Economics and*

Statistics, American Journal of Agricultural Economics, Journal of Environmental Economics and Management, and *Journal of Urban Economics.*

Allen Blackman is a senior fellow at Resources for the Future, research fellow at the Environment for Development Center for Central America, and adjunct professor at Georgetown University Public Policy Institute. His recent research has focused on pollution control and forest conservation in Latin America. He is editor of *Small Firms and the Environment in Developing Countries* (RFF Press) and has published articles in scholarly journals such as the *American Journal of Agricultural Economics, Journal of Environmental Economics and Management,* and *World Development.*

Martin Börjesson is a Ph.D. candidate at the Department of Energy and Environment, Chalmers University of Technology, Sweden. His research is in energy systems analysis and modeling, and his main focus is on system aspects of low-carbon alternatives in road transport. Recent publications include contributions to *Critical Issues in Environmental Taxation, Volume VII* and *Energy Policy* (journal).

Jing Cao is assistant professor at the School of Economics and Management, Tsinghua University. She received her Ph.D. in public policy at John F. Kennedy School of Government, Harvard University. She is affiliated with the Environment for Development initiative (EFD) – China Center, Tsinghua University Center for China in the World Economy (CCWE), and Tsinghua University – National Institute for Fiscal Studies. Her current research includes analyzing environmental tax reform in China using computable general equilibrium model and micro-level household survey data, economics of climate change, firm behavior and environmental regulation, and integrated top-down and bottom-up modeling on energy modeling.

Emanuel Carlsson is currently working with Consumer Price Index and Purchasing Power Parities for Statistics Sweden. He is a former research assistant for Thomas Sterner at the University of Gothenburg and has co-authored an article on the distributional effects of the Swedish fuel tax, published in the journal of the Swedish economic association, *Ekonomisk Debatt.*

Margaret Chitiga is the Executive Director for the Economic Performance and Development Programme of the Human Sciences Research Council in South Africa. Her recent research involves investigating the impact of policies and shocks on welfare. She is an associate editor for the *Journal of Environment and Development Economics* and is co-editor for the Centre for Environmental Economics and Policy in Africa (CEEPA)'s working paper series. Her recent publications include an article on "Oil prices and the South African economy: A macro-meso-micro analysis," published in *Energy Policy.*

Ashokankur Datta is a graduate student at the Indian Statistical Institute, New Delhi. His recent research has focused on the role of research and development in climate policy and distributional issues of community forestry.

Sanaz Ettehad is currently involved in an ongoing project in International Union for Conservation of Nature (IUCN). Her recent research has focused on global biodiversity finance. She recently completed a research project focusing on estimation of gasoline demand elasticities and distributional effects of fuel taxation in Iran with supervision by Thomas Sterner. Previously, she has had a short work experience in Iran's Research Institute of Petroleum.

Liyousew GebreMedhin is a junior research fellow and data manager at the Environmental Economics Policy Forum for Ethiopia, the EfD initiative center in Ethiopia and hosted by Ethiopian Development Research Institute. His recent research interest focuses on participation in off-farm employment, risk preferences, weather variability, local social networks, and shocks. He has produced and co-authored working papers on gender and social exclusion; benefit incidence of public spending on health care and education; and ability to pay for health insurance.

Lawrence H. Goulder is the Shuzo Nishihara Professor in Environmental and Resource Economics and director of the Stanford Environmental and Energy Policy Analysis Center at Stanford University. Goulder's research examines the environmental and economic impacts of regional, national, and international environmental policies, including policies to deal with climate change and pollution from power plants and automobiles. At Stanford he teaches undergraduate and graduate courses in environmental economics and policy.

Ana Laura Lozada Guzmán is head of research of advertising effectiveness in Havas Media Company. Her recent research has focused in the economic instruments to improve the air quality of Mexico City. She recently served as researcher in the National Institute of Ecology where currently she gives consultant services.

Roger H. von Haefen is associate professor in the Department of Agricultural and Resource Economics and associate director of the Center for Environmental and Resource Economic Policy at North Carolina State University. He is also a faculty research fellow at the National Bureau of Economic Research. His research interests span environmental, transportation, economic measurement, and applied econometric topics.

Kevin A. Hassett is a senior fellow and director of Economic Policy Studies at the American Enterprise Institute for Public Policy Research. Before joining AEI, Mr. Hassett was a senior economist at the Board of Governors of the

Federal Reserve System and an associate professor of economics and finance at the Graduate School of Business of Columbia University, as well as a policy consultant to the Treasury Department during the George H. W. Bush and Clinton administrations. Mr. Hassett writes a weekly column for *Bloomberg*.

Mark R. Jacobsen is an assistant professor of economics at the University of California, San Diego and a faculty research fellow at the National Bureau of Economic Research. His research focuses on environmental regulation and taxes, with particular attention to the automobile sector. His work includes publications in the *American Economic Review* and the *Journal of Public Economics*.

Kangni Kpodar is an economist in the African Department of the International Monetary Fund (IMF) and invited lecturer at the Centre for Studies and Research in International Development (CERDI). His recent research has focused on distributional effects of oil price changes, and financial development and poverty issues in developing countries. He co-authored a paper on CGE modeling of the impact of oil price changes on household expenditures in Mali, which was recently published in the *Journal of African Economies*.

Ramos Mabugu is head of the Research and Policy Division at the Financial and Fiscal Commission of South Africa. Most of Ramos's recent work is on the application of computable general equilibrium models to issues of macroeconomics, fiscal policy, and poverty. Ramos has taught economic modeling courses on two occasions at the Ecological and Environmental Economics Programme at the Abdus Salam International Centre for Theoretical Physics. He has also taught at various universities including the University of Pretoria in South Africa. Ramos is currently a member of the Editorial Board of *Environment and Development Economics Journal.*

Aparna Mathur is a resident scholar on AEI's economic policy studies team. She received her Ph.D. in Economics in 2005 from the University of Maryland. She has published in several academic journals including the *Energy Journal, Spatial Economic Analysis,* and *Small Business Economics.* A recent co-authored paper exploring the incidence of a carbon tax on households is forthcoming in an National Bureau of Economic Research book titled *The Design and Implementation of Climate Policy.* She is currently working on the topic of how well the bankruptcy system works for small businesses.

John K. Mduma is a senior lecturer in economics at the University of Dar es Salaam. His research area is poverty and equality analysis. In 2009–2010, he was the chief technical adviser in the preparation of phase two of the National Strategy for Growth and Reduction of Poverty for Tanzania. He was involved as a consultant in the preparation and production of the first Zanzibar Human

Development Report. His recent publications include "Application of Point Density Estimation in Assessing Changes in Polarization in Tanzania between 1991 and 2001." He is also co-authoring a book on growth and distribution in Tanzania: 2000 to 2009.

Alemu Mekonnen received his Ph.D. from the University of Gothenburg, Sweden. He is currently an assistant professor of economics at Addis Ababa University and coordinator of the Environmental Economics Policy Forum for Ethiopia at the Ethiopian Development Research Institute. His research focuses on energy, forestry, and poverty in Ethiopia and, more recently, climate change issues. His journal publications include the *Journal of Development Economics, Land Economics* and *Environment and Development Economics.*

Gilbert E. Metcalf is a professor of economics at Tufts University and a research associate at the National Bureau of Economic Research and MIT's Joint Program on the Science and Policy of Global Change. His current research focuses on energy and climate change policy evaluation and design. Metcalf serves as a contributing lead author for the upcoming IPCC Fifth Assessment Report. His book publications include the edited volumes *U.S. Energy Tax Policy* and *Behavioral and Distributional Effects of Environmental Policy* (with Carlo Carraro).

Adolf F. Mkenda is a senior lecturer of economics and the head of the department of economics at the University of Dar es Salaam. He is also the chairman of the Academic Board of the Collaborative Masters of Arts Program in Economics in Sub Saharan Africa under the African Economic Research Consortium, and is a fellow of Environment for Development Initiative (EfD) in Tanzania. He recently served as the lead consultant in the preparation and production of the first Zanzibar Human Development Report as well as in the preparation of the second Zanzibar Strategy for Growth and Reduction of Poverty. He is finalizing a book on growth and distribution in Tanzania: 2000 to 2009, which he is co authoring with John Mduma, Eliab Luvanda, and Remidius Rutihinda.

John M. Mutua is a researcher at the Kenya Institute for Public Policy Research and Analysis (KIPPRA), and junior research fellow at the EfD Centre in Kenya. He is also a Ph.D. candidate at the School of Economics, University of Nairobi. His recent research has focused mainly on energy, transport and environmental fiscal policy. He has published in the Handbook of Environmental Taxation: The Critical Issues in Environmental Taxation Volume VII, the OECD Working Papers series, and KIPPRA Discussion among others.

Mr. Wilhem Ngasamiaku is an assistant lecturer at the University of Dar es Salaam where he teaches economic theory, public sector economics, and mathematics for economists. He also participates in various research and consulting works. He has worked as a consultant at the World Bank Tanzania Country Office as an assistant research analyst in the Poverty Reduction and Economic Management (PREM). He also served as government consultant for monitoring and evaluating the performance of the National Strategy for Growth and Reduction of Poverty.

Rebecca Osakwe is currently a Ph.D. student with research interests in environmental and natural resource economics. She recently co-authored an article published in *Energy Policy*.

Budy P. Resosudarmo is an associate professor at the Arndt-Corden Division of Economics at the Crawford School of Economics and Government, The Australian National University (ANU). His research interests include determining the economy-wide impact of environmental policies, analyzing the impact of fiscal decentralization on regional economies, and understanding the political economy of natural resource utilization. His book publications include *The Impact of Environmental Policies on a Developing Economy: An Application to Indonesia*.

Elizabeth J. Z. Robinson is an associate professor of environmental economics at Gothenburg University in Sweden and a research associate with Environment for Development Tanzania. She is currently a visiting researcher with the International Food Policy Research Institute in Ghana where she is addressing agricultural value chain development in Africa. Her recent research focusing on the interface between people and protected areas has been published in the *Journal of Environmental Economics* and *Management and Land Economics*.

Milan Ščasný is research fellow and head of the Environmental Economics and Sociology Unit in Charles University Prague, Environment Center. Current research focuses on the valuation of non-market goods, particularly of health risks, external costs, consumer behaviour, and distributional aspects of environmental regulation. He has been involved in almost twenty European research projects, collaborated with OECD on household consumption projects, and provided consultancy for the Czech Ministry of the Environment. He has co-edited several books, including *Modelling of Consumer Behaviour and Wealth Distribution*.

Thomas Sterner is a professor of environmental economics at the University of Gothenburg in Sweden and a university fellow at Resources for the Future, Washington DC. He teaches various advanced economics courses both at the undergraduate and graduate levels in Gothenburg and in other universities. He has advised more than 20 Ph.D. students and built up a research group consisting of four full professors and a total of more than a dozen senior researchers in

Gothenburg. His primary research areas include issues concerning environment, resources, poverty, and development. A former president of the European Association of Environmental and Resource Economists, he has published a dozen books, including *Policy Instruments for Environmental and Natural Resource Management*, and more than 60 articles. Much of his work is focused on the design of policy instruments to deal with climate change and other environmental threats to the ecosystems on which we depend.

Sarah E. West is associate professor of economics at Macalester College in Saint Paul, Minnesota. Her research focuses on estimating behavioral responses to policies for the control of vehicle pollution, including taxes on gasoline, subsidies to clean vehicles, and Corporate Average Fuel Economy standards. She co-edited the book *Environmental Issues in Latin America and the Caribbean* and has published articles in *American Economic Review, Journal of Environmental Economics and Management, Journal of Public Economics, Journal of Transport Economics and Policy, National Tax Journal*, and *Regional Science and Urban Economics*.

Roberton C. Williams III is an associate professor in the Department of Agricultural and Resource Economics at the University of Maryland – College Park. He is senior fellow and director of academic programs at Resources for the Future, a research associate at the National Bureau of Economic Research, and a co-editor of the *Journal of Public Economics*. Professor Williams' research focuses on taxation and environmental regulation, and has been published in journals such as the *Journal of Political Economy, American Economic Review, Journal of Public Economics*, and *Journal of Environmental Economics and Management*.

Arief Anshory Yusuf is director of the Center for Economics and Development Studies, Padjadjaran University, Indonesia. His research interests lie in natural resource and environmental economics, poverty and inequality, and other economic development issues. He has done extensive work on general equilibrium models for Southeast Asian countries such as Indonesia, Thailand, and Lao PDR. He co-authored a book on green accounting and sustainable development in Indonesia and is now secretary general of Indonesian Regional Science Association.

Emmanuel Ziramba is a professor of economics at the University of South Africa. His recent research has focused on the relationships between energy consumption, economic growth and environmental pollution. He has published in journals such as *Energy Economics* and *Energy Policy*.

Foreword

BY DON FULLERTON, UNIVERSITY OF ILLINOIS AT
URBANA-CHAMPAIGN AND NBER

This ambitious book does not just summarize existing literature on distributional effects of fuel taxes. Rather, it contributes original research and teaches us something new! We may have thought we knew that fuel taxes were regressive, because fuel expenditures constitute a relatively large fraction of low-income family budgets. But we learn here that this view is wrong, or at least way too simple.

The book starts with a useful introduction that explains why this topic is important for the resolution of climate issues and that introduces the reader to some methodological issues. The book then includes three chapters that demonstrate alternative methodologies and therefore provide distinct results for just for one country, the United States. It then provides a large number of chapters with original research on a variety of other countries including various countries of Europe as well as Costa Rica, Mexico, China, India, Indonesia, Ethiopia, Ghana, Kenya, Mali, South Africa, Tanzania, and Iran. But it is not a random collection, as considerable effort is undertaken to maintain the comparability among these different research studies and to include important countries.

The distinguishing characteristic of this book is its consideration of what might be called "the heterogeneity of heterogeneity". We probably already thought about the heterogeneity, or differences, among families with different levels of income. But this book also discusses the heterogeneity among different countries, and the heterogeneity among different studies that use different data and different methodologies to estimate different measures of distributional effects for different policies!

Let me clarify that point with just a few examples. A gasoline tax in the U.S. might be regressive, because low-income families own fuel-inefficient cars and must drive further to get to work. They spend a high fraction of income on fuel. First, however, that conclusion depends on methodology. It is based on the use of annual income to categorize families, whereas the use of permanent income shows the gas tax might really be progressive. Second, it depends on how regressivity and progressivity are measured. Third, it depends on the policy being analyzed, since some policy reforms may explicitly return tax revenue to poor families and make the overall reform more progressive. Fourth, it may depend on the choice among partial equilibrium models, input-output models, and computable general equilibrium models.

Finally, of course, it depends on the country. In a rich country the gas tax might be regressive if low-income families spend a high fraction of income on fuel. But in many developing countries, the poor do not own cars at all. When only the rich own cars, then it is perhaps not surprising that a gasoline tax turns out to be quite progressive. On the other hand, the poor use public transport and other services that embody fuels. The research here then takes care to check that this and similar factors do not reverse the results that fuel taxes are actually progressive in important countries like India or China.

The book ends with a carefully crafted meta-analysis that draws interesting conclusions both from the viewpoint of methodology and for environmental policy making. Thus, I found that reading this book was an eye-opening experience, and I recommend it highly.

Preface

Being a dedicated environmentalist as well as a serious economist who takes my discipline very seriously, I have long been convinced that faulty energy prices are at the heart of many energy-related environmental problems. I have been arguing for incorporating energy externalities into consumer prices of energy since the late 1970s. I have lectured and debated this issue in various countries and usually felt fairly self-confident. Externalities should be internalized. Here was a really clear-cut case of a large externality threatening ecosystems and the answer was taxation (or possibly permit trade). These solutions were not always popular (in fact often very strongly unpopular) but I focused my attention on trying to convince my audience how much more expensive and irrational it would be to use regulation.

Over time the distributional argument has however troubled me. It is such a good argument! It is so particularly effective against environmentalists who are usually "good-hearted" people and don't want to hurt anyone, certainly not the poor. Hence the "what about the poor?" tends to kill the conversation or at least throw a very damp rag on the fire. The escalating gravity of climate change however made me think that something has to be done about fuel use from the transport sector. It is not necessarily going to be easier to deal with industry where issues of competitivity are a big problem, it will not be easier with non-carbon gasses such as methane or nitrogen oxides from agriculture. In fact gasoline and diesel might be two of the clearest cases. Biofuels are not easily going to provide the whole answer. I started to think that "maybe we will have to accept some regressivity"—but then of course we have to know how regressive a policy is and we might have to think about compensation— knowing however that compensation risks making policies expensive and inefficient. This was the atmosphere in which I started the research on this book. Somehow the opponents of fuel taxes had even managed to sow doubt in my mind and I felt that this is an issue that we really need to understand.

I have had several strands to my research. Most of it is focused on policy instruments but one part on climate internationally and one part on environment and resource management in developing countries. We have over the last decade built up many research collaborations with the Environment for Development, EfD, and various regional networks for environmental economics in Latin America, Africa and Asia (LACEEP, CEEPA, SANDEE and EEPSEA). So I contacted my friends and we started to discuss the possibility of this joint research endeavor.

Ana Lozada was the first collaborator and together we did a study on Mexico. I want to thank Ana particularly, she was at the Instituto Nacional de Ecología and did all the data collection and most of the empirical work. Ashokankur Datta saw an early draft of this paper and was one of the first to contact me for collaboration concerning India where he did a very thorough study. Then I came into contact with Emmanuel Ziramba through the Centre for Environmental Economics and Policy in Africa, CEEPA, from where he got a grant to write a paper on fuel demand and equity aspects in South Africa. Shortly thereafter two Swedish students started to write a thesis under my supervision, Hanna Ahola and Emanuel Carlsson. Emanuel later continued as my assistant on this project, and broadened the scope of that paper to a number of countries in Europe. At the EAERE conference that we hosted in Gothenburg, 2008, we organized a couple of sessions on distributional impacts of environmental policy—one specifically on fuel taxation. This gave the initial momentum for the book and I then turned to our network—the Environment for Development Initiative, that is funded by Sida and solicited more chapters as well as including more work in many countries through personal contacts created through the conference and other fora.

I would like to thank many friends in the networks mentioned, Sida, the Swedish International Development Cooperation Agency that has funded both the networks, the EfD and our research and capacity building efforts throughout many years. I would also like to thank the Sustainable Transport Initiative at University of Gothenburg and Chalmers University of Technology, CLIPORE—Mistra's Climate Policy Research Programme and Forskningsrådet Formas for financial support to this endeavor. Personally, I would also like to thank a number of my coauthors and other friends who have gone far beyond the call of duty in giving me good advice and feedback concerning the whole book (in addition I have some thanks for individual chapters). These include all the authors of the book who served as referees and helped with many aspects but I would particularly like to mention: Allen Blackman, Jing Cao, Emmanuel Carlsson, Ashokankur Datta, Don Fullerton, Mark Jacobsen, Gilbert Metcalf, Elizabeth Robinson, Johan Woxenius and Sarah West: THANKS.

Finally I would also like to thank Selma Oliveira, Kalle and Gustav Sterner for excellent editorial assistance. I would also like to thank the following for permissions to reproduce printed material: Armin Wagner, U.S. National Oceanic and Atmospheric Administration, United Nations Environment Programme (UNEP), the Deutsche Gesellschaft für Technische Zusammenarbeit (GTZ), *The Journal of Environmental Economics and Management* (West, Sarah E. & Williams, R.C.Roberton III, 2004. "Estimates from a consumer demand system: implications for the incidence of environmental taxes," *Journal of Environmental Economics and Management*, Elsevier, vol. 47-3.) and *American Economic Review* (Bento, Antonio M., Lawrence H. Goulder, Mark R. Jacobsen, and Roger H. von Haefen. 2009. "Distributional and Efficiency Impacts of Increased US Gasoline Taxes." *American Economic Review*, 99-3). Finally, thanks to Don Reisman and the staff at RFF Press for good collaboration.

## CHAPTER 1

Introduction: Fuel Taxes, Climate, and Tax Incidence

Thomas Sterner

This book combines four interesting issues: climate, energy, transport, and fairness in the distribution of goods and income. It is not only written for the sake of intellectual curiosity: It is motivated by deep concern for climate change and development. In this unique moment, when not only the giants of Asia have grown for decades, and even African economies have seen reasonable growth for a decade and a half, the promise of reduced poverty seems attainable. But it is also a moment when we understand that all this could be threatened and undone by the adverse effects of climate change. As this is written, in 2010, recognition of climate change as a serious problem has become widespread. Though some uncertainty and disagreement still exist about just how serious the problem is, the broad picture painted by the Intergovernmental Panel on Climate Change (the IPCC) requiring the world to drastically cut its emissions of fossil carbon (and other climate gases) has wide support. In contrast, there is no agreement on policies at local, national, and international levels. In fact, one might almost say paralysis prevails. Negotiations are held, but disagreements about both goals and methods—and particularly about the distribution of burdens—are deep. Meanwhile, year after year emissions continue to increase.

Global negotiations are mired by multiple difficulties. Though the United Nations Framework Convention on Climate Change (UNFCCC) is almost two decades old, and the first commitment period of the Kyoto Protocol, adopted in 1997, expires in 2012, there is still no clarity on its continuation or on what will replace it. The implementation of policies at the national level is typically resisted by vested interests, by those who fear they will be hurt locally, and because of concerns over the impact of implementation before other countries do the same. Yet global implementation is not easy—particularly not if there are too few national examples to follow.

This book focuses on national tax policies that affect the prices of fuels used in the transport sector. These fuels are interesting for a number of reasons. They are fairly homogeneous (more so than fuels and technologies

1

used for industry or heating), which allows us to compare and aggregate effects among countries. Global emissions from the transport sector account for a large share—about a quarter—of global fossil carbon emissions, and this share has been growing (for example, in the EU, it has gone from 20% to 30% in the last 20 years). Transport fuels are probably less sensitive from the viewpoint of jobs and competitiveness than industrial fuels, thus job loss, competition, and carbon leakage are less acutely pressing issues. And finally, the world happens to have almost a natural experiment in this area since fuel taxation policy differs considerably across countries, allowing us to study the effects of this policy instrument.

Ultimately any effective policy has to imply that the costs of emissions rise. However, discussions on fuel taxation are quite acrimonious and meet considerable resistance in many countries. One of the prime arguments given against fuel taxation is concern for the poor and for income distribution effects. Since taxation of fossil fuel (or equivalent instruments) is bound to be an important part of climate policy, a better understanding of the validity of these arguments is needed.[1]

As a starting point, we need to understand the determinants of fossil fuel use[2] in relation to income. The relationship could take on many different functional forms at both micro and macro levels. The former is the topic of this book, but it is worth pausing here to consider the macro relationship also.

In Figure 1-1 we show a number of possible Engel curves depicting the relationship between fuel use and income. In each, as it is natural to assume, absolute fuel use increases with increasing income. But this relationship need not be linear ("A"); it could just as well be accelerating ("B") or decelerating ("C"). The Engel curve will be different if we use cross-sectional data showing people with different income at one moment in time or time-series data as people get more income. Clearly there is some—although tenuous—relationship between these relations and the income distribution effects between rich and poor people at any one moment in time that we are looking for. Let us as

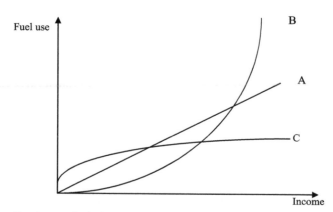

FIGURE 1-1 Engel curves for fuel use.

a thought experiment assume that high income earners always use a higher share of their income on fossil fuels (or emission intensive goods and services). If we aggregate (and assuming these relationships are stable enough to permit aggregation), this would yield an overall relationship for the country such as B, of accelerating emissions with rising income. This is the scenario the environmentalist fears the most, because it means that the environmental problem will get worse and worse as the world grows richer. On the other hand, the curve marked C in Figure 1-1 is indirectly related to the so-called environmental Kuznets curve (EKC) pattern that environmentalists hope for, whether natural or achieved as a result of suitable policy instruments such as making fossil fuels more expensive. C implies declining expenditure shares for fuel. This scenario could be explained by saturation, as more and more consumers already have a car, and by decreasing marginal utility of transportation. But this in turn suggests that there is a problem of regressivity, because at any given moment of time the poor have lower budget shares and are disproportionately affected by fuel taxes.

Many empirical estimates of fuel demand find that the income elasticity tends to be not far from unity, suggesting that the linear pattern A might be quite typical, and implying that fuel taxes would be neither regressive nor progressive but fairly neutral. However, some evidence also favors the Kuznets curve. But this is an empirical issue, and the results could vary from one time and place to another. Whether and how they do is the topic of this book. The introductory chapter will continue with the following section on climate policy in general followed by a section on the role played in such policy by fuel taxation. We then briefly discuss earlier studies before concluding with a discussion of how to measure progressivity.

Climate Challenges and Climate Policy

If a variable does not increase linearly with income, but follows a path such as the EKC, it is said to be decoupled from economic growth. Decoupling emissions from economic activity is generally a goal of environmental policy. The scale of decoupling needed in the next decades for climate gases is of unusual magnitude. If we are to continue global growth even at a moderate rate of, say, 3% to 4%, income will increase between four and seven times in the next 50 years. At the same time, it is not sufficient (in order to meet, for instance, the two degrees of temperature target) if carbon emissions stop increasing as in Figure 1-1: they actually need to be cut by 50% to 75%. This means that emission intensities have to be cut by about 90% to 95%. Carbon accumulates in the atmosphere because natural processes of absorption in the oceans and on land are insufficient to counteract current emission levels. We are thus essentially dealing with a stock pollutant, and ultimately we must reduce emissions of carbon very substantially. It is difficult to see

how these cuts could be achieved merely by behavioral changes or by technological changes on their own. Presumably we will need some combination,
but it is important to understand that both technology and lifestyle changes
are quite sensitive to prices. To handle the climate challenge, the world needs
substantially reduced emissions of carbon dioxide. One of the most efficient
instruments to reach this goal is high taxes on fossil fuels such as gasoline
and diesel. In Europe, these taxes are notably higher than in the U.S. That has
contributed considerably to lower emissions of greenhouse gases and even on
the atmospheric concentrations of carbon dioxide (Sterner 2007). Fuel taxes
might not originally have been designed for environmental purposes, but
their effect is surely environmental.[3] People discuss whether any sufficiently
powerful economic instruments are available but then fail to even see the
available evidence shown by the experience of fuel taxes in Europe, Japan, and
a few other countries—in fact a full-scale demonstration that these economic
instruments can be quite powerful.

Climate change is finally becoming accepted as an international priority.
The United Nations has set up the IPCC as one of the largest scientific evaluation efforts ever to provide analyses to support and inform the world community on the various aspects of the decisions that must be taken. The main focus
of the Kyoto Protocol was on stabilizing emissions from the industrialized
countries. It was intended as a compromise: designed to be acceptable to all,
it was supposedly better to take a small first step and get everyone on board.
However, Kyoto was not acceptable to all countries. It seems today that we
are witnessing a dramatic change in world opinion. This was catalyzed by the
Stern Review (2006), which was the first major, official report to give climate
change a prominent place among global problems, and its political backing
in the UK was one of the factors commanding attention. The Stern Review
considers the costs of unchecked climate change to be much larger than earlier
thought (not 1% of future GDP, but rather in the 10% domain). This analysis has met with considerable resistance, but it is fair to say that the general
consensus is shifting as the science of climate change becomes better known.

An overwhelming documentation of climate issues exists, and despite the
disagreements on some issues, there is in fact a fairly high degree of consensus
on a number of issues. This is not the place to review climate science,[4] but
a couple of facts are particularly important and worth repeating. First of
all, there has been an increase in mean global temperature of around 0.7 °C
during the last century (see Figure 1-2).

Secondly, there is an unprecedented increase in the carbon content of the
atmosphere. This is illustrated in Figure 1-3. Again, with all due respect to
the intricacies of the global carbon cycle into which we cannot delve here,
Figure 1-3 persuasively illustrates a number of important facts. First, there is
a connection between the carbon content and the climate regime (the troughs
of around 200 parts per million (ppm) coincide with ice ages, while the peaks
correspond to interglacial periods). Second, the natural range of variability

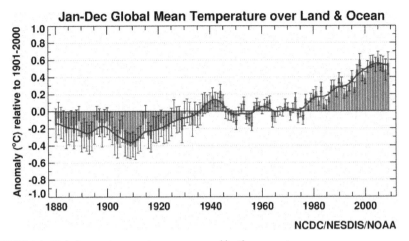

FIGURE 1-2 Global mean temperature as measured by thermometers.

Source: U.S. National Oceanic and Atmospheric Administration (NOAA); www.ncdc.noaa.gov/cmb-faq/anomalies.html

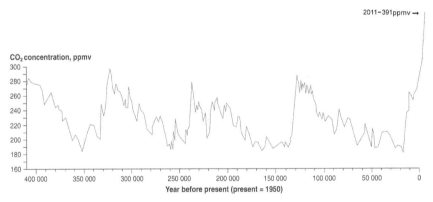

FIGURE 1-3 Carbon content of the atmosphere as estimated from the Vostok Ice core, and current values.

Source: United Nations Environment Programme (UNEP).

for the last half million years (and actually more) is around 200 to 300 ppm. Hence, third, the current value of 390 is far out of range. The carbon content of the atmosphere is now increasing—in fact accelerating (currently about 2 ppm/yr) in spite of the attention to this subject lately.

The carbon content of the atmosphere must be stabilized at some level, but there is considerable debate about which level. To reach targets such as 450 ppm (which is still very high compared to the experience of the last half a million years) will require radical reductions in carbon emissions: a cut in half before 2050 and almost complete phasing out within this century. Such dramatic cuts are needed because the carbon dioxide in the atmosphere is a stock pollutant, and the rate of absorption in the biosphere is low compared to emissions. The size of this challenge is all the more daunting given that population figures are still growing. Population is expected to increase to 10 billion people at the same time as incomes are expected to continue growing globally. China is today an average country in terms of emissions, which are about one ton of carbon per capita (almost 4 tons CO_2). If world emissions must decrease, then Chinese emissions must decrease. Even a lax target like 550 ppm only allows for a few years of growth before emissions peak and must then fall for the target to be met. This is staggering for a country like China, where incomes currently double every six or eight years and yet carbon emissions are to stay constant at best. To achieve such a dramatic decoupling of demand for fossil fuels from GDP is going to require considerable determination and very strong policy instruments.

The Role of Fuel Taxation in Dealing with Climate Change

Fuel demand is determined primarily by income and prices—in particular fuel prices, but also all other prices, such as for public transport and for vehicles. The effect of fuel price and other factors is usually expressed through elasticities, which show the percentage change in fuel consumption that would result from a 1% change in the exogenous variables.

There are literally hundreds of studies of fuel demand elasticities and a large number of surveys. Most studies use aggregate, yearly national statistics: either cross-sectional data for different countries, time-series for one country, or combined panel data. Other studies use data of monthly or quarterly frequency; still others use household data. Differences in country, time period, type of data, and model structure (particularly assumptions concerning time lags) are the main reasons for variation in the estimated elasticities. While a range of estimates is found, the consensus is that the long-run price elasticity is around –0.7 or –0.8, while the corresponding income elasticity is around unity. Goodwin (1992) finds that the time-series and cross-section methods broadly concur in giving long-run price elasticities of around –0.8. Dahl (1995) finds values from –0.7 and –1.0 and income elasticities between 1 and 1.4. Graham and Glaister (2002, 2004) find short-term price elasticities between –0.2 and –0.3, while long-run values go from –0.6 to –0.8. Long-run income elasticity is often somewhat higher than unity in rapidly developing countries (Dahl and Sterner 1991a, 1991b).

These surveys concur in finding a higher degree of *long-run* price sensitivity than many people appear to believe; on the other hand, the short-term elasticities tend, as mentioned, to be much lower (less than a third of the long-term values in the first year). That lag is problematic for politicians who are hurt immediately by any short-term discontent yet are not even likely to be in office to see the environmental benefit.

Practically all demand sensitivity studies focus on gasoline as opposed to diesel or other fuels. One reason might be that diesel is not only used heavily in transport (buses, trucks and cars) but also in non-transport machinery such as agricultural equipment, heating, light industry, and diesel generators with different pricing (often no tax for these uses) and also other explanatory factors from those for transport use. Another reason may be that estimations with several fuels would also have to deal with the different tax policies for the vehicles themselves; again, diesel and gasoline cars often face quite different taxes, and other instruments which tend to be complex and vary over time. As noted by Schipper et al. (1993), car fuel cannot be equated to gasoline use: gasoline can be used by mopeds, trucks, and other machines, leading to an overestimate of auto use. Meanwhile, many cars use diesel, resulting in under-estimation. The under- and over-estimation do not cancel out, however, because they are driven by very different processes and develop at different rates. One of the very few studies which explicitly sets out to estimate total *fuel* elasticities (thus including both diesel and gasoline and using correspondingly weighted fuel prices) is Johansson and Schipper (1997). They find overall fuel price elasticities of –0.7, suggesting that there is no large difference between the two fuels.

Price elasticities have important implications, because big differences among countries in fuel taxation in turn lead to big differences in final consumer price. (Other sources of price difference—efficiency, profit margins, costs at the gas stations—are fairly small.) The motorist also bears a number of additional taxes and fees levied directly on vehicles (registration fees, yearly taxes, vehicle sales taxes) and on road use (tolls, congestion fees, etc.). Also, a number of other policies vary between countries concerning the way in which public transport is financed, taxed, and subsidized. These policies all have effects on fuel consumption, but they are complicated to compare or even quantify.

Table 1-1 shows the average tax on gasoline (unleaded premium unless otherwise stated). The tax is in US cents converted by market exchange rates.[5] As we can see in Table 1-1, the simple average for the Western European countries selected is 119 cents per liter, which is very high compared to the U.S. and many other non-European countries. Countries such as Japan and Australia are intermediate. Two trends are notable: taxes within Western European countries have been converging over the last twenty years, whereas the spread between Western European and other countries has been increasing over time (since 2005, the average in this table for the West European countries went up by 35%, while the other countries' average rose by less than 20%).

The main reason why fuel prices vary between countries is differences in taxation. Other reasons include the costs of handling, quality-related and refinery costs, product transportation, and distribution costs and so forth, but these account only for a small share of the variation. Figure 1-4 gives an overview of the retail prices of gasoline in more than 170 countries. The picture for diesel is—broadly speaking—similar, and it is also available from the same source.

TABLE 1-1 Gasoline taxes in US cents/liter in selected countries, 2008

Western Europe	Gasoline Tax	Non European	Gasoline Tax
Italy	116	Japan	59
UK	123	Australia	43
Netherlands	139	New Zealand	51
France	121	Canada	32
Belgium	125	Mexico	10
Germany	128	USA	13
Finland	129	**Average**	**35**
Portugal	122		
Sweden	120		
Spain	90		
Austria	100		
Average	**119**		

Source: International Energy Agency (IEA) 2009; Unleaded premium gasoline (regular gasoline for Japan).

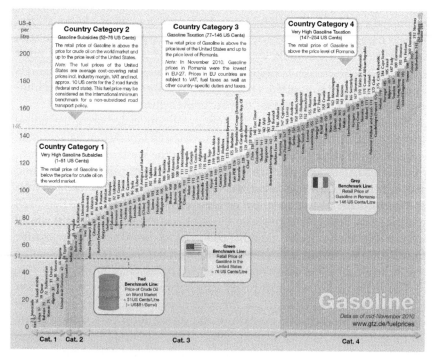

FIGURE 1-4 Retail price of gasoline in selected countries.

Source: www.gtz.de/fuelprices.

From a climate viewpoint, a particularly interesting comparison is between the U.S. and Europe. The very low fuel taxes in the U.S. compared even to the lowest tax rates in Europe are clearly related to higher fuel use—although the relationship is mainly apparent in the very long run. Thus for instance the average per capita consumption of gasoline in the U.S. is more than four times higher than in the UK or several other European countries. Such data suggest that if the EU had followed a similar tax policy to that in the U.S., aggregate carbon emissions would have been substantially higher. We can get an idea of the order of magnitude by using the average elasticities mentioned above to calculate the equilibrium gasoline consumption for each country with lower or higher prices. This is done in Sterner (2007), which calculates the effect if all OECD countries had had taxes as low as the OECD country with the lowest taxes (the U.S.). The differences in gasoline demand (and thus emissions)[6] analyzed are really very large. Part of this, however, is explained by an increased use of diesel in Europe, the inclusion of which reduces the dramatic differences when a comparison is made for total fuel consumption per capita between the U.S. and Europe. The difference is, however, still big. Table 1-2 (from Sterner 2007) shows the dramatic effect of fuel tax policies in Europe. To put this in perspective, total global emissions of carbon are on the order of 6 to 7 gigatons (Gton, or a billion metric tons). Roughly one of these Gtons (1.13 Gtons of fuel is roughly 1 Gton of carbon) comes from the OECD transport sector. This would have been about 30% higher in the absence of European (and Japanese) fuel taxes. If all the OECD countries—including, notably, the U.S.—had had equally high fuel taxes as the highest in Europe, the OECD total would instead have been at least 30% lower. These data strongly suggest that this is one of the few policies that actually has had a measurable effect on the carbon content of the atmosphere.[7] Note however that this is an approximation, and that carbon leakage to the benefit of the low tax countries is not considered. Presumably the high taxes in some countries reduced the world price of oil, thus encouraging consumption in the U.S. and other low tax countries, and partly offsetting the reduction in use. We have, however, not attempted to calculate this effect.

Similar comparisons could be made for developing countries with high or moderate fuel taxes compared to those (particularly the oil exporting countries) that do not tax, or tax very lightly (and in some cases even subsidize) domestic

Table 1-2 The effects on total OECD fuel use of high or low taxes

	Real	Hypothetical	
		UK prices	US prices
Fuel use (Mtons)	1,130	715	1,500
Percentage		−36%	+30%

Note: Sum of diesel and gasoline use in Ktons together with calculations (based on individual country estimates) for total fuel use if the countries had had the prices that correspond to the lowest or highest in the OECD area respectively.

consumption, as can be seen in Table 1-1 for Mexico. This price-setting is common, as shown by (for instance) Kosmo (1987) or Sterner (1989b and 1989c): oil exporters among the developing countries have typically had very low domestic prices. Such price-setting is thought of as a way to share the rent with the population; it is a fairly popular policy and often thought to promote industrialization and jobs (see also Larsen and Shah 1992 or IMF 2009).

The Popularity of Fuel Taxation: Casual Observation and Earlier Studies

One might think that the demonstration of how to make economic growth and sustainability compatible would meet with some interest and sympathy, but in my own experience that is not the typical reaction. I have been calculating fuel demand elasticities and studying the pricing and taxation of fuel in different countries for several decades. Although it is straightforward economics, the idea of continuously rising gas prices is typically viewed as unpleasant. The general idea of "taxing bads" may be accepted in the abstract, but when it comes to fuel (or electricity for that matter), the support tends to evaporate. We can observe much the same reaction in a whole range of countries which, though it may be at its strongest in the U.S., also applies, for example, to very low income countries in Africa. In fact, while we were working on this book, which includes a chapter on Ghana, there was indeed a sizable increase in fuel prices there. What followed was an immediate (and rather typical) uproar and outcry: "What is the government doing? They promised to care for the poor, and now they are raising the price of fuel!"

Let us start in the U.S., where the acrimony is most heated: The last time the gas tax was raised through a bill in Congress was in 1993 when Bill Clinton finally managed to push through a tiny tax that increased the price of gas by just 4.3 cents per gallon—little more than a cent per liter—and even that was not indexed to increase with inflation. In spite of a big debate on how to reduce public deficits at both the federal and state level, there is little talk of using such an obvious instrument as a gas tax that is needed anyway to reach even moderate climate goals. In 2010 the *Harvard Political Review* (*HPR*) featured an article that tried to understand this resistance. Its author, Will Rafey, states:

> *The modesty of the increase should not be surprising: since 1993, no prominent American politician has seriously supported a major increase in the gas tax. Virtually everyone agrees that supporting the gas tax is political suicide. As Michael Cragg, an energy consultant at The Brattle Group, told the* HPR, *"It's hard to see in this political environment how you'd get a gas tax passed."*

According to the *HPR*, the gas taxes collected in the U.S. are now so low they do not even cover road maintenance; additional funds for such maintenance must be taken from the general budget. Rafey also points to the apparent contradiction of a policy (of higher taxation to cover not only road building costs but also externalities) that enjoys such broad support from economists almost regardless of their political conviction but at the same time has so little political support in the general population. According to Rafey there is a whole range of explanations, including the influence of interest groups, the rural/urban divide, and the power of the anti-tax movement in general, which is also related to the short electoral cycle (which disfavors policies with short-term political costs and long-run benefits).

My experience from seminars and lectures in the U.S. is one of witnessing incredulity. The audience is skeptical that it is possible to raise fuel taxes. Sometimes they do not even believe how expensive fuel is in Europe. It seems unthinkable, and so the audience suspects some mistake with exchange rates and conversion from liters to gallons. When U.S. politicians from Bush to (Hillary) Clinton call for lower gas taxes they are merely responding to a very strong U.S. opinion. According to one recent survey, 86%[8] of Americans oppose a proposal to increase gasoline taxes by 50 cents a gallon, while only 8% are in favor. Yet 50 cents is still only a fraction of the gap that would be needed to bring U.S. and EU prices in line.

In September 2000, Europe too was rocked by fuel protests.[9] The most direct cause seems to have been rising international crude prices and a stronger dollar. This, combined with taxes that rise in absolute terms auto-matically with the price (as the Value Added Tax, or VAT, does), meant a rapid escalation in fuel prices. At least some partial "responsibility" fell upon politicians who in many cases were social democrats, many times in coalition with green parties. The protests were initiated by truckers and farmers, entre-preneurs with little sympathy for the red-green government coalitions, but they definitely had broad popular support. The movement appears to have started in France, where fishermen already in August had blockaded harbors and managed to extract concessions of lower fuel taxes. This was followed by farmers and truck owners—but such is political culture in France, whose farmers have, after all, successfully hijacked the whole EU common agri-cultural policy. It was more surprising that protests spread as they did to other major and minor European countries, becoming quite a pan-European movement. Suddenly former Lionel Jospin was joined by fellow socialist leaders such as Tony Blair in the UK and saw massive two-digit popularity losses in the polls. The same happened in Germany, where suddenly the scan-dalized Christian Democratic Union (CDU) caught up in the polls with the social democrats. Maybe this was all the more surprising given that Europe is so classically tax tolerant, and that the gas taxes which are so dramatically high by U.S. standards had not really been a major political issue earlier. It appears that the combination of a well-organized and highly motivated—although quite small in number—group of truckers, against a backdrop of a large passive popular support was very effective. Today, a Facebook page dedicated to "Keep Fuel Below £1.00 per litre" is a sign that this is an important issue of our time.

Even Sweden, with its tradition of high taxes and rather high tax tolerance, has had its share of protests. A few trucks were allowed to blockade entire refineries and fuel depots without much resistance. A few years later an "anti-gas-tax-appeal"[10] was started, and it claims to have collected 300,000 names in three months, over a million names in less than a year, and a total today (five years later) of 1.8 million signatures for their demands: lower gas tax, removed VAT on the actual fuel excise tax, and reduced VAT rates on fuel (from the standard 25% down to the 6% currently applicable to public transport). In a country of less than 9 million inhabitants, these are very large numbers. According to the group's webpage and their book on the fuel protest movement, leading politicians such as finance ministers of both the moderate and social democratic parties have refused to listen to them and are instead captive to small "car-hating" parties.

Non-economists sometimes support the idea of reducing emissions and fuel use but prefer to achieve these goals some other way: by telling car companies to make smaller and more energy efficient cars, by telling people to car pool, or through better organized public transport. Economists tend to think of the fuel tax as the best way of reaching an appropriate mix of all these sub-goals because it allows the market to find the least cost incentive-compatible combination. The risk of only building more energy-efficient cars and more buses so long as fuel is cheap is that no one will use these improvements. Economists also tend to think in terms of relative prices and balanced budgets. If fuel taxes are raised in lieu of other taxes so that public budgets are not affected then it is only the relative prices that are affected. The effect is very different from the situation when world oil producers raise their prices, which implies a much larger real loss in purchasing power and welfare for the country.

A common critique against fuel taxes uses the argument that fuel taxes will be unfair to the poor, to rural societies, or to certain groups in society. The critique has been so vehement that even I had the impression that most studies must actually support this hypothesis. Some early studies do speak of regressivity: The U.S. is a country of very high incomes and where even the poorer people have cars—in fact it is the poor who tend to own the older, more energy-inefficient cars, and it is the poor who have to drive the largest distance to work in a country which has very little public transport. So my first thought when writing this introduction was that I would review all these early studies that found regressivity, at least for the U.S. In fact the literature is much more balanced. Although many early studies are indeed for the U.S. (or some other high income countries), the picture is quite mixed with prominent early studies—by, for instance, Poterba (1991)—arguing that the tax is not very regressive.

Many researchers believe that taxes should be compared with a household's long-term or permanent income rather than its annual income. Measuring the tax burden relative to permanent income provides an estimate of household's ability to bear a tax over a lifetime. Annual income, by contrast, could substantially underestimate (or overestimate) the long-term ability of some households to pay a tax. For instance, households with retired workers may have small annual incomes but large savings. Moreover, households with people who are early in their careers may have low current incomes but expect substantially higher income in the future.

Studies measuring tax burden can use either income or expenditures to estimate incidence. The degree to which the gasoline tax burdens households within different income categories depends on whether income approach or expenditure approach is used. Poterba (1989 and 1991) used the expenditure approach and included households that own vehicles as well as those that do not own vehicles in his analysis. He found that low-income households spend less of their budget on gas than middle-income households. This suggests that a gas tax is less regressive than other studies would suggest. Walls and Hanson (1999) use annual income and a measure of lifetime income (based on expenditures) to analyze the distributional effects of vehicle pollution control policies. Results using annual income show all polices to be regressive across all income groups, while results using lifetime income are like Poterba's (1991) results, more progressive.

Poterba's (1991) findings provide some interesting results. According to his results, out of the 10 income deciles of U.S. households in 1986 the bottom four deciles had gasoline spending as a share of income of over 6.08%, with the lowest decile having 11.44% gasoline spending as percent of total income, as opposed to 2.4% for the uppermost income decile. (Each decile contains 10 percent of U.S. households, ranked on the basis of their income.) However, when he compared the result for total spending, results were much different. The lowest decile had only a 3.7% share of gasoline and motor oil spending as percent of total spending (where total spending included imputed values of rents and automobiles), while the uppermost income decile (here deciles are ranked on the basis of their total annual spending) had a 3.2% share of gasoline and motor oil spending out of total spending. Chernick and Reschovsky (1997) found similar results by using average family gasoline spending as a percentage of average family income for a period of 11 years, instead of annual income. According to their result the bottom income decile spent only 3.9% on gasoline, while the uppermost income decile spent 2.5%.

Fullerton (2011) has a very useful typology of six different types of distributional effects from climate policy, each of which may or may not be regressive: (1) the direct effect of higher prices of carbon-intensive products, (2) changes in relative returns to factors like labor, capital, and resources, (3) allocation of scarcity rents from a restricted number of permits—or in this case, the use of tax proceeds, (4) distribution of the benefits from improvements in environmental quality, (5) temporary effects during the transition, and (6) capitalization of all those effects into prices of land, corporate stock, or house

values. This book will focus on (1). Item (3)—the use of the "scarcity rent"—
is very important, although in our context it is fuel taxes rather than permit
values. Item (6) will also be discussed, while the others largely fall outside
the scope of the book—or, like item (4), will be mentioned but not analyzed
quantitatively. This does however serve to remind us just how complicated
distributional analyses can be.

Improving the appropriate measure of income is but one line of research.
There is a long list of further refinements that can be made if we want to judge
the progressivity or regressivity of a tax. We can take indirect resource use into
account: not only direct fuel use but indirect use is important, and thus we can
include public transportation—or indeed all transportation—since all goods
are transported. In fact, all goods are used in the production of other goods, so
general equilibrium effects are important. Adaptation to changing prices may
be different in different strata, and this too can be taken into account. Effects
on other factor markets also exist: the value of used cars, houses, and even
labor market skills can be differentially impacted by changes in fuel taxation.
(For instance, gas-guzzling cars would be expected to lose more in value.) If
ownership of these assets varies by social class, then there will be distributional
effects on these markets. Similarly, the pollution itself may have distributional
effects. In this book, not less than three chapters will use various sophisticated
models and datasets to analyze various aspects of the distribution of incidence
of fuel taxation in the U.S.

Though there is now quite a literature on these issues in the U.S., fairly little
has been written on other countries. Santos and Catchesides (2005) find that the
budget share of road fuel consistently declines with income for households in
the UK—implying regressivity. However, again this result depends crucially on the
use of income or expenditure approach. Steininger et al. (2007), using a general
equilibrium model that includes all feedback mechanisms in the economy, find
that such policies are overall progressive across the whole of the income distribution for Austria. For low income countries[11] much fewer studies exist, but the
issues are of course at least as important. It is of particular importance in rapidly
growing mega-economies such as China and India, whose carbon emissions will
be decisive for the future path of climate change. Ultimately, this is an empirical
issue. To judge the distributional consequences of fuel taxation in different countries is the main purpose of this book, which explores the distributional effects
in a large selection of both rich and poor countries with varying economic and
political policies. In addition to the U.S., we have sought to include a range of
large and small European, Asian, Latin American, and African economies.

Measuring Progressivity

Whichever methodological approach is preferred, to determine the burden a
tax places on richer and poorer households a researcher needs a measure of

progressivity. A very simple test of whether a given tax is progressive or regressive is available by comparing whether the budget shares for the consumption of that particular good are higher or lower among high- and low-income earners. For example, suppose a low-income person has a budget share for some item—say, food—of 40%, while a rich person only spends 10% of his income on food. Then a tax that increased the price of food by 10% would take 4% from the low-income person but only 1% of the rich person's income; such a tax would seem regressive. Conversely, if the poor have a low budget share for some good (like a luxury good) then a tax on that good is progressive. We can therefore illustrate the effect of a tax by showing the budget shares for different income groups. In terms of the Engel curves in Figure 1-1, consumption pattern A is proportional, and an excise tax on that good would be neutral as the budget shares are constant. Case B shows a consumption share that is growing with income—taxation would be progressive. If the consumption share falls with income as in C, then the poor are more affected by a tax, and the tax is regressive. We can therefore see that the conventional assumption of an "inverted-U" type Kuznets curve would imply tax regressivity at least for high incomes—yet the first part of the curve could in fact still be progressive.

The above definitions of regressivity and progressivity are simple as long as the budget shares are monotonically increasing or decreasing. The situation is more complex if a particular good is consumed heavily by middle classes but less by both the rich and the poor. A tax on this good would then hurt the middle class compared to the rich—a regressive effect—but it would also hurt the middle class compared to the poor—a progressive effect. Politically this may be very interesting as it stands, but in case we want a single summary measure to say if a tax is on balance either progressive or regressive, we need some manner of devising a summary index.

In this book we use the Suits and Kakwani indices for this purpose. These are the two most common indices used to measure the progressivity of taxes (see Suits 1977, or Kakwani 1977a). Both are inspired by, and analogous to, the Gini coefficient, since they are a simple summation or integration over the whole income distribution without using any explicit welfare weights. The Suits index varies from +1, which is extreme progressivity—the entire tax burden is borne by members of the highest income —to –1 at the extreme of regressivity, when the whole tax burden is borne by members of the lowest income bracket. This index allows us to compare between different taxes on the basis of progressivity. A flat or proportional tax has a Suits index of 0.

Figure 1-5 depicts one example of a concentration curve. (Note the similarity with the Lorenz curve, which depicts accumulated income against population.) Each income group's percentage share of the tax paid (out of the total for all groups) is accumulated on the vertical axis, and its percentage share of income or total expenditures on the horizontal. The case of a completely proportional (neutral) tax is depicted by the 45° line from the lower left corner to the upper right. The curve shown is for a progressive tax, while the

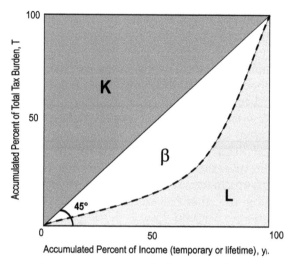

FIGURE 1-5 Example of a concentration curve with the areas used to calculate the Suits index.

corresponding curve for a regressive tax would lie above the diagonal. The intuition for this is that a regressive tax will take a larger proportion away from a low income decile than their share of income—thus making the overall (after tax) income distribution less even. The size of the Suits index is given by the share of the area between the concentration curve and the 45° line. This again is fully analogous to the Gini index, which is measured in the same way but using the Lorenz curve.

Even if the concentration curve is progressive in one part but regressive in the rest, thus crossing the proportional line, by calculating the Suits coefficient, a summary statistic measuring the degree of deviation from proportionality, we can determine whether such a tax is overall proportional, progressive, or regressive. This summary coefficient can therefore readily be used to compare distributional effects for different taxes. However, like all Gini-type measures, such indices are simple geometric constructs with no particular welfare economic interpretation—and in fact some properties are quite unintuitive when one looks at the finer details. (In this case, a given deviation from the diagonal has the same value irrespective of whether we are dealing with a rich or poor decile.) Despite such strange properties, Gini coefficients are quite popular.

The Suits coefficient, here denoted by S, is the share in the area under the 45° line of the white area, β, between the concentration curve and the diagonal, and is thus equal to $(1 - L/K)$ where K is the black area. If we write income, y, and tax burden, T, with a range from 1% to 100%, then S is defined as follows (see Suits 1977).

$$S = 1 - \frac{L}{K} = 2\int_0^1 (y - T(y))\,dy \qquad (1.1)$$

In the case of regressivity, $L > K$ and $-1 \leq S < 0$, while proportionality implies $S = 0$ and $L = K$. A progressive tax yields $L < K$ and $0 < S \leq 1$. Because our data is arranged in income deciles $i = 1 \ldots 10$, we use an approximation for the integral of L, suggested in Suits (1977) (see Equation 1.2).

$$S \approx 1 - \frac{\sum_{i=1}^{10} \frac{1}{2}\left[T(y_i) + T(y_{i-1})\right](y_i - y_{i-1})}{5000} \tag{1.2}$$

The Suits coefficient is thus a summary statistic or global progressivity index, as distinct from indices of structural progressivity or indices of local progressivity which measure the degree of progressivity at a given income level. An alternative is the elasticity of tax liability to pre-tax income (see Lambert 2001). Since the Suits index does not discriminate between departures from proportionality at the low, high, or middle end of the distribution, it is best interpreted together with a graph or table that describes the fuller distributional picture.

The Kakwani index (Kakwani 1977a) is a closely related alternative to the Suits. However, as Lambert (2001, *176*) points out, while the Suits index is limited between -1 and 1, the bounds for the Kakwani index depends on the income inequality. Furthermore, the few previous studies in the field that use an index have chosen the Suits index as measure (e.g., Chernick and Reschovsky 1997, Metcalf 1999, Walls and Hanson 1999, West and Williams 2004);[12] for the sake of comparison many of this book's chapters do the same. In other chapters the two indices are compared.

Although the Kakwani index also builds on concentration curves, there is a big difference: instead of using share of income as its horizontal axis, it has share of population, p, ranked from poor to rich. The vertical axis has cumulative shares of both income and tax. In Figure 1-6, $T(p)$ shows the cumulative share of taxes, while the Lorenz curve is the concentration curve showing the cumulative share of income $y(p)$.

Kakwani's index builds on the area B, *between* the concentration curve and the Lorenz curve in Figure 1-6. For positive values, the Lorenz curve is higher than the concentration curve, meaning that at each income level the cumulative population share has a higher share of cumulative income than its share of taxes it must pay, which is what we would expect from a progressive tax. Note that this again means that the particular tax is "progressive" with respect to the income distribution: the after-tax income distribution will thus be even more level than the pre-tax income distribution. If on the other hand this area is negative (that is, if the Lorenz curve is below the concentration curve) it means that the cumulative population share for any given income level must pay a tax share higher than its cumulative income share, suggesting that the tax is regressive. Using the formula for finding the area under a curve, the concentration index is given as:

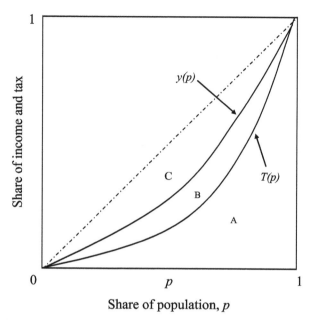

FIGURE 1-6 Concentration and Lorenz curves to calculate Kakwani's index.

$$K = 2\int_0^1 \big[y(p) - T(p) \big] dp \tag{1.3}$$

It is now straightforward to show the relationship between the Kakwani index and the Suits index. The horizontal axis of Figure 1-5 depicts y, which is the cumulative share of income ranked in the ascending order. This is in fact a transformation of the Lorenz, $y(p)$, in Figure 1-6, which is the cumulative share of income related to the cumulative share of the population ordered in the ascending order of income. Once we look at the cumulative share of income in this way, we immediately see that the relative concentration curve of tax under the Suits approach is just a transform of the concentration curve of tax under the Kakwani approach. That is:

$$y = y(p) \Rightarrow T(y) = T\big(y(p)\big) \tag{1.4}$$

Further, we can immediately see from Equation 1.4 that

$$dy = y'(p) dp$$

which means that the Suits index can be rewritten as

$$S = 2\int_0^1 \big(y - T(y) \big) dy = 2\int_0^1 \big[y(p) - T\big(y(p)\big) \big] y'(p) dp \tag{1.5}$$

Equations 1.3 and 1.5 show that $y'(p)$ (the marginal income tax for one more person—or slope of the Lorenz curve) constitutes the factor that differentiates the Kakwani index from the Suits index. The area β in Figure 1-6 is thus a transform of the area β in Figure 1-5. As shown by Formby et al. (1981), the two indices give different magnitudes of progressivity for the same tax structure and may actually even give conflicting verdicts on whether the tax structure is progressive or regressive (in cases when the curves cross the 45° line, illustrating progressivity in one part of the distribution and regressivity in another). Formby et al. state that there is no obvious way to choose between the two, but we can go a step further and say that the Suits index unnecessarily gives greater weight in its calculation to deciles with higher income (precisely through the factor $y'(p)$ which is rising from decile 1 through to decile 10 by definition). This is normally not desirable from a welfare perspective, where we would either want equal weights or perhaps *higher* welfare weights for poor households. In this sense the Kakwani index is preferable from a welfare perspective.[13] In most cases, however, they will give similar information.

Case Studies

This book has evolved from years of interaction and thought concerning climate change and fuel taxation. I have given many lectures and written a number of articles on fuel taxation and often met the argument that fuel taxes are regressive. This argument did make me uncertain, but I had not been fully persuaded. Eventually, I decided to tackle the issue head on. Following this introduction are three chapters on the U.S., where much of the debate has been centered. These articles cast light on the issue from a number of different methodological viewpoints. For instance, they deal with how to measure regressivity, how to define income, how to deal with capitalized effects in transition (for example, of used vehicles). After that the book turns to a whole array of different countries in Africa, Europe, the Americas, and Asia. The studies use some variety of methods and approaches, but a fairly high degree of methodological conformity has been imposed in order to make the book more readable and to allow for comparisons between countries. I will avoid repetition by not going into more detail: Chapter 19 provides an overview together with some comparative analysis and policy conclusions.

Acknowledgments

Very insightful assistance from Elizabeth Robinson is gratefully acknowledged. Thanks also for comments from Emanuel Carlsson, Don Fullerton, Mark Jacobsen, John Mduma, Gilbert Metcalf, Adolf Mkenda, Milan Ščasný, Daniel Slunge, and Michele Valsecchi.

Notes

1   This is particularly true because many other sources of emissions are likely to be much more difficult to deal with.

2   We are really interested in emissions and not fuel use. As a simplification we identify one with the other, although of course technology (such as carbon capture and storage) could modify this relationship.

3   The stated motives for gasoline taxes vary considerably. In some countries they are just a convenient tax base. In others they contribute to road building and maintenance, plus health effects. These vary geographically, and Parry and Small (2002) question their level for such purposes. Historically, climate externalities have played a small role (if any) in motivating gasoline taxes—but the taxes play a big role in reducing emissions of climate gases.

4   The IPCC is the natural source for those who want more information.

5   An alternative used by some researchers is to use purchasing power parities to provide an indicator of the *actual burden* the tax places on the representative motorist, but here this would imply some double-counting of the income effect—once directly through income elasticities, and once indirectly.

6   Emissions of carbon dioxide are closely tied to fuel use. The only real wedge between these figures would be achieved for biofuels but this factor is of little significance to date—at least compared to the differences in overall consumption levels.

7   The difference between the hypothetical use with high/low prices for total fuel is about 750 Mtons of fuel per year. A decade of such differences would correspond to emissions of roughly 25 billion tons of CO_2 or a carbon content of 3 ppm.

8   www.futurepundit.com/archives/005987.html (accessed May 17, 2011).

9   The section on fuel protests in 2000 is based on several newspaper sources at the time. See, for instance: Nyström, *Göteborgs-Posten* Sept. 19, 2000, p, 6; Holmberg, *Dagens Nyheter* Sept. 23, 2000, p. 20; http://en.wikipedia.org/wiki/Fuel_protests_in_the_United_Kingdom#Consequences (accessed May 17, 2011).

10 www.bensinskatteuppror.se/media/boken.pdf (accessed May 17, 2011).

11 If we broaden our outlook to environmental policies in general, these are often considered to be regressive in developed countries (OECD 1994, Atkinson et al. 1995), but for developing countries we do not have enough studies to say in general much about the distributional consequences of environmental policy (see, e.g., Yusuf and Resosudarmo 2008).

12 Walls and Hanson (1999) and West (2004) studies a tax on vehicle miles travelled (VMT). These results should be very similar to a tax per quantity of fuel consumed.

13 Neither index fully satisfies Atkinson's social welfare axioms. Sykes et al. (1987) have shown that "conclusions about comparative tax progressivity reached by use of the tax scale invariant measures of Suits and Kakwani are, at times, unsupportable and demonstrably invalid when evaluated from the perspective of social welfare." Another possible approach to assessing tax progressivity is proposed by Yitzhaki and Slemrod (1991). It builds on the notion of stochastic dominance of the concentration curves. The dominance approach has a reasonable welfare theoretic foundation but will only give a decisive verdict if one curve is really dominant throughout the whole income range. I thank Adolf Mkenda, author of Chapter 15 on Tanzania, for pointing this out to me.

## CHAPTER 2

The Consumer Burden of a Carbon Tax on Gasoline

KEVIN A. HASSETT, APARNA MATHUR, AND
GILBERT E. METCALF

G asoline is one of the major fuels consumed in the United States and the main product refined from crude oil. Consumption in 2007 was about 142 billion gallons—an average of about 390 million gallons per day—and the equivalent of about 61% of all the energy used for transportation, 44% of all petroleum consumption, and 17% of total U.S. energy consumption.[1] About 47 barrels of gasoline are produced in U.S. refineries from every 100 barrels of oil refined to make numerous petroleum products. While gasoline is produced year-round, extra volumes are made and imported to meet higher demand in the summer. Gasoline is delivered from oil refineries mainly through pipelines to an extensive distribution chain serving about 167,500 retail gasoline stations in the United States.

Most gasoline is used in cars and light trucks. It also fuels boats, recreational vehicles, and farm, construction, and landscaping equipment. A major concern with the use of gasoline today is in the context of climate change. The use of gasoline in transportation results in carbon dioxide (CO_2) emissions, which have been increasing at a rapid pace since the 1990s. In 2007, total CO_2 emissions stood at 6,022 million metric tons (MMT), an increase of more than 17% since 1990. The vast majority of carbon dioxide emissions come from the combustion of fossil fuels such as petroleum, coal, and natural gas, with petroleum accounting for nearly 43% of all emissions from these energy sources (EIA 2008).

The Energy Information Administration (EIA) further provides a breakdown of energy use by end-use sectors. In 2007, the transportation sector was the largest source of emissions compared with the residential, commercial, and industrial sectors, accounting for about 34% of all CO_2 emissions (EIA 2008).

In discussions over how best to address climate change issues, essentially two market-based approaches are being considered: a carbon tax and a cap-and-trade system. In this chapter, we focus on the effect of a carbon tax on gasoline to reduce CO_2 emissions, though our results essentially carry through for a cap-and-trade program as well.[2] A carbon tax is essentially a market-based instrument that creates a cost to emissions by directly taxing the carbon content of fuels. In the case of gasoline, this is essentially a tax on petroleum.

A consequence of a carbon tax (or a cap-and-trade program to cut emissions) is the subsequent rise in energy prices (as well as prices of energy-intensive goods and services) which would be passed on from the firm to the consumers in the form of higher prices for gasoline and other commodities. Therefore a major concern with any such program is that the burden of the costs arising from such a policy will fall disproportionately on low-income households. In other words, the policy will be regressive. For instance, a recent Congressional Budget Office (CBO) paper (2007) estimates that the price increases resulting from a 15% cut in CO_2 emissions through a cap-and-trade program would cost the average household approximately $1,600 (in 2006 dollars) and about 3.3% of income for households in the bottom quintile of the population (CBO 2007). In Hassett et al. (2009), we estimate that the cost of a carbon tax on the lowest decile of the population would be about 3.74% of income, over four times the cost on the highest income decile.

In this chapter, we measure to what extent a carbon tax on gasoline is regressive in a lifetime income framework. Note that our analysis involves imposing a carbon tax on all fossil fuels (coal, petroleum, and natural gas) and then assessing the price increases for all consumer goods that would occur as a result of the tax. While our detailed methodology for computing the price increases is discussed in our earlier paper (Hassett et al. 2009), we focus in this chapter only on the price increase for gasoline, an essential transport fuel and the burden of this price increase on households at different income levels.[3] It is also worth noting that other externalities in addition to global warming may require differential taxation of energy sources. To the extent that current gasoline taxes are Pigovian with regard to other factors such as congestion, then it may be that gasoline taxes may continue to exist even after the imposition of a carbon tax.[4] Accordingly, a separate analysis of the distributional impact of changes in the tax treatment of gasoline may be of some use to policymakers. Therefore, we present results for the incidence of current state gasoline taxes both across income deciles as well as regions.

Lifetime income measures are considered here, since in several cases measures based on current or annual income may provide a misleading picture of actual household incomes and may overstate regressivity. We discuss this in detail in the section to follow. Further, to capture intertemporal patterns, we look at three different years—1987, 1997, and 2003—to see how the incidence would change had a carbon tax been in effect in these three time periods.

Our results suggest that in general, gasoline taxes (both carbon taxes as well as state taxes) appear more regressive when income is used as a measure of economic welfare than when consumption (current or lifetime) is used to measure incidence. Studying the intertemporal distribution, we find that between 1987 and 2003 gasoline taxes have become marginally more regressive when calculated as a fraction of income.[5]

Gasoline taxes are also likely to have uneven regional effects depending upon the use of transportation in different regions. We report the average gasoline tax paid per household across regions and find that the regional variation is relatively modest.

In the next section, we explore different methods used to measure incidence and motivate the lifetime measure of consumption employed in Bull et al. (1994). Then we detail our data and methodology. The subsequent section presents results for the economic incidence of the tax. The final part concludes.

Measurement of Incidence

Tax incidence measures the ultimate impact of a tax on the welfare of members of society. The economic incidence of a tax may differ markedly from the statutory incidence due to price changes. For a carbon tax, the short-run economic incidence is likely to differ markedly from the statutory incidence. While the statutory incidence of an upstream tax on gasoline may be on the refinery owner, the economic incidence is likely to be on final consumers as the tax is shifted forward to consumers in the form of higher prices. Measuring the incidence of a tax requires numerous assumptions, and we begin the analysis by setting out our assumptions and methodology.

First, we must determine the appropriate unit of observation, which could be an individual or a household. For this study, we use the household as a unit. Second, we must choose the appropriate time frame of analysis. As we discussed above, the choice of the time frame for the analysis is extremely important. Early tax incidence analysis used current income as the base; that is, it compared the tax liability over a short period to income earned over that period. Following Friedman (1957) and the permanent income hypothesis, there was a realization that consumption decisions are made over a longer time horizon. Hence income should be measured as the present discounted value of lifetime earnings and inheritances. Failing to do so creates substantial measurement problems, particularly at the low end of the income distribution. For example, elderly people drawing down their savings in retirement will look poor when, in fact, they may be comfortably well off in a lifetime context. In other words, many low-income people are not necessarily poor.[6] Caspersen and Metcalf (1993) report cross tabulations on income and consumption that show a large fraction of households in consumption deciles substantially above their income deciles.

Poterba (1989) follows the approach of using current consumption as a proxy for permanent income, since if consumer behavior is consistent with the permanent income hypothesis, then consumers would set current consumption proportional to permanent income. However, as we mentioned earlier, using data from the 1987 Consumer Expenditure Survey, Bull et al. (1994) show that consumption, instead of being smooth, closely tracks current income over the life cycle. Moreover, energy consumption also shows a marked lifetime pattern.

This could be problematic for the incidence measurement. Suppose that as people grow old their energy consumption becomes a larger share of their total consumption, and suppose, as well, that over a lifetime the energy tax has a proportional incidence; then, using current consumption to measure lifetime income, the energy tax would appear regressive.

As an alternative to current consumption, we use an adjusted lifetime measure for consumption that is intended to correct for long-run predictable swings in behavior. This measure was first employed in Bull et al. (1994). Ideally, a lifetime measure of incidence would be constructed by taking the ratio of lifetime energy taxes to lifetime earnings. Unfortunately, the lack of any sufficiently long longitudinal panel data set precludes such an approach.[7]

To proxy for lifetime consumption, we therefore use the age profile of people sampled in a particular year by the Consumer Expenditure Survey (a more detailed description of the calculation is presented in Bull et al. 1994). In particular, we first classify people into different subsamples based on their educational level, since the pattern of income and consumption will be quite different for people with vastly different human capital stocks. For each subsample, we then calculate a "typical" path of consumption through the averages for the age groups. For a given person in the subsample we know the ratio of their current consumption to the average for their age group. We then compute their lifetime consumption by multiplying this ratio by the present value of the typical lifetime path.

For example, suppose an individual is a 35-year-old Ph.D. whose energy consumption is 80% of the average for her age and education group. Let's say the present discounted value of total lifetime energy consumption for a person with a Ph.D. is $80,000. Then for this individual, the imputed lifetime energy consumption is $64,000.

This procedure allows the age profile of each variable to be different. This flexibility helps to control for any confounding effects on the incidence calculation that predictable lifetime patterns of consumption behavior introduce in the cross-section. For example, suppose an alternative lifetime correction method was used where the share of consumption received at age 35 was used as the correction method. If 5% of consumption occurs on average at age 35 for a person in a given educational class, then that person's imputed lifetime consumption is 20 times their current consumption. Suppose that the lifetime incidence correction did not renormalize for each variable studied, but rather used the same correction factor for energy consumption. Then one would multiply current energy consumption by 20 to impute lifetime energy consumption. But if in reality, 10% of energy consumption is spent on average at age 35 for a person in a given educational class, then that person's imputed lifetime energy consumption should be ten times their current energy consumption. Failing to renormalize the incidence correction for each of the variables studied in this example would incorrectly double lifetime energy consumption, biasing incidence results.

Using consumption as the base for measuring income also addresses the problem of transitory income shocks. For instance, a transitory negative shock to income may push the recipient into a lower income decile, while leaving their energy consumption unchanged. In this case, the ratio of energy taxes to income would be higher than it would be under a correct lifetime measure. Similarly, an upward shock to income may push the recipient into a higher income decile, while leaving their energy consumption unchanged. Here, the ratio of energy taxes to income is lower than it would be under a correct lifetime measure. The combination of these effects would lead an income-based lifetime incidence correction to be biased toward regressivity. When lifetime incidence is measured against consumption, however, such transitory effects are less likely to lead to bias, since energy consumption and total consumption are likely to react together to income shocks, if they react at all.

The final issue in an incidence analysis is the allocation of the tax burden between consumers and producers. Taxes on energy can be passed forward into higher consumer prices or backward in the form of lower returns to factors of supply (capital, labor, and resource owners). Our approach implicitly assumes that consumers bear the burden of the tax.

Considerable theoretical work has been carried out on the incidence of energy taxes in general and of carbon taxes in particular. A number of large-scale general equilibrium models (CGE models) suggest that in the short to medium run, the burden of a carbon tax will be mostly passed forward into higher consumer prices (see, for example, Bovenberg and Goulder 2001; Metcalf et al. 2008).

This result falls out of the models because the demand for carbon-intensive fossil fuels is relatively inelastic in the short run, while the supply, which is determined globally, is relatively elastic. In the longer run, as consumers find substitutes for fossil fuels, demand would likely be more elastic, and more of the burden would be shifted back to the producers. To the extent that owners of natural resources and capital have higher incomes, our results will then tend to overstate the regressivity of a carbon tax.

While the identification of the full long-run effects of a carbon tax requires a dynamic CGE model, our approach has certain advantages. Most CGE models rely on a number of assumptions relating to the functional form of the production and demand functions and the parameters defining the elasticity of substitution between factors of production as well as the relevant demand elasticities. Our approach, in contrast, is fairly transparent and likely yields an accurate estimate of the short run impact of a tax, which in itself may be an upper bound on the regressivity of the tax.

Another advantage of our approach is that we are able to work with far more disaggregated data than the typical CGE model. In general, these models allow for a limited number of industrial sectors and consumption goods for ease of exposition and computability. However, we are able to work with more than 50 industry and commodity groups and more than 40 consumer goods. We can explore regional variation in the incidence of the tax as well.[8]

Our analytic approach assumes no consumer behavioral response. Consumer substitution away from more carbon-intensive products will contribute to an erosion of the carbon tax base. The burden for consumers, however, will not be reduced as much as tax collections will fall. Firms incur costs to shift away from carbon-intensive inputs, costs that will be passed forward to consumers. Consumers also will engage in welfare-reducing activities as they shift their consumption activities to avoid paying the full carbon tax. Although the burden impacts reported here do not take account of the range of economic responses to the tax, the impacts provide a reasonable first approximation of the welfare impacts of a carbon tax.

Methodology and Data

For purposes of our analysis, we consider the effect of a carbon tax set at a rate of $15 per metric ton of carbon dioxide assuming it were in effect in three different years: 1987, 1997, and 2003.[9] This allows us to see how changing consumption patterns over time influence the distribution of the tax. Because we are considering a carbon tax in different years, we deflate the tax rate to keep it constant in year 2005 dollars. Using the CPI deflator, the tax rates we consider are $8.73 in 1987, $12.33 in 1997, and $14.13 in 2003. The incidence calculations require two types of data. First, to assess the impact of the carbon tax on industry prices and subsequently on prices of consumer goods, we use the Input-Output matrices provided by the U.S. Bureau of Economic Analysis. Second, once we have the predicted price increases for the consumer goods, we need to assess incidence at the household level. For this, we used data from the U.S. Bureau of Labor Statistics Consumer Expenditure Survey for various years. Note that the carbon tax is essentially applied to coal, petroleum, and natural gas, and we obtain price increases for all consumer goods. These price increases were the focus of our previous paper (Hassett et al. 2009) and are reported here in Appendix Table 2-A1. In this chapter, we focus on the gasoline price increase as a result of the tax and estimate the tax burden on households. As the table shows, the estimated gasoline price increase is approximately 9.67% in 1987, 7.64% in 1997, and 7.73% in 2003. In this section, we discuss how we combined the two different data sets to calculate the incidence at the household level.

Energy related emissions of CO_2 were 4,821 million tons in 1987, 5,422 million tons in 1997, and 5,800 million tons in 2003. Given the tax rates and ignoring initial reductions in emissions, the tax would raise $42.1 billion in 1987, $66.9 billion in 1997, and $82.0 billion in 2003.[10]

We assume the tax is levied on coal at the mine mouth, natural gas at the well head, and on petroleum products at the refinery. Imported fossil fuels are also subject to the tax. As noted above, we assume in all cases that the tax is passed forward to consumers in the form of higher energy prices. Metcalf (2007) estimates that a tax of $15 per metric ton of CO_2 applied to average fuel prices in 2005 would nearly double the price of coal, assuming the tax is fully passed forward. Petroleum products would increase in price by nearly 13% and natural gas by just under 7%. The tax is also passed on indirectly to other industries that use these energy sources as inputs.

The procedure for evaluating the effect of a carbon tax as it is passed through the economy is discussed in detail in Fullerton (1995) and Metcalf (1999). We provide a summary of the methodology in this chapter's Appendix. The starting point for the analysis is the use of Input-Output (I-O) matrices available from the Bureau of Economic Analysis. In particular, we use the Summary Make and Use matrices from the I-O tables for 1987, 1997, and 2003. The Make matrix shows how much each industry makes of each commodity; the Use matrix shows how much of each commodity is used by each industry. Using these two matrices for each year, we derive an industry-by-industry transactions matrix that enables us to trace the use of inputs by one industry by all other industries. Various adding-up identities along with assumptions about production and trade allow the accounts to be manipulated to trace through the impact of price changes (and taxes) in one industry on the products of all other industries in the economy.

For each year, we cluster the industry groups provided in the I-O tables into 60 categories. For 2003, we separate out aggregate mining into two separate groups—mining and coal mining—using the split provided in the 2002 benchmark I-O files. We do a similar split to break out electricity and natural gas from other utilities. This was not a problem for the 1987 and 1997 benchmark I-O files, where these splits already existed.

Once we obtained the effect of the tax on prices of consumer goods, we used data from the Consumer Expenditure Survey (CEX) to compute energy taxes paid by each household in the survey. The CEX contains data on household income and expenditures for numerous consumption goods. We combine commodities to work with 42 categories of personal consumption items. Having computed the average price increase for each industry using the I-O tables, we translate those price increases into corresponding price increases for these consumer items. This is also discussed in detail in the Appendix, where we provide tables showing the recorded price increases in each year for each consumer item as a result of the tax.

Results

Carbon Tax Burden of Gasoline

Tables 2-1A and 2-1B present our results for incidence using annual income as our measure of economic welfare. We have grouped households by annual income and sorted the households into ten income deciles from the poorest

10% of the population to the richest 10%. Confirming conventional wisdom, the gasoline tax is quite regressive when measured relative to current income for all three years. The burden in the lowest decile in 1997 and 2003, for example, is over four times the burden in the top decile when measured as a fraction of annual income.

Table 2-1A shows that the overall burden distribution changes marginally across the 16-year period when annual income is used to rank households. The burden has increased marginally in the bottom decile but declined for all other deciles between 1987 and 2003. For the top income household, the burden of the tax falls from about 0.23% of income to about 0.15% of income. On average, taxes paid as a fraction of income have declined over the 16-year period. This in part reflects the greater energy efficiency of the economy. Aggregate energy intensity in the United States (measured as energy consumption relative to real GDP) fell by 23% between 1987 and 2003.[11]

The overall picture that emerges from Table 2-1A is that had a carbon tax been in effect in 1987, 1997, and 2003, the gasoline burden of the tax would have looked quite regressive using annual income as a measure of household well-being. Using an annual income approach, the regressivity is increasing slightly over this time period.

As an interesting aside, we show the total incidence of the carbon tax across income deciles in 2003 in Table 2-1B. To study whether there are any qualitative or quantitative differences in the gasoline portion of the tax relative to the overall carbon tax, we present Figure 2-1. Figure 2-1 essentially scales the burden for each income group by the average across all income groups. For instance, the average carbon tax across all income groups is about 1.88%. Therefore, for the bottom decile (and all other deciles) we plot the ratio of their burden scaled by 1.88. This should show whether the gasoline burden is

TABLE 2-1A Distribution of Carbon Tax Burden on Gasoline: Annual Income

Decile	1987	1997	2003
Bottom	0.69	0.66	0.70
Second	0.57	0.51	0.49
Third	0.58	0.56	0.47
Fourth	0.62	0.47	0.45
Fifth	0.48	0.41	0.40
Sixth	0.44	0.35	0.35
Seventh	0.44	0.32	0.29
Eighth	0.36	0.30	0.26
Ninth	0.31	0.23	0.22
Top	0.23	0.16	0.14

Source: Authors' calculations. The table reports the within decile average ratio of gasoline tax burdens to income.

TABLE 2-1B Distribution of Total Carbon Tax Burden 2003: Annual Income

Decile	2003
Bottom	3.74
Second	3.06
Third	2.36
Fourth	2.06
Fifth	1.76
Sixth	1.53
Seventh	1.3
Eighth	1.23
Ninth	1.01
Top	0.81

Source: Authors' calculations. The table reports the within decile average ratio of carbon tax burdens to income.

relatively higher or lower than the carbon tax burden for each income group. One interesting pattern that emerges is that the relative carbon tax burden seems to be higher than the average and the gasoline burden lower than the average for the bottom income deciles. The picture is reversed at the top, where the gasoline burden is higher and the carbon tax burden is lower than the average for all incomes. Therefore, richer households pay a larger fraction of the gasoline tax than they pay of the total carbon tax.

Turning to the measures of incidence using consumption as a proxy for lifetime income, the results change dramatically. Table 2-2 shows the distribution of the gasoline tax in the three years when households are sorted by current

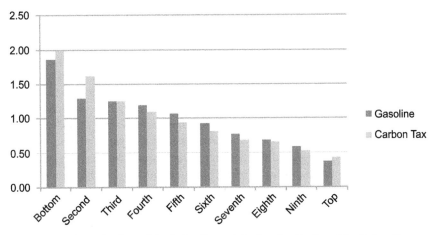

FIGURE 2-1 Relative Burdens of the Carbon Tax and the Gasoline Portion of the Carbon Tax across Income Deciles, 2003.

TABLE 2-2 Distribution of Carbon Tax Burden on Gasoline: Current Consumption.

Decile	1987	1997	2003
Bottom	0.38	0.29	0.30
Second	0.41	0.29	0.31
Third	0.40	0.32	0.29
Fourth	0.42	0.32	0.32
Fifth	0.41	0.29	0.28
Sixth	0.40	0.29	0.25
Seventh	0.35	0.27	0.26
Eighth	0.34	0.24	0.23
Ninth	0.32	0.21	0.20
Top	0.25	0.17	0.15

Source: Authors' calculations. The table reports the within decile average ratio of carbon tax burdens on gasoline to current consumption.

consumption. Now we find that the total tax is less regressive, with the ratio of average taxes paid by the bottom and the top varying from about 1.5 to 2.0 across the three years. The primary force driving this result is the tendency for consumption to be more evenly distributed than income, especially when one looks at the lower brackets.[12]

This result was similarly reported in Bull et al. (1994), who found that the lifetime calculation changed the results because the proportion of energy in total consumption (or ratio of energy consumption to income) varied significantly over the life cycle, with elderly low income individuals, in particular, having relatively large current energy consumption.

Turning to the measures of average tax burdens using lifetime consumption (Table 2-3), we see that this correction flattens the distribution even more. The incidence of the total tax is nearly proportional, with the ratio of burden in the lowest to top decile varying from 1.2 to 1.3 across the three years.

In summary, incidence calculations based upon annual income imply much steeper regressivity than do calculations based upon lifetime income measures. Moreover, the inter-temporal variation in incidence is reduced substantially using measures based on lifetime consumption rather than those using income. We suspect this occurs in large part because transitory income shocks exacerbate the apparent regressivity of the tax when measured against income.

We next turn to a regional analysis of the incidence of the tax. Policymakers are often concerned that a tax might disproportionately burden one region or part of the country at the expense of another. To measure the geographic burden of the tax, we group households by region and measure their average tax rate using weighted averages of the tax burdens.[13] Results are shown in Table 2-4.[14] Variation in the average tax rates is modest. The maximum difference in the average rate across regions is close to 0.30 percentage points in

TABLE 2-3 Distribution of Carbon Tax Burden on Gasoline: Lifetime Consumption.

Decile	1987	1997	2003
Bottom	0.40	0.27	0.21
Second	0.38	0.28	0.24
Third	0.36	0.28	0.29
Fourth	0.39	0.29	0.23
Fifth	0.42	0.29	0.30
Sixth	0.40	0.30	0.26
Seventh	0.40	0.30	0.25
Eighth	0.42	0.32	0.23
Ninth	0.34	0.27	0.20
Top	0.30	0.22	0.17

Source: Authors' calculations. The table reports the within decile average ratio of carbon tax burdens on gasoline to lifetime consumption.

TABLE 2-4 Distribution of Carbon Tax Burden on Gasoline: Annual Income.

Region	1987	1997	2003
New England	0.30	0.29	0.27
Mid-Atlantic	0.35	0.29	0.25
South Atlantic	0.46	0.37	0.31
East South Central	0.45	0.44	0.38
East North Central	0.42	0.31	0.30
West North Central	0.49	0.32	0.30
West South Central	0.51	0.46	0.36
Mountain	0.46	0.40	0.40
Pacific	0.43	0.33	0.39

Source: Authors' calculations. The table reports the within region average ratio of carbon tax burdens on gasoline to income. Regions are defined in footnote 14.

1987 and 0.20 percentage points in 1997. A closer look at the data reveals that the high regional burden for the Mountain and Pacific regions, for instance, may be due to the relatively higher need for transportation to travel long distances, relative to other regions.

State Tax Burden of Gasoline

To calculate the incidence of current state taxes on gasoline, we collected information on gasoline prices (inclusive of tax) and taxes for the most recent year of our analysis, 2003. Information on these variables for earlier years was missing for several states for 1997 and was unavailable for 1987. However, the results for 2003 should give a clear picture of how environmental taxes on gasoline would affect households across income levels and regions. We obtained data on gasoline prices and taxes from the U.S. Department of Energy as well as the International Fuel Tax Association.

The average gas tax across all regions (unweighted) was approximately 20 cents per gallon in 2003, and the average gasoline price was approximately $1.51 per gallon. From this we calculated the proportionate increase in price induced by the gas tax by taking the ratio of the gasoline tax in each region to the gasoline price in that region. These calculations are shown in Appendix Table 2-A2. The average proportionate increase in price due to the gasoline tax was approximately 13% across regions. We then applied these price increases to the expenditure data in the CEX to calculate the incidence.

Tables 2-5 and 2-6 show the results of our calculations. The current state taxes on gasoline are marginally less regressive when measured as a fraction

TABLE 2-5 Distribution of Burden of State Gasoline Taxes: 2003.

Decile	Annual Income	Current Consumption	Lifetime Consumption
Bottom	0.92	0.37	0.33
Second	0.77	0.50	0.36
Third	0.74	0.48	0.42
Fourth	0.74	0.53	0.38
Fifth	0.65	0.47	0.50
Sixth	0.57	0.43	0.45
Seventh	0.50	0.42	0.46
Eighth	0.43	0.39	0.40
Ninth	0.37	0.35	0.34
Top	0.25	0.28	0.29

Source: Authors' calculations. The table reports the within decile average ratio of gasoline tax burdens to income.

TABLE 2-6 Regional Distribution of Burden of State Gasoline Taxes: 2003.

Region	Annual Income
New England	0.51
Mid-Atlantic	0.48
South Atlantic	0.47
East South Central	0.61
East North Central	0.63
West North Central	0.52
West South Central	0.65
Mountain	0.70
Pacific	0.60

Source: Authors' calculations. The table reports the within region average ratio of gasoline tax burdens to income. Regions are defined in footnote 14.

of income than carbon taxes on gasoline would be. The burden on the lower income groups is nearly four times that on the top. The picture improves as we use current consumption and lifetime consumption (income) to measure the incidence. With lifetime incomes, the burden on the lowest decile is about 0.33% of income and that on the top about 0.29% of income. Turning to the regional incidence, the burden is fairly modest and evenly distributed, averaging about 0.56% of income.

Conclusion

This chapter measures the incidence of carbon taxes on gasoline using a lifetime incidence framework. We analyze the household burden of a $15 per metric ton tax on CO_2 in constant 2005 dollars at three different points in time. In particular, the focus is on the impact of such a tax on gasoline prices and the subsequent burden on consumers. The burden is measured ranking households by current income, current consumption, and lifetime consumption as the basis for the incidence measures. The methodology involves first working with the economy-wide Input-Output tables from the Bureau of Economic Analysis to assess how the $15 tax would affect the industrial sector, in particular the prices of energy goods and other industrial goods in which these energy goods serve as inputs. We then use this information to calculate the increase in prices of consumer goods as a result of the tax. Once we obtain the price increase in 42 categories of consumer goods, we calculate the burden of the tax on households using consumption data from the Consumer Expenditure Survey.

As the chapter discusses, energy taxes have different incidence effects across the life cycle. Therefore, it is important to measure the burden of taxes in terms of lifetime incidence, not just their burden in a given year. To take account of the lifetime incidence, we use two proxies. First we use current consumption following the work of Poterba (1989). Second we use lifetime-corrected consumption introduced in Bull et al. (1994) and explained in detail in the Appendix to that paper.

Our results suggest that when the total lifetime effect of a carbon tax is taken into account, the regressivity of the tax decreases. This is particularly true when we use lifetime-corrected consumption to rank households, rather than current consumption. In addition to looking at the economic incidence of the tax, we studied the incidence of the tax across regions. These data show that the variation across regions is fairly modest with marginally higher values driven by the relatively higher use of gasoline in the Mountain and Pacific regions.

These results carry through to our analysis of current state taxes on gasoline consumption. The tax is fairly regressive when using current income to measure incidence but less regressive using lifetime income measures. Further, the regional burden of the tax is fairly evenly distributed.

Appendix 2-1. A Lifetime-Corrected Incidence Measure

This section presents our derivation of the lifetime-corrected incidence measure. The starting point is the assumption that the expected consumption and income streams for people with vastly different human capital stocks are likely to be very different. Hence we first classify people into different subsamples based on their education levels. Within each education class, we then calculate average consumption for people in different age groups. The "typical" path of lifetime consumption can therefore be proxied by the average consumption for each age group within an education class, starting from the youngest to the oldest. Using this path of lifetime consumption, we then derive the present value of lifetime consumption for each education class.

Once we have the present value of lifetime consumption for each education class, we can use this to derive the lifetime consumption for any individual in the sample. For instance, suppose that an individual is a 35-year-old Ph.D. whose current consumption is 80% of the average for her age and education group. If the present value of lifetime consumption for a Ph.D. is $1 million, then her lifetime consumption is calculated as 80% of $1 million, or $800,000. We can use a similar technique to derive lifetime measures of income and direct and indirect taxes for every individual in the sample. Below, we present the mathematical derivation of our lifetime measures.

For each observation in the CEX sample, we know the age and education level. We first assign each observation to a particular education group, e, where e goes from 1 to 11 (1 = no schooling, 11 = doctorate). Then within the education group, we calculate the average of consumption for different age groups, a. We include all individuals above age 20. Let $\overline{C}^a(e)$ be the average consumption for all individuals in education group e and age a. (We can similarly define $\overline{Y}^a(e)$, $\overline{D}^a(e)$ and $\overline{I}^a(e)$, respectively for income, direct taxes, and indirect taxes).

Let $\hat{C}(e)$ denote the present value of lifetime consumption for all individuals in the same education group. This is derived as the present discounted value of average consumption across all age groups within the same education group, as shown below.

$$\hat{C}(e) = \sum_{a=a_o}^{A} \left(\frac{1}{1+r}\right)^{(a-a_o)} \overline{C}^a(e)$$

where $a_o = 20$, A is the highest age group and $r = 5\%$

To compute the lifetime consumption for each individual, we take the ratio of their actual consumption, $C_i^a(e)$, to the average for their age and education group and then multiply this ratio by the computed present value of lifetime consumption for their education class. Therefore,

TABLE 2-A1 Consumer Goods Price Increases as a Result of the Carbon Tax.

	CEX categories	1987	1997	2003
1	food at home	0.57%	0.65%	0.70%
2	food at restaurants	0.47%	0.56%	0.58%
3	food at work	0.61%	0.75%	0.86%
4	tobacco	0.54%	0.60%	0.67%
5	alcohol at home*	0.47%	0.56%	0.58%
6	alcohol on premises**	0.47%	0.56%	0.58%
7	clothes	0.53%	0.52%	0.40%
8	clothing services	0.75%	0.38%	0.41%
9	jewelry	0.54%	0.45%	0.43%
10	toiletries	0.87%	0.85%	0.72%
11	health and beauty	0.70%	0.38%	0.42%
12	tenant occupied non-farm dwellings	0.14%	0.21%	0.31%
13	other dwelling rentals	0.65%	0.41%	0.42%
14	furnishings	0.64%	0.62%	0.55%
15	household supplies	0.78%	0.77%	0.71%
16	electricity	10.18%	13.15%	12.55%
17	natural gas	16.87%	16.61%	12.28%
18	water	1.20%	0.73%	0.63%
19	home heating oil	7.67%	10.33%	9.56%
20	telephone	0.20%	0.21%	0.26%
21	domestic services	0.70%	0.41%	0.49%
22	health	0.44%	0.37%	0.39%
23	business services	0.14%	0.21%	0.50%
24	life insurance	0.28%	0.21%	0.31%
25	automobile purchases	0.67%	0.59%	0.90%
26	automobile parts	0.70%	0.64%	0.65%
27	automobile services	0.75%	0.34%	0.40%
28	gasoline	9.67%	7.64%	7.73%
29	tolls	0.65%	0.30%	0.64%
30	automobile insurance	0.14%	0.21%	0.31%
31	mass transit	0.95%	0.70%	0.90%
32	other transit	0.96%	0.50%	0.62%
33	air transportation	1.93%	1.82%	1.86%
34	books	0.43%	0.32%	0.34%
35	magazines	0.45%	0.31%	0.49%
36	recreation and sports equipment	0.52%	0.56%	0.42%
37	other recreation services	0.60%	0.36%	0.51%
38	gambling	0.39%	0.28%	0.31%
39	higher education	0.56%	0.27%	0.30%
40	nursery, primary and secondary education	0.60%	0.33%	0.34%
41	other education services	0.62%	0.26%	0.30%
42	charity	0.74%	0.43%	0.41%

Notes: * Values for alcohol have been set equal to the value for food at restaurants in each year.

      ** These price increases are calculated using a $15 per metric ton carbon dioxide tax.

TABLE 2-A2 Regional Gasoline Taxes as a Fraction of Regional Gasoline Prices (2003).

	Regions	Tax Inclusive Price (cents)	Tax (cents)	Tax as a Share of Final Price
1	New England	155	22	0.143
2	Mid-Atlantic	150	22	0.144
3	South Atlantic	141	16	0.117
4	East South Central	143	17	0.122
5	East North Central	151	24	0.161
6	West North Central	148	20	0.134
7	West South Central	142	20	0.138
8	Mountain	158	21	0.133
9	Pacific	168	20	0.116

$$\widehat{C}_i^a(e) = \frac{C_i^a(e)}{\overline{C}^a(e)} \widehat{C}(e)$$

We can compute similar measures for income ($\widehat{Y}_i^a(e)$), direct ($\widehat{D}_i^a(e)$), and indirect taxes ($\widehat{I}_i^a(e)$).

Finally, we can compute the lifetime-corrected incidence of direct carbon taxes as a proportion of consumption as follows;

$$\lambda_{D_i}^a(e) = \frac{\widehat{D}_i^a(e)}{\widehat{C}_i^a(e)}$$

Similar measures can be computed for the lifetime-corrected incidence of indirect and total carbon taxes.

To compute the incidence across income or consumption groups, we can repartition the sample into deciles of lifetime income or consumption and then take the average across each decile of the incidence for members of that decile.

Notes

1   The rest is about 24% diesel, 8% jet fuel, and 2% natural gas.
2   For our paper on the distributional consequences of cap-and-trade, see Hassett et al. (2009), "The Consumer Burden of a Cap-and-Trade Program," AEI Working Paper #144.
3   Note that this experiment is different than simply raising the gasoline tax. Coal and natural gas are used (indirectly through electricity) in the production of gasoline. A carbon tax raises the price of these fuels while a gasoline tax would not (except to the extent that higher gasoline prices raise the price of coal and natural gas).

4   Parry and Small (2005) for a discussion of why a portion of gasoline taxes may be considered Pigovian.

5   This refers only to the carbon tax induced burden of the gasoline tax, since our analysis for state taxes refers only to 2003.

6   Pechman (1985) realized that income data for the low-income groups suffered from substantial income mismeasurement. Since then, the approach adopted by him and several others, including us in this chapter, is to discard the bottom half of the lowest decile, i.e., to only look at the bottom 5% to 10% of the bottom decile, rather than the entire 10%.

7   The Panel Study of Income Dynamics (PSID) has good data on income but lacks detailed data on consumption, which we would require. Other authors have used it to study the lifetime incidence of alcohol and cigarette taxes (Lyon and Schwab 1995) and gasoline taxes (Chernick and Reschovsky 1992).

8   For those wishing to use our research as a jumping off point for more disaggregated CGE modeling, all data and programs are available at www.aei.org/carbontax.

9   The effect of a different tax rate on consumer goods prices can be approximated by appropriately grossing up (or down) Appendix Table 2-A1.

10 An analysis by the Energy Information Administration suggests that a $15 tax on CO_2 would reduce emissions by about 5% in the short run (see EIA 2006).

11 See Metcalf (2008) for an analysis of the determinants of changes in energy intensity.

12 This relationship is well known in the literature. Krueger and Perri (2002) for example, document this fact.

13 As with the distributional tables across income, we drop the bottom 5% of the income distribution from the analysis before carrying out the regional analysis.

14 The states in each region are as follows: New England: Maine, New Hampshire, Vermont, Massachusetts, Connecticut, Rhode Island; Mid-Atlantic: New Jersey, New York, Pennsylvania; South Atlantic: West Virginia, Virginia, North Carolina, South Carolina, Georgia, Florida, District of Columbia, Maryland, Delaware; East South Central: Kentucky, Tennessee, Missouri, Alabama, Mississippi; East North Central: Wisconsin, Illinois, Michigan, Indiana, Ohio; West North Central: North Dakota, South Dakota, Nebraska, Kansas, Minnesota, Iowa; West South Central: Texas, Oklahoma, Arkansas, Louisiana; Mountain: Montana, Idaho, Wyoming, Nevada, Utah, Colorado, Arizona, New Mexico; Pacific: California, Oregon, Washington, Alaska, Hawaii.

CHAPTER 3

Distributional and Efficiency Impacts of Increased U.S. Gasoline Taxes[*]

Antonio M. Bento, Lawrence H. Goulder,
Mark R. Jacobsen, and Roger H. von Haefen

For several reasons, reducing automobile-based gasoline consumption is a major U.S. public policy issue. Gasoline use generates environmental externalities. In 2004, approximately 22% of U.S. emissions of carbon dioxide—the principal anthropogenically sourced "greenhouse gas" contributing to global climate change—derived from gasoline use. Other environmental externalities from gasoline combustion include the impacts from emissions of several "local" air pollutants such as carbon monoxide, nitrogen oxides, and volatile organic compounds. Reduced gasoline use could lead to improved air quality and associated benefits to health.[1,2] In addition, gasoline consumption accounts for 44% of the U.S. demand for crude oil, and the nation's dependence on crude oil makes the U.S. vulnerable to changes in world oil prices emanating from disruptions in the world oil market. Some analyses claim that this vulnerability is not accounted for in individual consumption decisions and thus represents another externality from gasoline consumption.[3] The various externalities provide a potential rationale for public policy oriented toward gasoline consumption.

Recently, analysts and policymakers have called for new or more stringent policies to curb gasoline consumption. Under the Obama administration U.S. fuel economy standards are being raised from their 2010 level of 27.5 miles per gallon for cars and 23.5 miles per gallon for light trucks to a new combined standard of 35 miles per gallon effective in 2016. The 2005 Energy Bill includes tax credits for households purchasing relatively fuel-efficient vehicles such as hybrid cars. The California State Assembly enacted AB 1493, which mandates carbon dioxide emissions that would require significant improvements in automobile fuel economy. Other proposals include subsidies to retirements of older (gas-guzzling) vehicles and increments to the federal gasoline tax.[4]

This chapter examines the gas tax option, employing an econometrically-based multi-market simulation model to evaluate the policy's efficiency and distributional implications. We investigate the impacts of increased U.S.

gasoline taxes on fuel consumption, relating these impacts to changes in fleet composition (shifts to higher mileage automobiles) and vehicle miles traveled (VMT). We also evaluate the economy-wide costs of higher gasoline taxes and explore how the costs are distributed across households that differ by income, region of residence, race, and other characteristics. We consider how the distribution of impacts depends on the ways revenues from the tax are returned to the private sector.

Some prior studies have examined the impact of gasoline taxes by estimating the demand for gasoline as a function of gasoline price and household income. For example, Jerry A. Hausman and Whitney K. Newey (1995) use household-level data on gasoline consumption to estimate deadweight loss from gasoline taxes, while Sarah E. West and Roberton C. Williams III (2005; see Chapter 4) use such data to assess the distributional impacts of gasoline taxes and the optimal gasoline tax.

Other studies infer the demand for gasoline from automobile choice and utilization models. For example, James Berkovec (1985), Fred L. Mannering and Clifford Winston (1985), Kenneth E. Train (1986), and West (2004) estimate the household's discrete automobile purchase decision and its continuous choice of VMT. Following Jeffrey A. Dubin and Daniel L. McFadden (1984), these authors account for the connections between these two choices, although the cross-equation restrictions implied by a unified structural model of behavior are not imposed.

A third set of studies focuses on supply-side phenomena—in particular, the impacts of policies on new car production and the composition of the automobile fleet and the associated effect on gasoline consumption. In contrast with the previously mentioned studies, this third set considers explicitly the imperfectly competitive nature of the new car market and the pricing behavior of new car producers. For example, Steven T. Berry, James Levinsohn and Ariel Pakes (1995), Pinelopi K. Goldberg (1998), and David H. Austin and Terry M. Dinan (2005) develop models of the new car market that combine supply decisions by imperfectly competitive producers with discrete demand choices by households. The latter two studies explore impacts of automobile policies on the new car market. Goldberg (1998) and Andrew M. Kleit (2004) analyze tighter CAFE standards; Austin and Dinan (2005) examine CAFE standards and a gasoline tax increase.

The present study differs from earlier work in several ways. First, in contrast with nearly all prior work,[5] this analysis considers supply and equilibrium not only in the new car market but in the used car and scrap markets as well. The wider scope helps provide a more complete picture of the impact of a gasoline tax. In addition, addressing the equilibrium in all three car markets enables us to capture important dynamic effects. Higher gasoline taxes are likely to cause an increase in the share of relatively fuel-efficient cars among new cars sold. The extent to which the fuel-efficiency of the *overall* (new and used car) fleet improves will depend on the rate at which the newer, more efficient cars replace older cars. This depends on the relative size of the stocks of new and used cars and the rate at which older cars are taken out of operation (scrapped). By considering the new, used, and scrapped car markets, the model is able to consider the dynamics of changes in fleet composition and related short- and long-run impacts on gasoline consumption. As in Goldberg (1995), Berry et al. (1995), Amil K. Petrin (2002), and Austin and Dinan (2005), we consider the imperfectly competitive nature of the new car market. However, in contrast with these studies, we connect this market to the used and scrap markets. This allows us to consider how policies affect the entire fleet of cars and associated demands for gasoline.

A second major difference from earlier work is the model's ability to capture distributional effects. The model considers over 20,000 households that differ in terms of income, family size, employment status (working or retired), region of residence, and ethnic background. This enables us to trace distributional impacts in several important dimensions. All household demands stem from a consistent utility maximization framework, enabling us to measure distributional impacts in terms of theoretically sound welfare indexes. Prior studies have examined distributional effects by focusing on how gasoline expenditure shares differ across income groups.[6] In contrast, the present model considers not only the expenditure-side impacts but also the ways that the government's disposition of gas tax revenue influences the distribution of policy impacts.

Finally, the model differs in its econometric approach to estimating consumer demand for automobiles, VMT, and gasoline. Berkovec (1985), Mannering and Winston (1985), Goldberg (1998), and West (2004) account for the connections between the automobile purchase and use (VMT) decisions by employing sequential, two-step estimators. Their approach accounts for correlations between the discrete and continuous choice margins but ignores the cross-equation restrictions implied by a unified behavior model. In contrast, we adopt a full-information, one-step structural approach that simultaneously estimates these choice dimensions within a utility-theoretic framework that permits us to recover sound welfare estimates.[7] In addition, we assume that all parameters entering preferences vary randomly across households. Random coefficients allow us to account for correlations in the unobservable factors influencing a household's discrete car choice and continuous VMT demand while simultaneously allowing for more plausible substitution patterns among automobiles (McFadden and Train 2000; Bunch et al. 1996).

The rest of the chapter is organized as follows. The next section describes the equilibrium simulation model. We then present outlines of the model's data sources, with emphasis on the data employed to estimate household demands for vehicles and travel. The next section presents our approach for estimating households' automobile purchase and driving decisions. The subsequent section presents and interprets results from simulations of a range of gasoline tax policies. The final section offers conclusions.

Model Structure

The economic agents in the model are households, producers of new cars, used car suppliers, and scrap firms. The model considers the car-ownership and VMT decisions of 20,429 households. The ownership and VMT decisions are made simultaneously in accordance with utility maximization.

The model distinguishes cars according to age, class, and manufacturer. Table 3-1 displays the different car categories, which comprise 350 distinct cars, of which 284 appear in our data set and simulation.[8]

The used-car market equates the supply of used cars remaining after scrapping with the demand for ownership of those cars. Producers of new cars decide on new car prices in accordance with Bertrand (price) competition. These producers consider households' demand functions in determining optimal pricing. Price markups reflect the various price elasticities of demand for cars as well as the regulatory constraints posed by existing CAFE standards.

The model solves for a sequence of market equilibria at one-year intervals. Car vintages are updated each year, so that last year's new cars become one-year-old cars, last year's one-year-old cars become two-year-old cars, etc. Once a car is scrapped, it cannot re-enter the used car market. Characteristics of given models of new cars change through time (as described further in the simulation results section). In particular, producers change the fuel economy of new models in a manner consistent with profit maximization.

Household Demands

Households obtain utility from car ownership and use, as well as from consumption of other commodities. The utility from driving depends on characteristics of the automobile as well as VMT. Each household has exogenous

TABLE 3-1 Included Car Types.

Classes	Age categories	Manufacturers
Compact	New cars	Ford
Luxury compact	1–2 years old	Chrysler
Midsize	3–6 years old	General Motors
Fullsize	7–11 years old	Honda
Luxury mid/fullsize	12–18 years old	Toyota
Small SUV		Other Asian
Large SUV		European
Small truck		
Large truck		
Minivan		

income; most households also are endowed with cars. If a household has a car endowment, it chooses whether to hold or relinquish (sell or scrap) that car; if it relinquishes the car it also decides whether to purchase a different car (new or used). If a household does not have a car endowment, it chooses whether to purchase a car.

If household i owns car j, its utility can be expressed by:

$$U_{ij} = U_{ij}(z_j, M_i, x_i) \tag{3.1}$$

where z_j is a vector of characteristics of car j and M_i and x_i respectively refer to household i's VMT and its consumption of the outside (or Hicksian composite) good. The household's utility conditional on choosing car j can be expressed through the following indirect utility function:

$$V_{ij} = V_{ij}' + \mu_i \varepsilon_{ij} \tag{3.2}$$

with

$$V_{ij}' = V_{ij}' \left(y_i - r_{ij}, p_{ij}^M, p_{ix}, z_j, z_i, z_{ij} \right) \tag{3.3}$$

where

y_i = income to household i
r_{ij} = rental price of car j to household i
p_{ij}^M = per-mile operating cost
p_{ix} = price of the outside good, x
z_i = vector of characteristics of household i
z_{ij} = vector of characteristics of household i, interacted with characteristics of car j

Household income y_i is devoted toward purchasing a car (or cars),[9] car operation, and the purchase of the outside good. We treat car purchases as rentals, so that payments are spread over many years. The household budget constraint can then be written as:

$$y_i = r_{ij} + p_{ij}^M M_i + p_{ix} x_i \tag{3.4}$$

If a household owns a vehicle, the stream of rental income from that vehicle is included in its income. A household that chooses to retain its existing car effectively makes a rental payment equal to its implicit rental income from that car. Income also includes the household's share of profits to new car producers, government transfers, and capital gains or losses resulting from changes in automobile prices.[10] The government transfer component of income includes revenue from the gasoline tax and adjusts as policy changes.

The operating cost p_{ij}^M includes the fuel cost (including gasoline taxes) as well as maintenance and variable insurance costs. The rental price r_{ij} accounts for depreciation, registration fees and fixed insurance costs. As indicated in expression (3.2) above, indirect utility includes the random component $\mu_i \varepsilon_{ij}$, where ε has a type I extreme-value distribution (following the econometric model) and μ is a scale parameter. We assume the household chooses the vehicle (or vehicles) yielding the highest conditional utility, given V' and the random error. The probability that a given car j maximizes utility for household i is:

$$\exp\left(\frac{V_{ij}'}{\mu_i}\right) \bigg/ \sum_k \exp\left(\frac{V_{ik}'}{\mu_i}\right) \tag{3.5}$$

The indirect utility function V_{ij} can be differentiated following Roy's identity to yield the optimal choice of miles traveled, M_{ij}, conditional on the purchase of car j. Aggregate automobile and VMT demand are the sum of these micro decisions. In specifying aggregate demand for automobiles, we treat each individual in our sample as a representative of a subpopulation of like individuals and sum up the probabilities. Similarly for aggregate VMT demand, we sum up each individual's probability-weighted VMT demand for each car.

Supply of New Cars

Each of the seven producers in the model sets prices for its fleet of automobiles to maximize profits, given the prices set by its competitors and subject to fleet fuel economy constraints. Thus we assume Bertrand competition. Producers face less than perfectly elastic demands for their cars: that is, two new cars of the same class can sell at different prices if produced by different firms.

The producer problem accounts for the presence of CAFE standards. These standards require that each manufacturer's fleet-wide average fuel economy be above a certain level in each of two general categories of cars: "light trucks" and "passenger cars." The classes in the passenger car category are non-luxury compact, non-luxury midsize, non-luxury fullsize, luxury compact, and luxury midsize/fullsize. Those in the light truck category are small truck, large truck, small SUV, large SUV/van and minivan.[11]

In the following, the subscript k refers to the cars made by a particular manufacturer. The boldface vector p includes prices of the cars made by all seven manufacturers.[12] T and C denote the sets of cars (for a given manufacturer) in the light truck and passenger car categories, respectively. $\bar{e}_T$ and $\bar{e}_C$ refer to the efficiency requirements for light trucks and passenger cars and e_k is the fuel economy of car k. A given producer chooses a vector of prices, p_k, and a vector of individual-model fuel economies, e_k, to maximize profit:

$$\max_{\{p_k, e_k\}} \sum_k \left(p_k - c_k(e_k)\right) \cdot q_k(p, e) \tag{3.6}$$

subject to:

$$\frac{\sum_{k \in C} q_k}{\sum_{k \in C} \frac{q_k}{e_k}} \geq \overline{e}_C \text{ and } \frac{\sum_{k \in T} q_k}{\sum_{k \in T} \frac{q_k}{e_k}} \geq \overline{e}_T$$

where p_k and c_k refer to the purchase price and marginal cost, respectively, of a particular car and q_k is the demand as a function of all prices.[13] For any given model k, marginal cost is a function of e_k, the chosen level of fuel economy for that car. Each producer's solution to (3.6) determines the quantities of vehicles sold in each class. Producers can alter fuel economy and the mix of vehicles but cannot introduce new vehicle classes, exit existing vehicle classes, or alter attributes like weight and horsepower that determine class.

The solution to (3.6) requires a demand function (which is given within the model by the sum of individual demands from (3.5)) and a cost function. To identify the cost function parameters, we employ data on automobile markups, prices and quantities sold, along with our estimated household demand elasticities for different automobiles. The relationship between production cost and fuel economy is taken from engineering estimates of the incremental costs of fuel economy from the National Research Council (2002). These relationships pose the technological and cost constraints under which producers in the model choose optimal levels of fuel economy. Details are provided in the Appendix (available online at www.aeaweb.org/aer/data/june09/20071186_app.pdf).

We must solve the constrained optimization problem for all of the firms simultaneously since the residual demand curve faced by a given firm depends on the prices set by the others. The solution method is discussed below.

Used Car and Scrap Markets

The Used Car Market

In the model, "used car" refers to all cars that are neither new nor scrapped. The available supply of used cars of a particular vintage (i.e., model year) is the total stock of that vintage operating in the previous year less those that are scrapped. The total supply of all used cars in the current period is the aggregate supply from the previous period net of scrappage, plus an increment to the supply representing the cars that were new in the previous period. Let ℓ refer to a given manufacturer and class of vehicle. For each manufacture-class category ℓ, the quantity of used cars evolves according to:

$$q^U_{\ell,t+1} = (1 - \theta_\ell) q^U_{\ell,t} + q^N_{\ell,t} \tag{3.7}$$

where $q^U_{\ell,t}$ and $q^N_{\ell,t}$ refer to the quantity of used and new cars of the manufacturer-class combination ℓ available in year t and θ_ℓ represents the average probability that used cars of type ℓ are scrapped. This scrap rate will depend on the car's expected resale value if kept in operation. We discuss the specification of θ_ℓ in the next section.

Each used car type, or age-manufacturer-class combination, has a different rental price. The model determines the set of rental prices that clears the used car market, that is, that causes every car to be sold. Household demands for used cars come from household demands computed as in (3.5). Since the demand for a given used car will depend on the rental prices of all used cars (and on new car prices), all used car rental prices need to be solved simultaneously.

The Scrap Market

We assume that households will scrap a car when the scrap value exceeds the resale value. However, each car (age-manufacturer-class combination) in our model actually represents a group of cars of varying quality and value, some of which may fall under the cutoff for scrapping even if the average car in the group does not. To allow for scrapping of some cars of a given type, we assign a scrap probability to each car. The scrap decision depends on p_j, the purchase price or resale value of a used car. Today's purchase price is the discounted sum of future rental prices, adjusted for the possibility that a car will be scrapped (and earning no rental price) before reaching each progressively older age. The household has myopic expectations: it assumes that future rental values will be the same as the current-period rental values of older vintages of the same vehicle type. Changes in the gasoline tax affect scrap decisions through their effects on purchase prices.[14] When this value changes as a result of a change in the gasoline tax, so does the probability of scrap.

In terms of the resale value for each used car, the scrap probability θ_j is modeled simply as:

$$\theta_j = b_j \cdot \left(p_j\right)^{\eta_j} \tag{3.8}$$

where b_j is a scale parameter used for calibration and η_j is the elasticity controlling the change in scrap probability as the price of the car changes. Scrap rates increase with car age and are calibrated to 0.05, 0.06, 0.09, and 0.20 for the four categories of used cars. These values are derived from the distribution of car age in the data (see Appendix for details).

Solution Method

To solve the model, we must obtain the full vector of new and used car rental prices for a particular year that satisfies the following conditions: (3.1) every available (not scrapped) used car has a buyer (or retainer) and (3.2) for every new car producer, the first-order conditions for constrained profit maximization are satisfied.[15] Note that the second requirement is a function of all prices, not just new-car prices. We determine overall demands for a given car by aggregating across households their probability-weighted demands for that car.

The solution method embeds the used car problem within the broader problem of solving for both used and new car prices. Specifically, we solve for the used car prices that satisfy requirement (3.1), conditional on a set of posited prices for the new cars. We then adjust the new car prices in an attempt to meet condition (3.2) and solve again for used car prices that meet requirement (3.1) conditional on the adjusted new car prices. We repeat this procedure until conditions (3.1) and (3.2) are met within a desired level of accuracy.[16]

The government's revenue from gasoline taxes is returned to households according to the various "recycling" methods described in the simulation results section. Government revenues and transfers are mutually dependent: the level of transfers affects household demands and government revenues, while the level of revenues determines the transfer level consistent with the government's budget constraint. Thus, solving the model also requires that we determine the equilibrium level of government revenue and transfers. The overall solution is a set of prices for each car that simultaneously clears all markets and an aggregate transfer level that equals the government's revenues from the gasoline tax. To solve the multidimensional system we use Broyden's method, a derivative-based quasi-Newton search algorithm.

Data

Our data set has two main components: 1) a random sample of U.S. households' automobile ownership choices from the 2001 National Household Travel Survey (NHTS) and 2) new and used automobile price and non-price characteristics from *Ward's Automotive Yearbook*, The National Automobile Dealer's Association (NADA) *Used Car Guide*, and the Department of Energy (DOE) fueleconomy.org website. By merging these two types of information, we obtain an unusually rich data set, one that allows us to consider household choices among a wide range of new and used cars and that permits us to distinguish households along many important dimensions. In this chapter's Appendix, we offer details on how we merged the data sets and constructed needed variables.

The NHTS Sample

The 2001 NHTS consists of 26,038 households living in urban and rural areas of the United States. With the help of Department of Transportation staff, we obtained the confidential NHTS data files containing relevant data for our analysis. For each household we have information on income, automobile holdings (by make, model, and year) and VMT. In addition, we have data on the household's demographic characteristics (including household size, composition, gender, education, and employment status) and geographical identifiers (including the state, metropolitan statistical area, and zip code of residence).

After cleaning the data our final sample consists of 20,429 households from the original 26,038. Table 3-2 presents major demographic statistics of our final sample.

The Automobile Sample

The 1983–2002 *Ward's Automobile Yearbook* provided most of the car and truck characteristics used in our analysis. Automobile characteristics include horsepower, weight, length, height, width, wheelbase, and city and highway miles per gallon (MPG) by make, model, and year for all cars and trucks sold during this period. We obtained information on car and truck prices from the National Automobile Dealer's Association (NADA), which publishes this information in the monthly *NADA Used Car Guide*. We used price information from the April 2001 and 2002 editions of the *Car Guide*, which we obtained in electronic format. Each edition contained the manufacturer's suggested retail price and current resale price (a weighted average of recent transaction prices) for all new and used cars and trucks dating back to 1983. As indicated in the Appendix, we calculated depreciation based on changes in prices for a given car over the 2001–2002 period.

Combining information from the *Ward's* and NADA data sets yielded a vector of prices and various automobile characteristics for roughly 4,500 automobiles distinguished by manufacturer, model, and year. We aggregated these data into the seven manufacturer categories, ten class categories, and five age categories in Table 3-1. We used a weighted geometric mean formula to aggregate price and non-price characteristics within each make, class, and age category, where the weights were proportional to the holdings frequencies in the *NHTS*.

Table 3-3 displays statistics on miles per gallon, horsepower, and rental price from our data. The data show significant MPG differences across classes and age categories. A new compact, for example, is 1.48 times more efficient than a large SUV. The newest compacts yield 1.47 more miles per gallon than those in the oldest age category. In contrast, the newest midsize and large SUVs are less fuel-efficient than the older models. As for horsepower, most of the increases apply to compacts and fullsize cars. Average horsepower of compacts increased 60%, and average horsepower of fullsize cars rose 75%. Differences in rental price are most substantial for new cars, due to the particularly rapid depreciation of new luxury vehicles. Older cars have much lower rental prices and these prices are more similar across classes.

TABLE 3-2 Sample Demographic Statistics from the 2001 NHTS: 20,429 Observations.

Variable	Mean (SD)
Household size	2.490 (1.34)
# of adults ≥ 18 years old	1.861 (0.69)
# of adults ≥ 65 years old	0.380 (0.67)
# of children ≤ 2 years old	0.096 (0.32)
# of children 3–6 years old	0.136 (0.41)
# of children 7–11 years old	0.185 (0.49)
# of children 12–17 years old	0.211 (0.54)
# of workers	1.272 (0.95)
# of females	1.033 (0.52)
Average age among adults (≥ 18)	49.56 (16.8)
Household income (2001 $s)	56,621 (43,276)

Household breakdown:	Percentage
1 male adult, no children, not retired	5.71
1 female adult, no children, not retired	7.88
1 adult, no children, retired	10.3
2+ adults w/average age ≤ 35, no children, not retired	7.10
2+ adults w/average age > 35 and ≤ 50, no children, not retired	8.43
2+ adults w/average age > 50, no children, not retired	9.04
2+ adults w/average age ≤ 67, no children, retired	9.29
2+ adults w/average age > 67, no children, retired	8.47
1+ adults w/youngest child < 3 years old	8.69
1+ adults w/youngest child 3–6 years old	7.65
1+ adults w/youngest child 7–11 years old	8.64
1+ adults w/youngest child 12–17 years old	8.85
White household respondent[a]	85.6
Black household respondent	7.62
Hispanic household respondent	6.25
Asian household respondent	2.17
Adults with high school diplomas	89.4
Adults with four-year college degrees	30.5
Resident of MSA[b] < 250k	7.62
Resident of MSA 250–500k	8.22
Resident of MSA 500k–1m	8.30
Resident of MSA 1–3m	22.2
Resident of MSA > 3m	32.5
Non-resident of MSA	21.1
Household income ≤$25,000	22.8
Household income ≤$50,000 and >$25,000	33.3
Household income ≤$75,000 and >$50,000	19.8
Household income >$75,000	24.1

[a] The white, black, Hispanic, and Asian percentages sum to more than 100 percent because some respondents have multicultural backgrounds. [b] MSA = metropolitan statistical area.

TABLE 3-3 Automobile Characteristics.

Characteristic	Compact	Luxury Compact	Midsize	Fullsize	Luxury Mid/Full	Small SUV	Large SUV/Van	Trucks and Minivans	Total
Miles Per Gallon[a]									
All Car Ages:	29.73	24.18	27.16	25.57	23.65	23.75	20.04	22.19	24.39
	(27.8, 35.6)	(22.2, 26.9)	(24.2, 31.0)	(22.6, 30.5)	(21.3, 25.0)	(17.8, 27.0)	(16.6, 26.8)	(16.8, 27.7)	(16.6, 35.6)
Model Years:									
2001–2002	30.29	24.47	26.90	25.61	23.70	24.17	19.08	21.51	24.15
	(28.0, 32.8)	(22.9, 26.9)	(24.2, 30.5)	(23.0, 28.0)	(23.0, 24.2)	(21.9, 26.4)	(17.2, 22.5)	(16.8, 25.9)	(16.8, 32.8)
1999–2000	30.32	24.45	27.29	25.79	23.86	23.80	18.21	22.07	24.18
	(28.1, 35.6)	(23.1, 26.8)	(25.1, 29.7)	(22.6, 28.0)	(23.0, 24.4)	(19.8, 27.0)	(16.7, 19.6)	(18.8, 26.3)	(16.7, 35.6)
1995–1998	30.02	24.24	27.50	25.51	24.29	23.44	19.60	22.01	24.44
	(28.4, 32.1)	(22.3, 26.4)	(25.4, 29.8)	(23.0, 27.8)	(23.3, 25.0)	(19.6, 26.3)	(16.6, 23.7)	(17.7, 27.2)	(16.6, 32.1)
1990–1994	29.21	23.81	26.74	25.37	22.91	22.67	20.90	21.80	24.08
	(27.8, 30.4)	(22.2, 26.3)	(25.2, 30.0)	(23.5, 28.8)	(21.3, 24.0)	(17.8, 24.9)	(17.2, 26.0)	(17.6, 26.0)	(17.2, 30.4)
1983–1989	28.82	23.94	27.38	25.56	23.23	24.84	22.88	23.75	25.14
	(28.2, 29.4)	(22.6, 26.1)	(24.3, 31.0)	(23.8, 30.5)	(22.1, 24.3)	(23.3, 26.3)	(18.1, 26.8)	(20.0, 27.7)	(18.1, 31.0)
Horsepower/100									
All Car Ages:	1.286	2.275	1.530	1.726	2.177	1.531	1.909	1.665	1.719
	(0.88, 1.78)	(1.56, 3.63)	(0.98, 1.96)	(0.86, 2.21)	(1.42, 2.81)	(1.02, 1.95)	(0.88, 2.59)	(0.94, 2.79)	(0.86, 3.63)
Model Years:									
2001–2002	1.526	2.621	1.787	2.123	2.463	1.763	2.391	2.023	2.036
	(1.34, 1.78)	(1.64, 3.63)	(1.65, 1.96)	(1.97, 2.21)	(2.13, 2.81)	(1.65, 1.95)	(2.15, 2.59)	(1.40, 2.79)	(1.34, 3.63)
1999–2000	1.454	2.488	1.682	1.917	2.376	1.648	2.271	1.920	1.932
	(1.23, 1.68)	(1.70, 3.45)	(1.58, 1.80)	(1.50, 2.07)	(2.10)	(1.45, 1.88)	(2.12, 2.52)	(1.34, 2.63)	(1.23, 3.45)

continued

TABLE 3-3 (*Cont.*)

Characteristic	Compact	Luxury Compact	Midsize	Fullsize	Luxury Mid/Full	Small SUV	Large SUV/Van	Trucks and Minivans	Total
1995–1998	1.342	2.414	1.597	1.835	2.237	1.554	2.024	1.633	1.773
	(1.09, 1.47)	(1.75, 3.38)	(1.47, 1.72)	(1.41, 2.07)	(2.01, 2.53)	(1.35, 1.83)	(1.86, 2.17)	(1.09, 2.06)	(1.09, 3.38)
1990–1994	1.152	2.075	1.418	1.469	1.952	1.467	1.476	1.430	1.516
	(1.05, 1.24)	(1.60, 2.54)	(1.28, 1.54)	(0.90, 1.74)	(1.83, 2.11)	(1.29, 1.59)	(0.90, 1.77)	(1.07, 1.78)	(0.90, 2.54)
1983–1989	0.955	1.777	1.166	1.212	1.637	1.164	1.244	1.243	1.270
	(0.88, 1.03)	(1.56, 2.15)	(0.98, 1.41)	(0.86, 1.36)	(1.42, 2.01)	(1.02, 1.27)	(0.88, 1.46)	(0.94, 1.51)	(0.86, 2.15)
Rental Price/1000									
All Car Ages:	2.570	5.959	2.749	3.029	5.680	3.141	4.289	3.149	3.681
	(0.38, 6.84)	(0.55, 26.6)	(0.38, 8.55)	(0.39, 8.67)	(0.45, 21.4)	(0.42, 7.81)	(0.43, 14.4)	(0.26, 8.32)	(0.26, 26.6)
Model Years:									
2001–2002	5.798	15.94	6.528	7.463	14.45	6.823	10.27	6.750	8.792
	(5.14, 6.84)	(7.23, 26.6)	(5.65, 8.55)	(6.84, 8.67)	(11.8, 21.4)	(6.12, 7.81)	(7.92, 14.4)	(4.78, 8.32)	(4.78, 26.6)
1999–2000	3.258	6.819	3.274	3.628	5.712	3.724	4.566	3.850	4.237
	(2.14, 4.24)	(3.74, 12.6)	(2.10, 4.72)	(3.13, 4.52)	(3.99, 8.69)	(3.11, 4.35)	(2.20, 7.69)	(2.91, 5.24)	(2.10, 12.6)
1995–1998	2.320	4.506	2.420	2.521	3.823	2.884	3.638	2.842	3.051
	(1.62, 3.27)	(2.59, 5.72)	(1.68, 3.18)	(2.06, 3.17)	(2.54, 5.61)	(2.20, 3.58)	(2.53, 5.66)	(1.94, 3.68)	(1.62, 5.72)
1990–1994	0.972	1.679	1.015	1.019	1.317	1.259	1.253	1.118	1.186
	(0.72, 1.29)	(1.11, 2.34)	(0.73, 1.33)	(0.75, 1.26)	(0.86, 1.79)	(0.98, 1.74)	(0.69, 2.04)	(0.74, 1.51)	(0.69, 2.34)
1983–1989	0.503	0.850	0.509	0.491	0.714	0.589	0.676	0.514	0.585
	(0.38, 0.67)	(0.55, 1.31)	(0.38, 0.67)	(0.39, 0.64)	(0.45, 1.21)	(0.42, 0.82)	(0.43, 1.31)	(0.26, 0.73)	(0.26, 1.31)

Notes: Minimum and maximum values reported in parentheses. The categories small truck, large truck, and minivan have been aggregated in this table.
[a] Weighted harmonic mean of EPA test miles per gallon estimates.

Calculation of Rental Prices and Per-Mile Operating Costs

Two important variables we must construct from our data are the automobile rental prices and per-mile operating costs (the "price per mile" variable in the section on model structure) for all 284 autos. The underlying inputs into these prices and costs differ by region as well as automobile type. For household i owning car j, the rental price is given by:

$$r_{ij} = D_j + 0.85I_{ij}^A + F_{ij} + R \cdot p_j \qquad (3.9)$$

where

$\quad D_j \quad$ = depreciation in the real value of car j
$\quad I_{ij}^A \quad$ = household i's annual insurance costs for car j
$\quad F_{ij} \quad$ = household i's automotive registration fees for car j
$\quad R \quad$ = real interest rate

Thus, the one-year rental price of a car is the sum of depreciation, insurance, and registration costs, plus the forgone real return on the principal value of the car.[17] For the real interest rate, R, we use a value of 3.89%, the 2001 average daily real rate on 30-year T-bills. We include insurance costs in both the rental price (associated with the choice of car) and the per-mile operating cost (associated with VMT). Representatives from State Farm Insurance suggested to us that roughly 85% of auto insurance premiums are fixed and independent of VMT. Hence, 85% of insurance costs appear in the rental price formula, while the remainder is allocated to operating costs.

The rental prices are included in the household utility function relative to the price of the outside good (cost of living) faced by each household. We incorporate a cost of living index for 363 distinct regions that, together with differences in insurance and registration fees, reflects variation across households in the effective rental price of vehicles.[18]

The per-mile operating cost, p_{ij}^M, is expressed by:

$$p_{ij}^M = \frac{p_i^{gas}}{MPG_j} + N_j + 0.15I_{ij}^M \qquad (3.10)$$

where

$\quad p_i^{gas} \quad$ = household i's per gallon price of gasoline
$\quad MPG_j \quad$ = miles per gallon for car j
$\quad N_j \quad$ = per-mile maintenance and repair costs for car j
$\quad I_{ij}^M \quad$ = household i's per-mile insurance costs for car j

The price of gasoline (and therefore operating cost) varies among households based on differences across 363 distinct regions of residence. The average after-tax gasoline price faced by households in 2001 ranged from $1.19 (Albany, Georgia) to $1.86 per gallon (San Francisco).

Estimation of Household Ownership and Utilization Decisions

The Econometric Model

Challenges

Two overarching concerns influenced our approach to estimating household automobile demand. The first was our desire to integrate consistently the car ownership and utilization decisions. Such integration is crucial for generating consistent estimates of welfare costs from gasoline taxes. The second concern arose from an important feature of the data: households frequently own more than one car. In the 2001 *NHTS*, 41.5% of households own zero or one car, another 43.6% own two cars, and the remaining 14.9% own three or more autos. This implies that many households have a potentially enormous number of auto bundles from which to choose. If, for example, there are J different cars and trucks and we consider only bundles consisting of no more than two cars, there are $1 + J + J(J + 1)/2$ bundles that households can potentially choose. With our automobile data set consisting of 284 composite cars and trucks, there are 40,755 distinct bundles that households might choose (and this large number ignores all bundles with three or more autos).[19]

As discussed in the chapter introduction, nearly all past efforts to integrate automobile ownership and utilization decisions have relied on reduced-form, sequentially-estimated models. Our structural approach estimates simultaneously the decisions on both margins. To account for different households owning different quantities of cars, we adopt a variation of Igal E. Hendel (1999) and Jean-Pierre H. Dubé's (2004) repeated discrete-continuous framework. In the context of automobile choice, the framework assumes that a household's ownership and utilization choices arise from separable choice occasions. On each choice occasion, the household makes a discrete choice of whether to own one of J automobiles. If an auto is chosen, the household conditionally decides how much to drive it during the year. To account for ownership of multiple automobiles, households have multiple choice occasions on which different automobile services may be demanded. Intuitively, different choice occasions in our framework correspond to different primary tasks or purposes for which households might demand automobile services (e.g., commuting to work, family travel, shopping excursions, or any combination thereof). We assume their number depends on the number of adults in a given household.[20]

Our approach to modeling automobile demand has advantages and drawbacks. Its main advantages are that it consistently links ownership and utilization decisions and reduces the dimension of the households' choice set on a given choice occasion to $J + 1$ alternatives (J autos and the no auto alternative).

The latter feature makes our approach econometrically tractable with our 284 composite auto data set. It also has the virtue of allowing for households to own several cars. A main drawback is that it does not allow for interaction effects among the fleet of autos held by households—for example, a four-person household's utility from holding a second minivan being less than holding a single minivan. To account for such interactions, one would need to regard bundles of automobiles, rather than individual cars, as the objects of choice. However, as suggested above, such an approach would require *substantially* more aggregation of cars beyond what we have pursued.[21] This would rule out significant product differentiation and thus severely limit our ability to account for the imperfectly competitive nature of the automobile industry. In addition, it would compel us to put a limit of two on the number of cars owned by any household, which would eliminate from our sample those households likely to be most affected by changes in gasoline taxes.

Specifics

Our repeated discrete-continuous model of automobile demand works as follows. Household i $(i = 1,\ldots,N)$ is assumed to have a fixed number of choice occasions, T_i. We let T_i equal the number of adults in each household plus one.[22] On choice occasion t, household i is assumed to have preferences for car j $(j=1,\ldots,J)$ that can be represented by the following conditional indirect utility function:

$$V_{itj} = V'_{ij} + \mu_i \varepsilon_{ijt}$$

$$V'_{ij} = \frac{-1}{\lambda_i} \exp\left(-\lambda_i \left(\frac{y_i / T_i - r_{ij}}{p_{ix}}\right)\right) - \frac{1}{\beta_{ij}} \exp\left(\alpha_{ij} + \beta_{ij} \frac{p_{ij}^M}{p_{ix}}\right) + \tau_{ij}$$

$$\alpha_{ij} = \tilde{\alpha}_i^\top z_{ij}^\alpha$$

$$\beta_{ij} = -\exp\left(\tilde{\beta}_i^\top z_{ij}^\beta\right) \qquad\qquad (3.11)$$

$$\lambda_i = \exp\left(\tilde{\lambda}_i^\top z_i^\lambda\right)$$

$$\tau_{ij} = \tilde{\tau}_i^\top z_{ij}^\lambda$$

$$\mu_i = \exp\left(\mu_i^\cdot\right)$$

where $(y_i, r_{ij}, p_{ij}^M, p_{ix})$ are household i's income, rental price for the jth auto, utilization (or VMT) price for the jth car, and the Hicksian composite commodity price, respectively; $(z_{ij}^\alpha, z_{ij}^\beta, z_{ij}^\tau)$ are alternative automobile characteristics (including make, age, and class dummies that control for unobserved attributes)[23] interacted with household demographics; z_i^λ contains just household characteristics; $(\tilde{\alpha}_i, \tilde{\beta}_i, \tilde{\lambda}_i, \tilde{\tau}_i, \mu_i^\cdot)$ are parameters that vary randomly across households; and ε_{ijt} contains additional unobserved heterogeneity that varies randomly across households, automobiles and choice occasions.[24] If the household instead decides not to rent a car (i.e., automobile 0), its conditional indirect utility function is:

$$V_{it0} = \frac{-1}{\lambda_i} \exp\left(-\lambda_i \left(\frac{y_i / T_i}{p_{ix}} \right) \right) + \phi_i^T z_i^\phi + \mu_i \varepsilon_{i0t} \tag{3.12}$$

where z_i^ϕ and ϕ_i are individual characteristics and parameters, respectively. The rational household is assumed to choose the alternative that maximizes its utility on each choice occasion. Assuming each ε_{ijt} $(j = 0, ..., J)$ can be treated as independent draws from the normalized type I extreme value distribution, the probability that individual i chooses alternative j on choice occasion t condition on the model's structural parameters is:

$$PR_{it}(j) = \frac{\exp\left(V_{ij}' / \mu_i \right)}{\sum_k \exp\left(V_{ik}' / \mu_i \right)} \tag{3.13}$$

Assuming the household chooses automobile j, Roy's identity implies that the household's conditional VMT demand is:

$$M_{itj} = \exp\left(\alpha_{ij} + \beta_{ij} \left(\frac{p_{ij}^M}{p_{ix}} \right) + \lambda_i \left(\frac{y_i / T_i - r_{ij}}{p_{ix}} \right) \right) \tag{3.14}$$

We assume the analyst imperfectly observes M_{itj} due to measurement error in our data.[25,26] The analyst observes $\tilde{M}_{itj} = M_{itj} + \eta_{itj}$, where η_{itj} is an independent draw from the normal distribution with mean zero and standard deviation $\sigma_i = \exp(\sigma_i^*)$.[27] The likelihood of observing $\tilde{M}_{itj}$ conditional on the model parameters is:

$$l\left(\tilde{M}_{itj} \mid j \text{ chosen}, j \neq 0 \right) = \frac{1}{(2\pi)^{1/2} \sigma_i} \exp\left(-\frac{1}{2} \left(\frac{\tilde{M}_{itj} - M_{itj}}{\sigma_i} \right)^2 \right) \tag{3.15}$$

Given our assumed structure, the full likelihood of household i's automobile demand conditional on $\delta = \left(\tilde{\alpha}_i, \tilde{\beta}_i, \tilde{\lambda}_i, \tilde{\tau}_i, \phi_i, \mu_i^*, \sigma_i^* \right)$ is then:

$$L_i = \prod_{t=1}^{T_i} \left[\prod_{j=0}^{J} PR_{it}(j)^{1_{itj}} \prod_{j=1}^{J} l\left(\tilde{M}_{itj} \mid j \text{ chosen} \right)^{1_{itj}} \right] \tag{3.16}$$

where 1_{itj} is an indicator function equal to one if car j is chosen on individual i's tth choice occasion and zero otherwise.

Estimation Strategy

Past econometric efforts to model vehicle ownership and derived VMT demand decisions have used variations of Dubin and McFadden's (1984) sequential estimation strategy that accounts for the induced selectivity bias in derived VMT demand with a Heckman-like (1979) correction factor. We employ a full-information estimation approach that accounts for correlations in the unobserved determinants of choice across discrete and continuous dimensions through random parameters (McFadden and Train 2000). Intuitively, random parameters allow unobserved variations in taste to influence automobile ownership decisions and VMT demand decisions. We allow all parameters, $\delta = \left(\tilde{\alpha}_i, \tilde{\beta}_i, \tilde{\lambda}_i, \tilde{\tau}_i, \phi_i, \mu_i^*, \sigma_i^* \right)$, to be distributed multivariate normal with mean $\bar{\delta}$ and variance-covariance matrix Σ_δ. This approach is more general than earlier random coefficient discrete-continuous applications (e.g., King 1980; Feng et al. 2005) that include only one random parameter. The more general specification offers a far richer degree of unobserved preference heterogeneity to influence household ownership and use decisions than previous applications.[28]

Given the nonlinear nature of our likelihood function, the large number of households and sites in our data set, and the potentially large number of parameters on which we wish to draw inference, classical estimation procedures such as maximum simulated likelihood (Gourieroux and Monfort 1996) would be exceptionally difficult, if not impossible, to implement. In light of these computational constraints, we adopt a Bayesian statistical perspective and employ a variation of Greg M. Allenby and Peter J. Lenk's (1994) Gibbs sampler estimation procedure that is less burdensome to implement in our application.[29]

The Bayesian framework assumes that the analyst has initial beliefs about the unknown parameters $(\bar{\delta}, \Sigma_\delta)$ that can be summarized by a prior probability distribution, $f(\bar{\delta}, \Sigma_\delta)$. When the analyst observes a set of choices x, she combines this choice information with the assumed data-generating process to form the likelihood of x conditional on alternative values of $(\bar{\delta}, \Sigma_\delta)$, $L(\mathbf{x} \mid \bar{\delta}, \Sigma_\delta)$. The analyst then updates her prior beliefs about the distribution of $(\bar{\delta}, \Sigma_\delta)$ to form a posterior distribution for $(\bar{\delta}, \Sigma_\delta)$ conditional on the data, $f(\bar{\delta}, \Sigma_\delta \mid \mathbf{x})$. By Bayes' rule, $f(\bar{\delta}, \Sigma_\delta \mid \mathbf{x})$ is proportional to the product of the prior distribution and likelihood, i.e., $f(\bar{\delta}, \Sigma_\delta \mid \mathbf{x}) = f(\bar{\delta}, \Sigma_\delta)L(\mathbf{x} \mid \bar{\delta}, \Sigma_\delta)/C$ where C is a constant. In general, $f(\bar{\delta}, \Sigma_\delta \mid \mathbf{x})$ will not have an analytical solution, and thus deriving inference about the moments and other relevant properties of $(\bar{\delta}, \Sigma_\delta)$ conditional on the data is difficult. However, Bayesian econometricians have developed a number of Markov Chain Monte Carlo (MCMC) procedures to simulate random samples from $f(\bar{\delta}, \Sigma_\delta \mid \mathbf{x})$ and in the process draw inference about the posterior distribution of $(\bar{\delta}, \Sigma_\delta)$.

Following Allenby and Lenk (1994), we specify diffuse priors for $(\bar{\delta}, \Sigma_\delta)$ and use a Gibbs sampler with an adaptive Metropolis-Hastings component to simulate from $f(\overline{\delta}, \Sigma_\delta \mid x)$. By decomposing the parameter space into disjoint sets and iteratively simulating from each set conditionally on the others, the Gibbs sampler generates simulations from the unconditional posterior distribution after a sufficiently long burn-in. The implementation details of the algorithm are described in the Appendix.

One further dimension of our estimation approach is worth noting. Because of the large number of households in our data set ($N = 20,429$) and our desire to account for differences in automobile demand across different household types (e.g., single males, two-adult households with and without children, retired couples), we stratified the sample into 12 groups based on demographic characteristics and estimated separate models within each strata. In addition to decomposing a computationally burdensome estimation problem on a large data set into a series of more manageable estimation problems on smaller data sets, stratification allows us to better account for observable and unobservable differences among households.

Empirical Results

For all 12 strata, we obtain precisely estimated posterior mean values for $(\overline{\delta}, \Sigma_\delta)$.[30] Many of the parameters that are common across the 12 strata vary in magnitude considerably, suggesting significant preference heterogeneity across the different subpopulations. We also find that the diagonal elements of Σ_δ are generally large, suggesting considerable preference heterogeneity within each stratum as well. The latter preference heterogeneity and the highly nonlinear structure of our preference function mean that the estimated parameters do not have a simple economic interpretation. Thus instead of focusing on the estimated parameters, we examine the various elasticities that they imply. We display these elasticities in Table 3-4, broken down by household and automobile types. Our cross-section estimation implies that these should be interpreted as long-run elasticities.

The first column of Table 3-4 reports the elasticity of gasoline use with respect to gasoline price. In the "All" and "By Household" panels, the elasticities allow for responses in both VMT and car choice (and associated fuel economy). In the "By Auto" panel, the elasticities are conditional on car choice. Across all households and cars, we obtain a mean estimate of –0.35. The estimated elasticities are larger for families with children and owners of trucks and SUVs. D. J. Graham and Stephen Glaister's (2002) survey of past studies indicates long-run elasticities in the U.S. ranging from –0.23 to –0.80. Kenneth A. Small and Kurt Van Dender's (2007) more recent state-level analysis produces a central estimate of –0.33.

The second column of the table shows the elasticity of gasoline use with respect to income. On average, we find estimates of around 0.76. The elasticity is highest for families with children and owners of new vehicles. Graham and Glaister report long-run estimates in the range of 1.1 to 1.3.[31]

TABLE 3-4 Posterior Mean Long-run Elasticity Estimates

	Elasticity of gasoline use wrt price[a]	*Elasticity of gasoline use wrt income*[a]	*Car ownership elasticity wrt rental price*	*VMT elasticity wrt operating cost*[a]
All	−0.35	0.76	−0.82	−0.74
By Household				
Retired	−0.32	0.61	−0.93	−0.69
Not retired, no children	−0.32	0.68	−0.72	−0.69
Not retired, with children	−0.39	0.96	−0.85	−0.83
By Auto				
By Class				
All Cars				
Compact	−0.27	0.83	−0.65	−0.59
Luxury compact	−0.30	0.78	−1.25	−0.64
Midsize	−0.28	0.74	−0.67	−0.60
Fullsize	−0.29	0.75	−0.73	−0.63
Luxury midsize/ fullsize	−0.30	0.79	−1.25	−0.63
Small SUV	−0.29	0.93	−0.73	−0.63
Large SUV/van	−0.32	0.88	−0.98	−0.69
Small truck	−0.34	0.78	−0.62	−0.72
Large truck	−0.31	0.79	−0.85	−0.66
Minivan	−0.31	0.85	−0.77	−0.65
New Cars				
Compact	−0.28	1.14	−1.44	−0.60
Luxury compact	−0.27	0.76	−3.14	−0.46
Midsize	−0.29	0.95	−1.58	−0.60
Fullsize	−0.29	1.04	−1.77	−0.61
Luxury midsize/ fullsize	−0.28	0.83	−3.04	−0.47
Small SUV	−0.26	1.86	−1.58	−0.55
Large SUV/van	−0.34	1.06	−2.30	−0.69
Small truck	−0.37	0.91	−1.32	−0.75
Large truck	−0.32	1.05	−1.69	−0.65
Minivan	−0.31	0.98	−1.67	−0.63
By Age				
New cars	−0.30	1.10	−1.97	−0.63
1–2 year old cars	−0.29	0.79	−1.01	−0.63
3–6 year old cars	−0.27	0.76	−0.73	−0.59
7–11 year old cars	−0.30	0.75	−0.28	−0.65
12–18 year old cars	−0.31	0.83	−0.13	−0.68

[a] Elasticities in the "By Auto" panel are conditional on car choice.

The third column reports car ownership elasticities with respect to the own rental price. For new cars, rental price elasticities should track purchase price elasticities if rental and purchase prices vary proportionally. Our results imply mean rental price elasticities of –0.88 for all vehicles and –1.97 for new vehicles only. Luxury cars, large SUVs, and large trucks, which have the highest rental prices, have the highest rental price elasticities among automobile classes.

Our estimated elasticities with respect to rental prices are smaller in absolute magnitude than those found in some studies, such as Berry et al. (1995), which obtained elasticities ranging from –3 to –4.5. A plausible explanation is that the objects of choice in our study are not individual make-models but automobile composites (i.e., make-model combinations aggregated by age, class, and make). This aggregation implies that we have only 59 new cars in our data set, not the 200 to 300 cars typically found in other applications. By collapsing make-models into composite cars, we reduce the number of channels for substitution.

Much of the work estimating these elasticities has focused exclusively on new vehicles (e.g., Berry et al. 1995 and Petrin 2002). These studies have generally employed multiple years of automobile sales data and controlled for the potential endogeneity of price arising from unobserved car characteristics.[32] In contrast, we have a single household-level cross-section and control for unobserved product characteristics through class, make, and age dummies that vary across household types.

A close comparison with data types and results from other studies suggests that our smaller elasticities might also reflect limitations from cross-sectional data and are not likely due to a failure to adequately control for unobserved car characteristics. Our elasticities are similar in magnitude to those reported in Berry et al. (2004) and Train and Winston (2007), studies that employ household-level cross-sectional data and control for price endogeneity with alternative specific constants and instrumental variables.[33] In addition, Goldberg (1995), using five years of household-level data and an identification strategy comparable to ours, finds elasticities for make-models that are similar in magnitude to Berry et al. (1995).[34] Combined, these results suggest that our estimates control sufficiently for endogeneity of price but may reflect limitations from the cross-section of data used.

The final column of Table 3-4 reports long-run VMT elasticities with respect to operating costs. Across all households and cars, the average elasticity is –0.74. This elasticity is lower for new cars than for older vehicles. Older cars are disproportionately owned by lower income households, who exhibit higher VMT elasticities. Because gasoline makes up slightly less than half of per mile operating costs, our average estimate implies an average VMT elasticity with respect to the price of gasoline of –0.34. In their survey, Graham and Glaister report that from prior studies the average estimate for this long-run elasticity is –0.30, while Small and Van Dender report an estimate closer to –0.1. Both pairs of authors note that existing estimates are quite sensitive to the data and modeling assumptions employed, and thus the caveats mentioned earlier concerning the limitations of cross-sectional data may apply here as well. Past applications that (like ours) use disaggregate household data to control for endogenous vehicle choice tend to find larger elasticities than aggregate time-series or panel data studies that combine household and commercial demand for highway VMT (Mannering 1986).

Simulation Results

Assumptions Underlying the Simulation Dynamics

The simulation model generates a time path of economic outcomes over ten years at one-year intervals. As mentioned, the model solves in each period for the market-clearing new and used car prices. We assume that household incomes grow at an annual rate of 1%. In all simulations, the pre-tax price of gasoline is $1.04 and is taken as exogenous and unchanging over time.[35]

Baseline Simulation

The baseline simulation offers a reference scenario with which we compare the outcomes from various gasoline tax policies. Consistent with historical trends, we assume in this simulation that automobile horsepower and weight increase at an annual rate of 5%. In our central case we calibrate the baseline fuel economy technology to the "Path 1" assumptions of the National Research Council (2002) regarding improvements in fuel economy: over a 10-year period, such improvements range from 11% for compacts to 20% for large SUVs. As part of a sensitivity analysis below, we adopt in the baseline the more optimistic NAS "Path 3" assumptions regarding growth in baseline fuel economy technology. In our policy simulations, producers adjust fuel economy away from these baseline technologies following equation (3.6) above.

Table 3-5 displays the equilibrium quantities of new and used cars under the baseline simulation. Our reference case overpredicts the size of the vehicle fleet by about 8%, ranging from 5% for midsize cars to 14% for small trucks.

Impacts of Gasoline Tax Increases under Alternative Recycling Methods

Here we present results from simulations of permanent increases in gasoline taxes. We start by focusing on the impacts of a tax increase of 25 cents per gallon (other tax increases are considered below) under the following alternative ways of recycling the additional revenues from the tax increase:

- *"Flat" recycling:* Revenues are returned in equal amounts to every household.
- *"Income-based" recycling:* Revenues are allocated to households according to each household's share of aggregate income.

TABLE 3-5 Baseline Fleet Composition.

	Year 1			Year 10		
	New	*Used*	*All cars in operation*	*New*	*Used*	*All cars in operation*
Class						
Compact	4.98	44.68	49.66	5.27	49.52	54.79
Lux compact	0.22	4.44	4.66	0.26	2.79	3.05
Midsize	2.63	27.58	30.21	2.82	27.30	30.12
Fullsize	1.32	16.32	17.64	1.49	14.64	16.13
Lux mid/full	0.32	8.30	8.62	0.39	4.67	5.06
Small SUV	1.32	10.65	11.97	1.41	12.99	14.40
Large SUV	1.10	15.93	17.02	1.30	12.92	14.23
Small truck	1.27	10.26	11.54	1.35	12.25	13.60
Large truck	2.17	19.83	22.00	2.42	22.16	24.58
Minivan	1.32	12.74	14.06	1.45	13.62	15.07
Total	16.65	170.73	187.39	18.15	172.87	191.03

Note: Units are millions of privately owned cars in operation.

- *"VMT-based" recycling*: Revenues are allocated lump-sum according to each household's share of aggregate vehicle miles traveled in the baseline.

Recycling could be accomplished by the government's mailing rebate checks to households on an annual basis. The shares of total revenues going to different households depend on baseline values and thus do not depend on behavioral responses to the gasoline tax.

Aggregate Impacts

Gasoline Consumption

Table 3-6 presents the impacts of this policy on gasoline consumption. In the short run (year 1), the percentage reduction is about 5.1% under flat and income-based recycling and about 4.5% under VMT-based recycling. Compared with other recycling methods, VMT-based recycling gives a larger share of gasoline tax revenue to car owners, who tend to have larger income elasticities of gasoline use than those who do not own cars. As a result, there is a larger offsetting income effect on gasoline use under VMT-based recycling than under other recycling methods and the overall reduction in gasoline consumption is smaller.

The percentage change in gasoline use is approximately equal to the percentage change in miles traveled (VMT) minus the percentage improvement in fuel economy (miles per gallon). Table 3-6 shows the contributions of these

TABLE 3-6 Change in Gasoline Use with 25-Cent Tax Increase.

Recycling Method	Flat		Income–based		VMT–based	
	Year 1	Year 10	Year 1	Year 10	Year 1	Year 10
Baseline gasoline use per household (gallons)	775.18	828.89	775.18	828.89	775.18	828.89
% change in gasoline use	−5.09	−4.99	−5.06	−5.07	−4.51	−4.40
% change in VMT	−5.01	−4.84	−4.98	−4.93	−4.43	−4.21
% change in VMT per car	−4.62	−4.37	−4.56	−4.38	−4.01	−3.69
% change in cars in operation	−0.41	−0.49	−0.44	−0.57	−0.44	−0.54
% change in overall MPG	0.08	0.16	0.08	0.15	0.09	0.20

two components. Most of the reduction in gasoline use comes from the reduction in VMT: the improvements in fleet-wide fuel economy are fairly small.[36]

In the short run, the major channel for improved aggregate fuel economy is an increase in the scrapping rate for vehicles with unusually low fuel economy. The augmented gasoline tax raises per-mile operating costs, which makes vehicles with low fuel economy relatively less desirable, causing their demand and prices to fall and their scrap rates to rise. In the first year of the policy, an additional 160,000 used large trucks and large SUVs are scrapped. Over the longer term, average fuel economy is influenced by changes in fleet composition attributable to increased relative sales of new cars that are more fuel-efficient and by price-induced increases in fuel economy of given models. The percent increase in fuel economy is larger in the long run, although fuel economy improvements still account for a small share of the overall reduction in gasoline consumption.

Table 3-7 summarizes the changes in fleet composition. On impact, the higher gasoline tax causes a shift away from cars (more cars are scrapped)

TABLE 3-7 Fleet Size and Composition

	Baseline[a]		25-cent gasoline tax increase[b]					
			Flat recycling		Income-based recycling		VMT-based recycling	
	Year 1	Year 10	Year 1	Year 10	Year 1	Year 10	Year 1	Year 10
Cars in operation:								
All	188.3	191.0	−0.41%	−0.49%	−0.44%	−0.57%	−0.44%	−0.54%
New	16.7	18.2	−1.00%	−0.08%	−1.12%	−0.38%	−0.93%	−0.07%
Used	171.6	172.8	−0.35%	−0.53%	−0.37%	−0.59%	−0.39%	−0.59%
Low MPG	75.9	78.9	−0.47%	−0.81%	−0.50%	−0.82%	−0.49%	−0.77%
High MPG	112.4	112.1	−0.37%	−0.26%	−0.40%	−0.39%	−0.40%	−0.38%

[a] Millions of cars. [b] Percent change relative to the baseline.

and, among cars that remain in operation, a shift toward used cars (which, as illustrated in Table 3-3, are on average more fuel-efficient). In the long run, the percentage reduction in new cars is smaller. This is the case because new cars become increasingly efficient relative to older cars as time passes, and the gasoline tax increase gives greater importance to fuel-economy.[37]

Several prior studies[38] suggest that the overall reduction in gasoline consumption should be larger in the long run than in the short run, since the fleet composition (fuel economy) channel requires considerable time to take effect. In fact our simulations indicate that, in percentage terms, the long-run reduction is smaller than the short-run reduction. This occurs because VMT per household falls by a smaller percentage in the long run than in the short run (see Table 3-6). This in turn stems from the fact that although in the long run there is a larger percentage reduction in the number of cars owned by the average household, there is a smaller percentage reduction in miles traveled per car.[39]

Table 3-6 shows that the 25-cent per gallon increase in the gasoline price leads to a reduction of 4.5% to 5% in the *equilibrium* demand for gasoline in the long run, or about a 0.2% reduction for each penny increase in the gasoline price. It is difficult to compare this result with other studies, since other studies do not consider market equilibrium for both new and used cars and do not consider time explicitly. However, it may be noted that Austin and Dinan (2005) report that a 30-cent per gallon increase in the gasoline tax would reduce gasoline consumption (by new cars) by 10% (cumulatively) over a 14-year period, or 0.3% reduction (cumulatively) for each penny increase.

Efficiency Costs

Table 3-8 displays the efficiency cost of gasoline tax increases of 10, 25, and 75 cents per gallon. This cost is the weighted sum of the negative of each household's equivalent variation, where a household's weight is proportional to its share of the total population. Here "cost" should be interpreted as a gross

TABLE 3-8 Revenue and Costs from 25-Cent Increase in Gasoline Tax (Results for Year 1).

Revenue recycling	Flat			Income-based			VMT-based		
Tax increase (cents)	10	25	75	10	25	75	10	25	75
Net tax revenue ($billion)	7.43	17.96	48.46	7.43	17.97	48.43	7.52	18.29	49.91
Efficiency cost[a]									
Total ($billion)	1.23	3.24	11.43	1.25	3.28	11.72	1.11	2.89	10.38
Per dollar of additional revenue	0.17	0.18	0.24	0.17	0.18	0.24	0.15	0.16	0.21
Per avoided gallon of gasoline consumed ($)	0.71	0.76	0.96	0.73	0.78	0.98	0.72	0.77	0.97

[a] Negative of the weighted sum of equivalent variations of each household.

measure, since it does not net out the environmental or national security benefits stemming from the policy change.

Under flat recycling, the (gross) cost per dollar raised is $0.16, $0.18 and $0.24, for gasoline tax increases of 10, 25, and 75 cents per gallon, respectively. The costs under the alternative recycling cases are not much different from those in the flat recycling case: the nature of recycling, though very important distributionally (as indicated below), does not much affect the aggregate costs. This result requires careful interpretation. Another choice in the recycling decision is whether to return revenues in lump-sum form or instead by way of cuts in the marginal rates of prior taxes such as income or sales taxes. Prior studies have shown that returning revenues through marginal rate reductions can significantly reduce policy costs, relative to lump-sum recycling.[40] Because our simulation model does not include prior taxes (except for taxes on gasoline), we can only consider recycling through lump-sum transfers and cannot contrast other aspects of recycling.[41]

Distributional Impacts

Effects across Income Groups

Figure 3-1 displays the impacts of a 25-cent gasoline tax increase on household income groups.[42] The distribution of impacts depends crucially on the nature of recycling. Under flat recycling, lower income groups experience a welfare improvement from the policy change, while higher income groups suffer a welfare loss. Here the lower income groups receive a share of the tax revenues that is considerably larger than their share of gasoline tax payments. While policy discussions often refer to the potential regressivity of a gasoline tax, these simulations indicate that flat recycling more than fully offsets this potential regressivity.[43]

Under income-based recycling the pattern of impacts is U-shaped. In this case the middle-income households experience the largest welfare loss. As indicated in Table 3-9, for these households the ratio of miles driven (or gasoline taxes paid) to income is highest; hence recycling based on income benefits these households less than other households. Only the very rich experience welfare gains under income-based recycling; these households have the lowest ratio of miles traveled (or gasoline tax paid) to income.

VMT-based recycling implies a fairly flat pattern of impacts across the income distribution, although the welfare losses are greater for higher-income households. In comparison with lower-income households, rich households drive more luxury cars, which are relatively less fuel-efficient. As a result, the ratio of gasoline taxes paid to VMT is especially large for richer households and these households benefit least from VMT-based recycling.

Table 3-10 decomposes the welfare impacts into the various contributing factors: the change in gasoline price, the transfer (rebate) of gasoline tax revenue, the net capital gain or loss associated with policy-induced changes in car prices, and changes in profit to new car producers. We have assumed that households

**Welfare Impacts across Household Income Groups
Under Alternative Revenue-Recycling Methods**

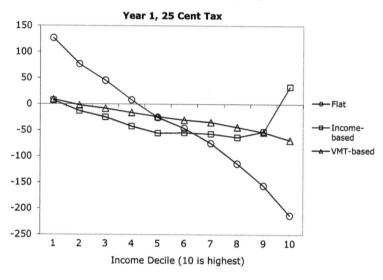

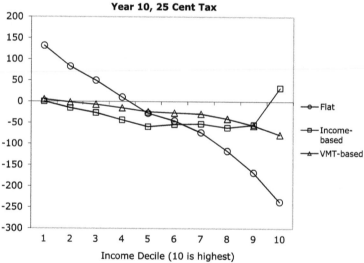

FIGURE 3-1 Welfare Impacts across Household Income Groups Under Alternative Revenue-Recycling Methods.

Note: Welfare impacts are in average price-adjusted 2001 dollars per household.

own shares of new car profits in proportion to their share of benchmark aggregate income. The table makes clear that changes in the gasoline price and the transfer are by far the most important sources of the household welfare impact. It also confirms that, depending on the type of recycling involved, the transfer may or may not offset the gasoline price impact to a particular household.

TABLE 3-9 Consumption, Mileage, and Car-ownership Patterns of Household Income Groups.[a]

Income Decile	Gasoline Consumption		Miles Traveled		Avg. Fuel Economy of Owned Vehicles[b]	Share of Economy's Light Trucks and SUVs
	Level (gallons)	Share of Total	Level (1000s)	Share of Total		
1	157.3	0.02	4.03	0.02	25.61	0.02
2	315.7	0.04	7.97	0.04	25.25	0.04
3	473.6	0.06	11.69	0.06	24.68	0.05
4	588.3	0.08	14.33	0.08	24.35	0.08
5	724.0	0.09	17.65	0.09	24.38	0.09
6	823.7	0.11	19.76	0.11	23.98	0.11
7	922.0	0.12	22.35	0.12	24.25	0.13
8	1060.8	0.14	25.46	0.14	24.00	0.15
9	1227.1	0.16	29.55	0.16	24.08	0.17
10	1459.8	0.19	35.28	0.19	24.17	0.17

[a] Predicted values from simulation model. [b] VMT-weighted.

TABLE 3-10 Decomposition of Welfare Impacts of 25-Cent Gasoline Tax Increase.

	Gasoline price	Transfer	Car prices	Producer profits	EV	EV as a percent of income
Flat Recycling						
Income						
<25	−84.36	157.58	2.62	−3.12	74.96	0.45%
25–50	−196.36	160.22	−0.43	−7.19	−51.87	−0.14%
50–75	−284.09	158.88	−3.16	−11.88	−154.50	−0.24%
>75	−334.45	160.29	−4.62	−19.11	−213.94	−0.21%
All	−176.02	159.04	0.04	−7.22	−29.73	−0.08%
Income-based Recycling						
Income						
<25	−83.90	68.33	2.90	−3.42	−13.75	−0.08%
25–50	−196.40	157.21	−0.40	−7.86	−55.45	−0.15%
50–75	−284.65	259.81	−3.40	−13.00	−55.33	−0.09%
>75	−336.04	417.87	−5.07	−20.90	39.99	0.04%
All	−176.06	157.83	0.10	−7.90	−31.48	−0.08%
VMT-based Recycling						
Income						
<25	−84.26	79.40	2.86	−2.80	−2.01	−0.01%
25–50	−197.37	181.56	−0.38	−6.44	−30.56	−0.08%
50–75	−285.89	261.01	−3.31	−10.64	−52.80	−0.08%
>75	−340.00	307.48	−4.92	−17.12	−69.87	−0.07%
All	−177.08	162.93	0.11	−6.46	−25.70	−0.07%

Note: Welfare effects are expressed in price-adjusted 2001 dollars.

Effects along Other Demographic Dimensions

Figure 3-2 shows VMT and policy impacts by race and income. The figure reveals two main results. First, income seems to be a more important determinant of welfare impact than race: greater variation in welfare impacts exists across income groups than across racial categories. This reflects the fact that much of the welfare impact is determined by VMT, and the differences in VMT across income groups are much larger than the VMT differences across racial groups, after controlling for income (Figure 3-2a). Second, low-income

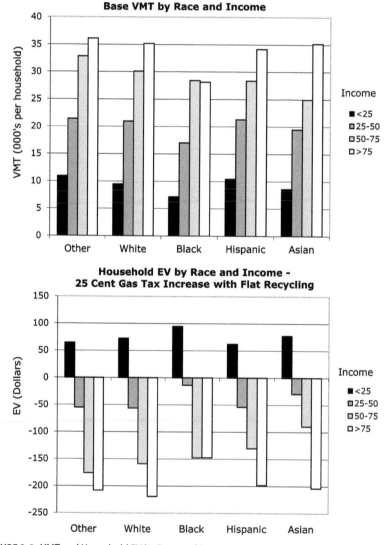

FIGURE 3-2 VMT and Household EV by Race and Income.

African-American households enjoy the largest gains from flat recycling, while high-income African-Americans experience the smallest losses. This is in keeping with the relatively small differences in VMT between low-income and higher-income African-American households.[44]

Figure 3-3 displays differences in welfare impacts across states.[45] The top map displays average VMT per household from the data. The bottom

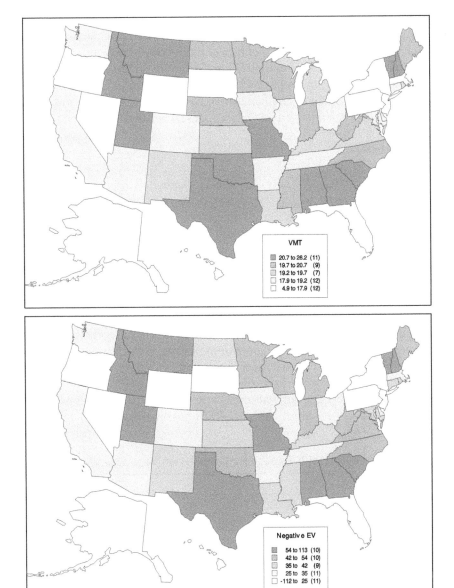

FIGURE 3-3 Average Household VMT and EV.

map exhibits the differences in average household welfare impact. The top and bottom maps are nearly identical, indicating that benchmark VMT is a strong predictor of the welfare impact. Benchmark VMT seems to be strongly correlated with population density. Several relatively densely populated states—New York, Pennsylvania, New Jersey, and Florida—experience the smallest average welfare impact, while many of the relatively sparsely populated states—Montana, Idaho, Utah, Oklahoma, Texas, Alabama, Georgia, and South Carolina—suffer the largest adverse impacts. However, population density does not perfectly correlate with benchmark VMT or the magnitude of the impact: some sparsely populated states—Wyoming and Nevada—nevertheless have low benchmark VMT and relatively small welfare impacts.

Table 3-11 shows how impacts differ depending on the employment status of the household. Retirees fare better than younger individuals, as they tend to drive less. Households with no children also do better, for the same reason.

Sensitivity Analysis

Here we consider the sensitivity of results to parameters affecting changes in fuel economy and scrapping. We also explore the extent to which responses to gasoline tax increases depend on the existence of the CAFE standard. Finally, we consider the welfare effects when revenue from the tax is not returned to households in any form (or used in any productive manner).

The impacts of gasoline tax increases could well be affected by the rate of technology change in automobiles over the next decade. One aspect of faster technological improvement would be speedier growth in the fuel economy of given car models in successive years.[46] To explore this possibility, we created an additional scenario allowing for easier improvements in fuel economy. Here we adopt the "Path 3" baseline assumptions from the National Research Council (2002) study. This scenario's lower costs of fuel-economy improvements imply larger improvements over time in the baseline.[47]

Table 3-12 shows the different implications of the two technology paths. In contrast with the central case, in which baseline model fuel economy

TABLE 3-11 Welfare Impact of 25-Cent Gasoline Tax Increase on Selected Household Groups.

	Year 1			Year 10		
Recycling	*Flat*	*Income-based*	*VMT-based*	*Flat*	*Income-based*	*VMT-based*
All	−29.73	−31.48	−25.70	−31.02	−32.64	−25.55
Retired	46.81	8.65	−15.02	52.78	11.98	−12.69
Not retired, no children	−26.27	−33.02	−21.55	−21.52	−28.77	−15.16
Not retired, with children	−88.01	−58.40	−37.81	−101.03	−68.64	−45.98

Note: Welfare effects are expressed in price-adjusted 2001 dollars.

TABLE 3-12 Impacts of Gasoline Taxes under Alternative Parameter Assumptions.

	Year 1		Year 10	
	Baseline	*25-cent tax increase*[a]	*Baseline*	*25-cent tax increase*[a]
Central Case				
Gasoline consumption (gallons/ household)	775.18	−5.09%	828.89	−4.99%
Aggregate VMT (000s miles/ household)	18.80	−5.01%	21.23	−4.84%
Avg. MPG (miles weighted)	24.26	0.082%	25.62	0.155%
Avg. EV (price-adjusted dollars)	–	−30.13	–	−31.28
Faster Fuel-Economy Improvements[b]				
Gasoline consumption	773.66	−5.07%	751.56	−4.48%
Aggregate VMT	18.83	−4.99%	22.25	−4.23%
Avg. MPG	24.34	0.080%	29.60	0.263%
Avg. EV	–	−29.67	–	−24.23
High Scrap Elasticity				
Gasoline consumption	775.18	−5.16%	828.89	−5.00%
Aggregate VMT	18.80	−5.08%	21.23	−4.86%
Avg. MPG	24.26	0.088%	25.62	0.154%
Avg. EV	–	−29.75	–	−30.93
No CAFE Standard				
Gasoline consumption	775.18	−5.25%	828.89	−6.21%
Aggregate VMT	18.80	−4.93%	21.23	−4.90%
Avg. MPG	24.26	0.339%	25.62	1.401%
Avg. EV	–	−29.28	–	−30.11
Gasoline Tax Revenue Not Recycled[c]				
Gasoline consumption	775.18	−5.49%	828.89	−5.62%
Aggregate VMT	18.80	−5.41%	21.23	−5.50%
Avg. MPG	24.26	0.084%	25.62	0.122%
Avg. EV	–	−218.07	–	−231.32

[a] Percent change relative to the baseline under the same parameter assumptions. [b] Percent increases in fuel economy over 10 years are: Compact 41, Lux compact 41, Midsize 52, Fullsize 58, Lux mid/full 55, Small SUV 54, Large SUV 65, Small truck 58, Large truck 59, Minivan 59. [c] More precisely, assume revenue is neither recycled to households nor used in any productive way.

improves between 11% (compacts) and 20% (large SUVs), under this alternative scenario the baseline improvements are more than twice as large (see note [b] to the table). In the baseline, by year 10 average household gasoline consumption is 751.6 gallons in the fast technology improvement case. This is about 9% lower than in the central case baseline. Fuel economy (miles per gallon) is about 15% higher in the fast-improvement case. Average VMT is also higher (by 5%), reflecting the lower per-mile cost of driving associated with higher fuel economy.

In the case with lower cost fuel economy improvements, the gasoline tax increase induces a smaller long-run percentage reduction in consumption than it does in the central case. This is because gasoline occupies a smaller share of the household budget in the baseline in this alternative scenario, implying a smaller income effect from the tax increase. The average long-run welfare impact (EV) is 22% smaller under the fast technology growth scenario, which is also consistent with gasoline's smaller budget share. Thus, the baseline time-profile of fuel economy significantly influences the welfare consequences of a gasoline tax. However, the differences across the two scenarios in baseline welfare are greater than the welfare impact within either of the two scenarios of introducing a 25-cent gasoline tax increase.

The third main row heading in the table reports results from a simulation in which we double the scrap elasticity η_j to –6.0 from its central value of –3.0. With this change, the gasoline tax causes a somewhat larger reduction in gasoline use in the short run, reflecting a higher scrapping rate: with the higher scrap elasticity, the policy change causes 22% more cars to be scrapped compared with the policy under the central case. While the higher scrap elasticity implies a larger policy impact on gasoline consumption in the short run, it has little influence on the policy impact in later years.

The fourth panel of Table 3-12 displays results in an alternative case where (counter to fact) there is no binding CAFE standard—neither on small cars nor on trucks. In the absence of this standard, the increase in gasoline taxes yields a significantly larger short- and long-run improvement in fuel economy compared with the case of a pre-existing binding CAFE standard. Correspondingly, there is a larger reduction in gasoline consumption. After 10 years, gasoline consumption is reduced by about 6.2%, as compared with 5.0% in the central case. When firms are not constrained by the CAFE standard, producers have greater incentives to change the composition of their car or truck fleets to meet the increased consumer demands for fuel economy that stem from higher fuel costs. In contrast, when firms are constrained by the CAFE standard, the increase in the gasoline tax leads to smaller changes in the composition and average fuel economy of their fleets of cars and trucks. The composition of a producer's car or truck fleet is largely determined by the CAFE standard. In the presence of the CAFE standard, an increase in the gasoline tax affects a car producer's fleet composition mainly by altering the relative sales of cars versus trucks.

The final panel models the scenario where gasoline tax revenues are not returned to households in any form. The welfare costs rise from $30 to $218 per household since the tax revenue is now lost in addition to the distortion. Gasoline consumption and VMT fall more than in the central case due to the larger income effects. Finally, as opposed to the case with "flat" revenue recycling, this policy is regressive over most of the income distribution.

Conclusions

This chapter has examined the impacts of gasoline tax increases with a model that considers jointly supply- and demand-side responses to policy changes. The model links the markets for new, used and scrapped vehicles and accounts for the imperfectly competitive nature of the automobile industry. Linking the three markets enables us to account for the penetration of the car fleet by new cars and thereby assess how the impacts of policy interventions evolve through time. We also address the considerable range of car choices in a high-dimensional discrete-continuous choice model. Parameters for the household demand side of the model stem from a one-step estimation procedure that integrates individual choices for car ownership and miles traveled, thereby yielding consistent welfare measures. Finally, we allow for the considerable heterogeneity among car owners, which enables us to explore the distributional impacts of policy changes along many important dimensions.

We find that each cent-per-gallon increase in the price of gasoline reduces the equilibrium gasoline consumption by about 0.2%. The reduction in demand mainly reflects reduced miles traveled by car owners; shifts in demand from low to high MPG vehicles appear much less important. Under a 25-cent gasoline tax increase the size of the vehicle fleet falls about 0.5%. The impacts on new and used car ownership differ substantially over time. In the first year of the policy, the reduction in vehicle ownership comes largely by way of a decline in new car purchases. However, the ratio of fuel economy of new to old vehicles increases over time, and the increased gasoline tax gives greater importance to fuel economy. As a result, the decline in new car ownership is attenuated over time; by year 10 the reduction in car ownership applies nearly uniformly to new and used vehicles.

The gasoline tax's marginal excess burden (excluding external benefits) per dollar of revenue raised ranges from about $0.15 for a 10-cent tax increase to $0.25 for a 75-cent increase. This efficiency cost is considerably lower than the estimates of the marginal external benefits from higher gasoline taxes,[48] suggesting that increases in the gasoline tax would be efficiency-improving. Taking account of revenue-recycling (and disregarding external benefits), the impact of a 25-cent gasoline tax increase on the average household is about $30 per year (2001 dollars).

The distributional impacts of the gasoline tax differ dramatically under the three revenue-recycling approaches we considered. Under flat recycling, the average household in each of the bottom four income deciles experiences a welfare gain from a gasoline tax increase. The gain to the average household in the lowest income decile would be equivalent to about $125. This suggests that a single-rebate-check approach to recycling would more than eliminate (for the average household within a given income group) the potential regressivity of a gasoline tax increase. On the other hand, if revenues are recycled in proportion to income, only very poor households (those in the lowest decile) and very rich households (those in the highest) stand to gain. The different impacts of the various recycling methods largely reflect differences across the income distribution in car use (VMT). However, household income does not perfectly correlate with VMT and other important determinants of the welfare impacts: controlling for income, we find significant differences in impacts across racial categories and regions of residence.

The framework presented here has considerable potential to address other automobile-related policies, including tightening of CAFE standards and subsidies to retirements of low-mileage (or high-polluting) automobiles. We plan to investigate these policies in future work, examining impacts not only on gasoline consumption but on automobile-generated pollution as well.

Two limitations in our model deserve mention. First, although the model allows the fuel economy of new cars to respond endogenously to gasoline tax changes, it does not distinguish the demands for some of the most fuel-efficient cars—namely, hybrids—from the demands for conventional fuel cars in the same vehicle class. Only recently have sales of hybrids become significant, and thus the data for isolating demands for such cars are quite limited. Nonetheless, in future work we hope to develop surveys that will enable us to consider specifically the demands for hybrid vehicles. In addition, the model abstracts from transactions costs (e.g., time costs, information-gathering costs, and transactions-related taxes) associated with the purchase or sale of new or used cars. These costs cause households to change their car holdings less frequently than they otherwise would. Although data limitations currently make it impossible to assess such costs in a rigorous manner, we believe it would be useful in the future to incorporate alternative assumptions about such costs within the estimation effort and to judge the implications of such costs for policy outcomes.

Acknowledgments

The authors are grateful to David Austin, Tim Bresnahan, Don Fullerton, Michael Greenstone, Ken Small, James Sweeney, Kenneth Train, Kurt Van Dender, Frank Wolak, and three anonymous referees for very helpful suggestions, and to Emeric Henry for his important contributions to the econometric efforts. We also thank the William and Flora Hewlett Foundation and Stanford's Precourt Institute for Energy Efficiency for financial support.

Notes

* This chapter is based on the following article: Bento, Antonio M., Lawrence H. Golder, Mark R. Jacobsen and Roger H. Von Haefen. 2009. Distributional and Efficiency Impacts of Increased US Gasoline Taxes. *American Economic Review* 99(3). We gratefully acknowledge permission from *American Economic Review* to reprint it.

1 Ian W. H. Parry and Kenneth A. Small (2005) and the National Research Council (2002) examine the various externalities from gasoline use and offer estimates of the overall marginal damages. The former study estimates the overall external cost from U.S. gasoline consumption (including effects relating to local pollution, climate change, congestion, and accidents) to be about 75 cents per gallon in year 2000 dollars. This suggests that U.S. taxes on gasoline are below the efficiency-maximizing level, since the federal tax plus average state tax totals 41 cents.

2 The extent of the health improvement from improved air quality depends on both the reduction in gasoline use and possible changes in pollution per gallon of gasoline used. Air districts currently in compliance with air pollution regulations under the 1990 Clean Air Act amendments might well respond to reductions in gasoline use by relaxing "tailpipe" emissions requirements, that is, on the allowable emissions per unit of fuel combusted. This would offset the air-quality and health improvements from reduced gasoline consumption.

3 See, for example, National Research Council (2002).

4 The general public appears to be growing increasingly supportive of stronger measures to curb gasoline use. A February 2006 *New York Times*/CBS News Poll found that a majority of Americans would support a higher gasoline tax if it reduced global warming or made the U.S. less dependent on foreign oil.

5 One exception is Berkovec (1985), who develops a model with interactions among these markets. His model assumes pure competition among auto producers, however.

6 See James M. Poterba (1989, 1991) for expenditure-based estimates of the incidence of gasoline taxes.

7 A difficulty with welfare measurement from two-step estimators is that each step yields a different set of estimates for the same parameters. Each set may have different welfare implications for the same policy. One-step estimators generate a single set of parameter estimates and therefore avoid this difficulty. To our knowledge, the only other automobile study to incorporate a one-step procedure is that of Ye Feng, Don Fullerton, and Li Gan (2005).

   Other studies have estimated the demand for automobiles separately from the demand for gasoline and VMT. Berry, Levinsohn and Pakes (2004) and Petrin (2002) focus on the demand for automobiles; Hausman and Newey (1995), Richard L. Schmalensee and Thomas M. Stoker (1999) and West and Williams (2005) concentrate on the demand for gasoline. Austin and Dinan (2005) obtain demand functions for cars by calibrating the parameters of their simulation model to be consistent with internal estimates by General Motors.

8 The number of distinct cars increases over time as some unique new models become old and enter the used car fleet.

9 In the section of this chapter on estimation of household ownership and utilization decisions we discuss how we allow for multiple car ownership.

10 If a household is endowed with one vehicle of type j entering the period, its gain is computed as:

$$\left(r_j' - r_j\right)\cdot\left(1-\theta_j\right)+\frac{1}{2}\left(r_j' - r_j\right)\left(\theta_j - \theta_j'\right)$$

where r_j and r_j' respectively denote the rental price of car j in the reference and policy-change cases and θ_j and θ_j' represent the probability of the car's being scrapped in the two cases. The first term represents the gain in the value of the car owned, while the second is an adjustment to the gain that accounts for the change in the probability that the car is scrapped.

11 We remove a small (fixed) fraction of the largest vehicles from CAFE in order to incorporate the fact that the very largest trucks and SUVs are exempt from CAFE standards.

12 The purchase price is the same as the present value of rental prices over the life of the car.

13 Our treatment ignores some complexities of the CAFE regulations. The actual regulations allow for intertemporal banking and borrowing: the standard can be exceeded in one year if the firm overcomplies in another. In addition, some manufacturers can and do elect to pay a fine rather than meet the standards and others are not in fact constrained by the standards. Mark R. Jacobsen (2007) addresses these issues.

14 Here it is relevant that we are simulating a permanent and constant change to the gasoline tax. If the policy involved government committing to a path of varying gasoline taxes in the future, for example, a more complex modeling of expected future prices might be called for.

15 Note that the calibration procedure is embedded in a baseline simulation, before the introduction of an increment to the gasoline tax. The values of calibrated parameters (determining new car supply and costs and used car scrap rates) are then saved and introduced into the policy simulation, solved as described in this section.

16 The oligopolistic structure of the new car market involves both multiple products and multiple producers. Under these conditions, theory leaves open the possibility of non-uniqueness. We have tested for non-uniqueness by randomizing starting values over a uniform distribution, and in these experiments the model has always converged to one solution.

17 If the household has purchased the car using a loan, this term can be equivalently interpreted as the interest payment on that loan.

18 Further details about the regional cost of living index are provided in the appendix; It varies by a factor of 1.77 across households.

19 Past transportation applications have addressed this dimensionality problem by randomly sampling from the full set of choice alternatives in estimation. As discussed in Train (1986), such an approach works only with restrictive fixed parameter logit and nested logit models. We cannot adopt this sampling approach in our model because, as described below, our model employs random coefficients to introduce correlations in the unobservables entering the discrete and continuous choice margins. Moreover, although the sampling approach solves the dimensionality problem in estimation, it does not solve the problem in a simulation model, where the full choice set would need to be employed to construct aggregate automobile demands.

20 There is some evidence in the non-market valuation literature that the specification of the number of choice occasions, as long as it is larger than the chosen number of goods, does not have significant effects on estimated welfare measures (von Haefen et al. 2005). Moreover, we do not expect that it has much if any effect on the relative efficiency rankings of policies.

21 Feng, Fullerton, and Gan's (2005) bundling approach aggregates all automobiles into one of two composites—cars and trucks.

22 The 2001 NHTS indicates that 11.1% of households have more automobiles than the number of adults. For the 1.84% of household with more autos than the number of adults plus one, we set the number of choice occasions equal to the number of held autos.

23 Berry et al. (1995, 2004) use alternative specific constants for every automobile to control for unobserved characteristics. Given the highly nonlinear-in-parameters structure of our conditional indirect utility functions, we could not estimate a model with a full set of alternative specific constants and instead adopted a more parsimonious specification with make, age, and class dummies as in Goldberg (1995).

24 The level of income in the budget constraint associated with each choice occasion is the household's income divided by the number of choice occasions. This assures that overall spending is consistent with the household's total income.

25 Because the 2001 NHTS survey elicited VMT in part by asking respondents to recall their past driving behavior, we believe it is appropriate to account explicitly for measurement error in reported VMT.

26 Our assumption that some disturbances capture preference heterogeneity while others pick up measurement error makes our model conceptually similar to the Gary Burtless and Jerry A. Hausman (1978) two-error discrete-continuous model that is frequently used in nonlinear budget constraint applications.

27 Following Dubin and McFadden (1984), past automobile applications assume some degree of correlation between η_{itj} and the type I extreme value errors in the discrete choice model. Similar to King (1980), we instead assume that these disturbances are independent and introduce correlations between the discrete and continuous choices through random parameters as described below.

28 For example, under our random coefficients specification a household that is relatively insensitive to utilization costs and horsepower when purchasing a car will likewise be relatively insensitive to these factors when driving it.

29 Although the Bayesian paradigm implies a very different interpretation for the estimated parameters relative to classical approaches, the Bernstein-von Mises theorem suggests that the posterior mean of Bayesian parameter estimates, interpreted within the classical framework, are asymptotically equivalent to their classical maximum likelihood counterparts. Following Train (2003), we interpret this result as suggesting that both approaches should generate qualitatively similar inference; thus the analyst's choice of which to use in practice can be driven by computational convenience.

30 Parameter estimates for each of the 12 strata are reported in this chapter's Appendix.

31 Although our estimated income elasticities are lower than in much of the previous literature, we note that our stratification of the sample allows parameters controlling income effects to vary among types of households, which may yield a more accurate estimation of income effects than in prior (mainly time-series) work.

32 For example, Berry et al. (1995) employ 20 years of market-level data for new make-model combinations and use alternative specific constants and instrumental variables to identify price effects.

33 Train and Winston (2007) use a cross-section of household level data involving 200 new cars and find average elasticities for new cars ranging from −1.7 to −3.2.

34 We explored the sensitivity of estimates to several alternative specifications and estimation strategies. In particular, we experimented with allowing the income

coefficient to vary across car classes and age groups and restricting a subset of parameters to be fixed across the sample. None of these alternatives generated elasticities significantly different from those in Table 3-4.

35 Pre-existing federal taxes are $0.185 and average state taxes are $0.225.

36 In our simulations, the fraction of the response coming from fleet fuel economy improvements (as compared to reduced VMT) is much smaller than in Austin and Dinan (2005) and Small and Van Dender (2007), who find that over half of the response comes from fuel economy rather than VMT. The fuel economy response can be divided into (1) changes in fleet composition and (2) improvements in the fuel economy of particular models. Our results differ from the Austin and Dinan's mainly because of differences in the magnitude of the second factor. Austin and Dinan calibrate the potential for technological improvements based on the cost estimates in NRC 2002. In contrast, we calibrate parameters determining the marginal costs of engineering improvements in a way that reconciles observed automobile choices with the assumptions of profit maximization and utility maximization. This yields marginal costs of improving fuel economy that are larger than the marginal costs implied by the NRC study.

37 This increasing relative efficiency of new cars applies both under the baseline price path and under the policy change. The baseline path is based on the National Research Council's (2002) "Path 1" assumptions on new car fuel economy.

38 Examples are Jean Agras and Duane Chapman (1999), Graham and Glaister (2002), and Olof Johansson and Lee Schipper (1997).

39 In the long run, the cost of gasoline represents a smaller fraction of per-mile operating cost, a reflection of improvements in fleet fuel economy.

40 See, for example, Lawrence H. Goulder et al. (1999) and Ian W. H. Parry and Wallace E. Oates (2000).

41 The absence of prior taxes can also affect policy costs. The direction of the bias from this omission depends on the extent to which the commodity receiving the tax increase (gasoline) is a complement or substitute for taxed factors of production such as labor and capital. Previous studies indicate, in particular, that if gasoline is an average substitute for leisure, the presence of prior taxes raises the costs of a gasoline tax (or of an increase in this tax). See, for example, Goulder and Williams (2003). On the other hand, if gasoline is a sufficiently weak substitute (or relatively strong complement) for these factors, then the pre-existing taxes imply lower costs from a gasoline tax. West and Williams' (2007) empirical estimates suggest that gasoline and leisure may be complements, which imply an upward bias in our model's estimate of the cost of a gasoline tax increase. Their study calculates the cost of a marginal increase in the gasoline tax to be about 26 cents, somewhat higher than the cost in our simulations.

42 The pattern of impacts across households is similar for the 10-cent and 75-cent gasoline tax increases.

43 West and Williams' (see Chapter 4) econometric study also finds that a gasoline tax with flat (or lump-sum) recycling is regressive.

44 Although not displayed, the same pattern emerges under other forms of recycling: differences in income account for more of the variation of welfare impacts than racial differences do, and the variation in impacts between high-income and low-income African-American households is relatively small compared to the variation for other households.

45   To generate the results in these figures, we first regressed the household welfare impacts (EVs) from the simulation on household characteristics and on the predicted baseline VMT and predicted baseline VMT squared. Next we used the coefficients from the regression, the same set of household characteristics, and household baseline VMT from the data (as opposed to predicted VMT) to get a new fitted value of EV for each household. We then aggregated this information by state.

46   Growth in fleet-wide fuel economy has been promoted by the increased production and sale of hybrid vehicles. In our model, hybrid vehicles are merged with conventional cars within given manufacturer-class combinations (e.g., Toyota compacts). We are considering splitting out hybrids in future work. To estimate demands for hybrids, we may need to supplement our revealed-preference data with stated-preference information, since hybrids were introduced in the automobile fleet in 2001, the year corresponding to our benchmark data. Today they represent about four percent of the compact car fleet.

47   The NRC study interprets the scenario involving faster fuel economy growth as due to technological advances that reduce producers' costs of supplying more fuel-efficient cars. Our simulations also express such a scenario. However, it should be noted that changes in the baseline time-profile of fuel economy could also reflect changes in household preferences. Our model cannot capture such demand-side changes, since we assume a stable utility function in our econometric estimation.

48   See, for example, Parry and Small (2005).

## CHAPTER 4

Estimates from a Consumer Demand System: Implications for the Incidence of Environmental Taxes[*]

SARAH E. WEST AND ROBERTON C. WILLIAMS III

D istributional effects represent an important concern in designing and evaluating environmental policy. Most studies suggest that environmental taxes tend to be at least mildly regressive, making such taxes less attractive options for policy. In this chapter, we consider the distributional effects of a gasoline tax increase. The gasoline tax's regressivity is often cited as one of the strongest arguments against increasing this tax. Thus, it represents a policy for which distributional analysis is particularly important.[1]

Our study makes two main contributions. First, we compare four different measures of the incidence of the tax, which differ in the implicit simplifying assumptions that they make. These are the equivalent variation, two consumer surplus measures (one that allows demand elasticities to vary across income groups and one that does not), and a measure that implicitly assumes that all demands are completely unresponsive to price.[2]

To our knowledge, ours is the first study to compare incidence estimates under these four measures. Among these measures, there is a tradeoff between accuracy and ease of implementation: measures that implicitly make more restrictive assumptions will be less accurate, but they also require less information to implement. Thus, it is useful to know how much error the implicit simplifying assumptions induce.

This is a particularly important issue for environmental taxes, because (for reasons that will be explained in the next section), the errors will be insignificant if the policy-induced price changes are small but will become important for sufficiently large price changes. Environmental taxes could imply just such large price changes; proposed carbon taxes would lead to double-digit (or larger) percentage increases in fossil fuel prices, while a gasoline tax equal to marginal damage would nearly double the price of gasoline.[3]

Second, this chapter examines three simple scenarios for gas tax revenue use: that it is discarded, that it is used to cut taxes on wage income, and that it is returned through a uniform lump-sum distribution. The first scenario implies that the net wage and lump-sum income for each household will remain constant: only the price of gasoline changes. The second scenario implies that net wages will rise, because the recycled revenue will reduce marginal tax rates. The third scenario implies that household lump-sum income will rise.[4]

Discarding the revenue is a useful benchmark, as it isolates the distributional effect of the gas tax itself. Most recent studies of second-best optimal environmental taxes assume that revenue is returned by lowering the tax on wage income (see, for example, West and Williams 2004 or Parry and Small 2002). Some of these studies also consider returning revenue via a lump-sum transfer (e.g., Parry and Williams 1999). This yields a more progressive distributional effect, though it is less efficient than using the revenue to lower the tax rate on wages.[5]

We use the 1996 through 1998 Consumer Expenditure Surveys, which provide detailed data on household expenditures including gasoline expenditures and on each individual's earnings and working hours. We merge these data with state-level price information from the American Chambers of Commerce Researchers' Association (ACCRA) cost-of-living index. We use the National Bureau of Economic Research's TAXSIM model to calculate marginal and average tax rates of each worker. The resulting data set thus includes quantities and after-tax prices for the three goods defined in our demand system: leisure, gasoline, and a composite of all other goods.

We use the Almost Ideal Demand System (AIDS) derived by Deaton and Muellbauer (1980). The advantages of this system are well-known: it gives a first-order approximation to any demand system, satisfies the axioms of choice exactly, and imposes neither separability nor homotheticity. We estimate our demand system separately for one-adult households and for households with one female and one male adult, and for different income quintiles within each group.

This data set and estimation approach are very similar to those used by West and Williams (2004), with the only major difference being that the present chapter estimates the demand system separately for each quintile. However, the two inquiries differ greatly in how they use the demand system estimates. The present chapter focuses on incidence, whereas West and Williams (2004) focuses exclusively on the calculation of optimal gas tax rates and does no incidence analysis.

We find that increasing the gas tax is regressive, but that this depends greatly on how the revenue is used; using the additional revenue to provide lump-sum transfers more than offsets the regressivity of the gas tax and thus, in this case, increasing the gas tax is somewhat progressive. Using the revenue to cut the tax on labor income yields an efficiency gain and reduces the regressivity of the tax, but this effect is not sufficient to make the change progressive overall.

Using an incidence measure that allows for demand responses is important for measuring both the overall burden and the distribution of that burden.

Under the measure that assumes no demand response, the measured tax burden is significantly higher for all income groups than under other measures, but this effect is much weaker for the top quintile than for the rest of the income distribution. Thus, our estimates suggest that studies that use a measure that ignores demand responses will substantially overstate the regressivity of the gas tax. Comparing the two consumer surplus measures shows that assuming away differences in demand elasticities across income groups makes relatively little difference for the estimated aggregate burden of the tax, but it makes the tax appear to be more regressive. Finally, the differences between the equivalent variation and consumer surplus measures are relatively small.

The next section formally describes the four incidence measures used in the chapter and how they differ. The subsequent sections describe the demand system, the data, and the estimation approach, followed by the presentation of the incidence estimates; the final part concludes.

Measures of Tax Incidence

An ideal measure of tax incidence would begin by calculating the general equilibrium changes in prices that would occur throughout the economy in response to the change in the tax rate and then calculate the effects of those price changes on households' welfare. The obvious effect of an increase in the gasoline tax is to raise the consumer price of gasoline, thus imposing a burden on gasoline consumers. But it also could lower the producer price of gasoline, imposing a burden on the owners of gas stations and gasoline producers, and perhaps in turn on workers in those industries through lower wages. It could also affect the prices of other goods that use gasoline as an intermediate input.

However, calculating such effects requires a great deal of information, most notably the demand and supply elasticities for all relevant industries, and the distribution of ownership of firms in those industries. Thus, for simplicity, many incidence studies (including all those cited in this section) assume that the supply of consumer goods is perfectly elastic. This implies that the imposition of a tax on a consumer good does not affect the producer price of that good, and thus that the entire burden of the tax falls on consumers. Similarly, studies commonly assume that the burden of labor taxes falls entirely on workers.[6] Together, these two assumptions mean that there is no incidence on firms. The present chapter makes both assumptions. In practice, of course, these assumptions do not hold, and thus our incidence estimates will differ somewhat from the true incidence of the tax.[7] Our goal here is not to produce a perfect estimate of the incidence of the gasoline tax, but rather to compare different measures of incidence.

Given a particular set of price changes, the question then is how to measure the effect on household welfare. We compare four different measures, all of which use the same demand system estimates but differ in the implicit simplifying assumptions they make about demand in calculating incidence. The first of the measures we use in this chapter is the equivalent variation, which is the most rigorous measure (along with the analogous compensating variation) of the dollar equivalent of the utility change resulting from a policy. It is defined as the lump-sum tax or transfer that would have the same effect on utility as the set of price and income changes resulting from the tax. This can be expressed as

$$V_h\left(\overline{I}_h + E_h, \overline{\mathbf{p}}_h\right) = V_h\left(I_h, \mathbf{p}_h\right) \tag{4.1}$$

where $\overline{I}$ and I are income before and after the imposition of the tax, respectively, $\overline{\mathbf{p}}$ and $\mathbf{p}$ are the corresponding price vectors, E is the equivalent variation, and h denotes the household. This measure implicitly accounts for cross-price effects through the cross-price derivatives of the indirect utility function. Thus, the equivalent variation for two simultaneous price changes will not generally be equal to the sum of the equivalent variations for the two changes taken separately. We are unaware of any studies of the gasoline tax that use equivalent variation or compensating variation to measure incidence. These measures have been used in studies of the incidence of other environmental taxes, however, including studies of the incidence of a carbon tax by Cornwell and Creedy (1997) and Brännlund and Nordström (2004).

To calculate the equivalent variation, however, requires an estimate of the indirect utility function, which is often unavailable. The second measure we consider, the change in consumer surplus, requires much less information. The change in consumer surplus is defined as change in the area under the demand curve (i.e., the integral of inverse demand minus price) over the quantity purchased, which reflects the effect of changing prices on utility, plus the change in full income. Assuming a constant-elasticity demand curve for each good yields

$$\Delta CS_h = \sum_k \left\{ \frac{\overline{x}_h^k \overline{p}_h^k}{\varepsilon_h^k + 1} \left[1 - \left(\frac{p_h^k}{\overline{p}_h^k} \right)^{\varepsilon_h^k + 1} \right] \right\} + I_h - \overline{I}_h \tag{4.2}$$

where ΔCS is the change in consumer surplus, $\overline{x}^k$ is the consumption of good k (before the tax change), and ε^k is the uncompensated own-price elasticity of demand for good k, which is allowed to vary across income quintiles. This measure is much simpler to implement than the equivalent variation: one needs to know only the spending on the taxed good before the imposition of the tax, the percentage change in price and change in income induced by the tax, and the own-price demand elasticity for any good whose price changes. West (2004) uses a consumer surplus measure similar to this to examine incidence of a gas tax, though it implicitly assumes a linear demand curve, rather than the constant-elasticity demand curve assumed in deriving equation (4.2).

This measure may differ from the equivalent variation for three reasons. First, income effects will lead to a difference between the equivalent variation and the change in consumer surplus; the equivalent variation is the area under the compensated demand curve (corresponding to the level of utility after the tax change), which will differ from the area under the uncompensated demand curve unless the income elasticity of demand is zero.[8]

Second, the formula in equation (4.2) assumes that the change in consumer surplus from a set of price changes is equal to the sum of the changes in consumer surplus from each of the price changes taken separately. This will be generally true only if all cross-price elasticities are equal to zero.[9]

Finally, the formula in equation (4.2) assumes a constant-elasticity demand curve. This will introduce some error for a large price change if the elasticity changes along the demand curve. Each of these three differences will be insignificant (relative to either measure) for sufficiently small changes in price, but will become important for sufficiently large changes.[10]

Our third measure is very similar, differing only in that it assumes that demand elasticities do not vary across households. West (2004) also uses a similar consumer surplus measure, again differing in that it implicitly assumes a linear demand curve rather than a constant elasticity. For this measure, we calculate the average demand elasticity for all households and use this average elasticity to calculate incidence. There are many reasons to think that demand elasticities might vary with income. For example, because poor households have smaller budgets, they may respond more to prices than the wealthy. On the other hand, if poor people have fewer transportation options, they may be less price-responsive. To the extent that demand elasticities vary with income, this third measure will overstate the incidence on income groups with relatively elastic demand and understate the incidence on groups with relatively inelastic demand.

Many incidence studies ignore demand responses altogether, simply calculating incidence as the sum over all goods of the price change for a given good times the household's consumption of that good (before the imposition of the tax change) plus the change in income resulting from the policy. This is our fourth measure of incidence. Under this measure, the incidence on household k is given by

$$\sum_h \left(\bar{p}_h^k - p_h^k \right) \bar{x}_h^k + I_h - \bar{I}_h \qquad (4.3)$$

Metcalf (1999) uses this approach to estimate the incidence of a range of environmental taxes, Poterba (1991) and West (2004) use it to estimate the incidence of the gas tax, and Walls and Hanson (1999) use it to estimate the incidence of vehicle emissions taxes and taxes on miles.[11]

Unless all demand elasticities are equal to zero, this measure will differ from both the equivalent variation and the change in consumer surplus, because it does not reflect any demand responses—cross-price or own-price. It will tend to overstate the burden of tax increases, because it ignores consumers' shift away from the newly taxed good (and, conversely, understates the reduction in burden from a tax decrease, because it ignores the shift toward the good after its price drops).[12]

Again, this difference will be insignificant relative to the burden of the tax for a sufficiently small change, but will become important for larger changes.

Finally, it should be noted that all of these measures ignore the external benefits of reduced gasoline consumption. Incorporating such benefits would reduce net tax burdens for all income groups and might have important distributional effects if the external benefits are unevenly distributed across income groups.

Demand System Estimation

To calculate the welfare measures discussed above, we need compensated and uncompensated own- and cross-price elasticities of demand. In this section, first we specify the demand system we use to obtain these elasticities. Second, we describe the data, variable derivation, and summary statistics. Last, we discuss the estimation technique and system estimation results.

Specification of the Demand System

We consider an Almost Ideal Demand System (AIDS) defined over gasoline, leisure, and a composite of all other goods. Again, the advantages of the AIDS are well-known: it gives an arbitrary first-order approximation to any demand system, satisfies the axioms of choice exactly, is simple to estimate, and does not assume that the utility function is separable or homothetic.

The AIDS provides these advantages given that it is derived from a preference structure characterized by the expenditure function proposed in Muellbauer (1975, 1976):

$$\log c(\mathbf{p}, u) = (1 - u)\log[a(\mathbf{p})] + u\log[b(\mathbf{p})] \qquad (4.4)$$

where u is utility, $\mathbf{p}$ is a vector of goods prices and the price of leisure (the wage), and the following equations are defined over n goods:

$$\log a(\mathbf{p}) = \alpha_0 + \sum_k \alpha_k \log p_k + \frac{1}{2}\sum_k\sum_j \gamma_{kj} \log p_k \log p_j \qquad (4.5)$$

and

$$\log b(\mathbf{p}) = \log a(\mathbf{p}) + \beta_0 \prod_k p_k^{\beta_k} \qquad (4.6)$$

Applying Shephard's Lemma, we obtain the demand equations for gasoline, leisure, and other goods for household h in their budget share forms:

$$s_{ih} = \alpha_i + \sum_j \gamma_{ij} \log p_{jh} + \beta_i \log\left(\frac{y_h}{P_h}\right) \tag{4.7}$$

$$i = gasoline, leisure, other goods; h = 1,...,H$$

where α, β, and γ are parameters to be estimated, and y_h is total income, the amount spent on gasoline, leisure, and other goods. Two-adult households have one leisure share equation for the household's male adult and another for the household's female adult.

The price index P_h is defined by:

$$\log P_h = \alpha_0 + \sum_k \alpha_k \log p_{kh} + \frac{1}{2}\sum_j \sum_k \gamma_{kj} \log p_{kh} \log p_{jh} \tag{4.8}$$

Use of the price index in (4.8) requires estimation of a nonlinear system of equations. Following a suggestion in Deaton and Muellbauer (1980), we approximate (4.8) by using Stone's (1954b) linear price index defined by:

$$\log P_h \equiv \sum_k s_{kh} \log p_{kh} \tag{4.9}$$

where s_{kh} are by-quintile average expenditure shares.[13]

Demand theory imposes several restrictions on the parameters of the model, including:

$$\sum_i \alpha_i = 1 \tag{4.10}$$

$$\sum_i \gamma_{ij} = 0 \tag{4.11}$$

$$\sum_i \beta_i = 0 \tag{4.12}$$

$$\gamma_{ij} = \gamma_{ji} \tag{4.13}$$

Provided that these restrictions hold, (4.7) represents a system of demand functions that add up to total income, are homogeneous of degree zero in prices and total income, and that satisfy Slutsky symmetry. Budget shares sum to one.

Data, Variable Derivation, and Summary Statistics

The 1996 through 1998 Consumer Expenditure Surveys (CEX) are the main components of our data. The CEX Family Interview files include the amount spent by each household on gasoline, total expenditures, and a wide variety of household income measures. For each household member, the Member Files include usual weekly work hours, occupation, the gross amount of last pay, the duration of the last pay period, and a variety of member income measures.

The CEX is a rotating panel survey. Each quarter, 20% of the sample is rotated out and replaced by new consumer units. We pool observations for households across quarters.

We estimate two demand systems: one for one-adult households and the other for two-adult households with one adult male and one adult female (where an adult is at least 18 years of age). We do not include households with adults over the age of 65.[14]

The twelve quarters in the 1996 through 1998 Consumer Expenditure Surveys have 4,659 one-adult households and 5,047 two-adult households under 65 with complete records of the variables needed here. Households appear in the data between one and four times; the total number of one-adult observations is 9,725, and the total number of two-adult observations is 11,034.

Total income (y_h in equation (4.7) above) equals the amount spent on gasoline, leisure, and all other goods. The CEX contains quarterly gasoline expenditure. Since it also contains hours worked per week, we divide quarterly gasoline expenditure by 13 to get weekly gas expenditure. To derive weekly leisure "expenditure," we need to specify each person's time endowment, the total number of hours available either to work or to consume leisure. The highest number of hours worked in the sample is 90 per week, so we set the weekly time endowment at 90 hours. We then subtract the number of hours worked per week from 90 to get weekly leisure hours.

To obtain the price of leisure (the wage) we first calculate the wage net of tax using state and federal effective tax rates generated from NBER's TAXSIM model.[15]

Since we do not observe wages for individuals who are not working, we follow Heckman (1979) to correct for this selectivity bias and obtain selectivity-corrected net wages. We explain this estimation in detail in this chapter's Appendix. We multiply the selectivity-corrected net wage by the number of hours of leisure per week to obtain weekly leisure expenditure.[16]

To calculate weekly spending on other goods, we first convert the CEX's measure of quarterly total expenditures into weekly total expenditures. Then we subtract weekly gasoline expenditure from total weekly expenditures to obtain spending on all other goods.

For gas prices and the price of other goods, we use the ACCRA cost-of-living index. This index compiles prices of many separate goods as well as overall price levels for approximately 300 cities in the United States. It is most widely used to calculate the difference in the overall cost of living between any two cities. It also lists for each quarter the average prices of regular, unleaded, national-brand gasoline. Since the CEX reports household residence by state, not city, we average the city prices within each state to obtain a state gasoline price and state price index for each calendar quarter. Then we assign a gas price and a price index to each CEX household based on state of residence and CEX quarter. We use the ACCRA price index divided by 100 as our price of other goods.

We calculate the price index (P_h) using (4.9). Finally, we calculate the log of real income term in $(\log(y_h/P_h))$ by dividing total income by the price index in (4.9).

Table 4-1 lists summary statistics for the demand system estimation sample, working households that consume gasoline. Both one- and two-adult households spend about 2% of their income on gasoline. One-adult

TABLE 4-1 Summary Statistics for Working Households with Non-zero Gas Consumption*

Variable	One-adult Households		Two-adult Households	
	Mean	*Standard Deviation*	*Mean*	*Standard Deviation*
Gasoline per Week (gallons)	13.78	10.97	24.67	16.07
One-adult Hours per Week	41.25	10.92	–	–
Two-adult Male Hours per Week	–	–	44.76	10.34
Two-adult Female Hours per Week	–	–	37.54	11.09
Gasoline Share of Expenditures	0.02	0.01	0.02	0.01
One-adult Leisure Share of Expenditures	0.49	0.11	–	–
Two-adult Male Leisure Share of Expenditures	–	–	0.29	0.09
Two-adult Female Leisure Share of Expenditures	–	–	0.26	0.07
Other Good Share of Expenditures	0.50	0.15	0.44	0.12
Gas Price ($)	1.19	0.12	1.19	0.11
Other Good Price (index)	1.04	0.10	1.04	0.11
One-adult Heckman-corrected Net Wage	8.31	2.36	–	–
Two-adult Male Heckman-corrected Net Wage ($)	–	–	11.02	3.26
Two-adult Female Heckman-corrected Net Wage ($)	–	–	8.60	2.20
$\ln(y/P)$	5.72	0.50	6.23	0.45
One-Adult Education: < High School Diploma (%)	6.8	–	–	–
One-Adult Education: High School Diploma (%)	23.3	–	–	–
One-Adult Education: > High School Diploma (%)	69.9	–	–	–
Two-Adult Male Education: < High School Diploma (%)	–	–	8.5	–
Two-Adult Male Education: High School Diploma (%)	–	–	27.9	–
Two-Adult Male Education: > High School Diploma (%)	–	–	63.6	–

TABLE 4-1 *(Cont.)*

Variable	One-adult Households		Two-adult Households	
	Mean	*Standard Deviation*	*Mean*	*Standard Deviation*
Two-Adult Female Education: < High School Diploma (%)	–	–	6.8	–
Two-Adult Female Education: High School Diploma (%)	–	–	26.9	–
Two-Adult Female Education: > High School Diploma (%)	–	–	66.4	–
Race of Household Head				
White (%)	82.6	–	87.5	–
Black (%)	13.7	–	8.8	–
Asian (%)	0.7	–	0.8	–
Other race (%)	3.0	–	2.9	–
Number of Children	0.41	0.87	1.16	1.18
Region				
Northeast (%)	13.3	–	13.3	–
Midwest (%)	24.5	–	25.8	–
South (%)	34.1	–	35.5	–
West (%)	28.1	–	25.4	–
Observations	6,553	–	7,162	–

* Of the one-adult households 46% are headed by males while 54% are headed by females.

households spend a bit less than 50% of their total income on leisure and the remainder on other goods. Two-adult households spend about 55% of their income on leisure and the remainder on other goods. The average selectivity-corrected net wage in the one-adult sample is $8.31 per hour. Men in the two-adult sample make $11.02 per hour while women make $8.60 per hour.[17]

System Estimation and Results

To incorporate the effect that household and individual specific characteristics have on demand, we add a vector of these characteristics, c_{rh}, to the constant terms in (4.7) so that:

$$\alpha_i = \zeta_{io} + \sum_r \zeta_{ir} c_{rh} \quad i = gasoline, leisure, other goods \quad (4.14)$$

where ζ_{io} and the ζ_{ir}'s are parameters to be estimated.

Some households have zero expenditure on gasoline. To correct for the potential selection bias that may arise, we first estimate a probit on the choice to consume or not consume gasoline. From that probit, we obtain inverse Mills ratios for gasoline, R_{gh} (see the Appendix for details on this estimation). Substituting equation (4.14) into equation (4.7), and adding R_{gh} and error term e_{ih} gives us the following equations for estimation:

$$s_{ih} = \zeta_{io} + \sum_r \zeta_{ir} c_{rh} + \sum_j \gamma_{ij} \log p_{jh} + \beta_i \log\left(\frac{y_h}{P_h}\right) + \theta_i R_{gh} + e_{ih} \qquad (4.15)$$

We divide the sample into quintiles based on expenditures (on gasoline and other goods) which we interpret as a proxy for lifetime income. We use an equivalence scale to adjust total expenditures on gasoline and other goods for different family sizes. These quintiles are intended to reflect individuals' standard of living, and a given level of total household income clearly represents a higher standard of living if that household only includes one person than if it includes several people—thus necessitating some adjustment for household size.[18]

Our equivalence scale weights adults and children equally but allows for economies of scale in consumption. Specifically, we divide total expenditures on gasoline and other goods by $(adults + children)^{.5}$. We pool one-adult and two-adult households, rank them together by equivalence-scale adjusted total expenditures on gasoline and other goods, and divide them into quintiles.[19]

We estimate the demand system defined by (4.15) separately for each quintile using working households that consume gasoline, separately for one-adult and two-adult households. We impose the restrictions in (4.10) through (4.13) and drop the equation for other goods. Each system includes member and household characteristics that may affect gasoline or leisure shares: the members' age, age squared, race, sex (in one-adult estimation only), education, and number of children. We also include state dummy variables to account for unmeasured state-level sources of variation.[20]

Because some regressors may be endogenous, we use instrumental variables techniques. The net wage, for example, may be endogenous for two reasons. First, the gross wage is determined by dividing earnings by hours of work, and both variables may be measured with some error. Thus, hours worked and wages may be correlated. Second, the marginal income tax rate depends on income. We therefore use the mean net wage by occupation by state, calculated separately for men and for women, to instrument for the net wage.[21]

In principle, the full econometric model, including all discrete and continuous choices, might be estimated using maximum likelihood procedures. While such an approach would produce consistent estimates, it would be difficult to implement. Since censoring occurs in both gasoline and leisure demand (and for either or both the male and female in two-adult households) we would need to evaluate multiple integrals in the likelihood function. Furthermore, such a procedure would probably be computationally intensive enough that it would be impractical, given that we need to bootstrap standard errors for our elasticity and incidence estimates.

We expect the error term to be correlated with the endogenous variables. In addition, since the equations in (4.15) are functions of the same explanatory variables, we also expect error terms among the equations to be correlated. We assume, however, that the expected values of the error terms equal zero ($E(e_{ih}) = 0$ for all i) and we estimate the demand system using three-stage least squares (3SLS).[22]

Table 4-2 presents compensated and uncompensated demand elasticities by quintile, evaluated using by-quintile means, and aggregate elasticities evaluated using full sample means.[23]

These elasticity estimates aggregate male and female labor for two-adult households and then aggregate the one-adult and two-adult households in each quintile. Gasoline own-price elasticities range from about –0.18 to –0.73 and fall in the span of estimates reported in gas demand survey articles (see Dahl 1986; Dahl and Sterner 1991a; Espey 1996). Upper income quintiles are less responsive to gas price changes than lower income quintiles.[24]

Table 4-2 also presents standard errors for the elasticity estimates. The standard errors were calculated based on 1,500 replications of a nonparametric bootstrap. Because observations for the same household for multiple quarters are not independent, observations are clustered by household in generating each bootstrap sample.[25]

Incidence Estimates

This section considers the incidence of a substantial increase in the gasoline tax, under a range of different assumptions about how the gas tax revenue is used and about how consumer demands react to the policy changes.

Modeling Incidence

We consider the effect of increasing the gasoline tax from its current level (which averages roughly 37 cents per gallon: the 18.4-cent federal gas tax plus the average state gas tax) to $1.39 per gallon.[26] This is the optimal second-best gas tax found by West and Williams (2004), based on demand system estimates similar to those in the present chapter, but with a representative agent model. That study in turn used an estimate of 83 cents for the marginal external damage per gallon of gasoline, which was taken from a survey by Parry and Small (2002). For simplicity, we assume that the tax is changed only for gasoline used directly by households—there is no change in the tax on gasoline used as an intermediate input.[27]

We compute incidence for ten representative individuals: a representative one-adult household and a representative two-adult household for each of the five quintiles. The price and quantity consumed of each good and the level of income for each one-adult representative household is equal to the

Sarah E. West and Roberton C. Williams III

TABLE 4-2 Estimated Elasticities by Quintile (Standard Errors in Parentheses).

Good		Compensated Elasticities			Uncompensated Elasticities		
		Gas Price	Wage	Other Good Price	Gas Price	Wage	Other Good Price
Quintile 1	Gasoline	−0.738	−0.049	0.787	−0.724	−0.335	1.117
		(0.754)	(0.517)	(0.955)	(0.753)	(0.547)	(0.980)
	Labor	−0.004	0.183	−0.179	0.040	−0.739	0.872
		(0.024)	(0.092)	(0.094)	(0.023)	(0.104)	(0.106)
	Other Good	0.033	0.164	−0.198	0.029	0.250	−0.305
		(0.040)	(0.087)	(0.099)	(0.040)	(0.094)	(0.106)
Quintile 2	Gasoline	−0.692	0.119	0.573	−0.689	0.051	0.652
		(0.222)	(0.145)	(0.232)	(0.222)	(0.153)	(0.236)
	Labor	−0.007	0.149	−0.142	0.034	−0.770	0.773
		(0.007)	(0.037)	(0.038)	(0.007)	(0.043)	(0.043)
	Other Good	0.026	0.140	−0.166	0.023	0.217	−0.244
		(0.011)	(0.039)	(0.041)	(0.011)	(0.043)	(0.045)
Quintile 3	Gasoline	−0.550	−0.293	0.843	−0.549	−0.316	0.864
		(0.209)	(0.131)	(0.206)	(0.209)	(0.141)	(0.215)
	Labor	0.012	0.136	−0.148	0.049	−0.726	0.750
		(0.006)	(0.028)	(0.029)	(0.006)	(0.036)	(0.035)
	Other Good	0.035	0.139	−0.174	0.030	0.258	−0.300
		(0.009)	(0.029)	(0.030)	(0.009)	(0.034)	(0.035)
Quintile 4	Gasoline	−0.450	−0.300	0.749	−0.448	−0.329	0.795
		(0.188)	(0.123)	(0.206)	(0.188)	(0.149)	(0.219)
	Labor	0.011	0.139	−0.151	0.043	−0.617	0.749
		(0.005)	(0.024)	(0.024)	(0.005)	(0.031)	(0.032)
	Other Good	0.026	0.124	−0.151	0.018	0.323	−0.385
		(0.007)	(0.021)	(0.022)	(0.007)	(0.025)	(0.028)
Quintile 5	Gasoline	−0.180	−0.146	0.326	−0.180	−0.136	0.304
		(0.187)	(0.158)	(0.245)	(0.187)	(0.157)	(0.245)
	Labor	0.005	0.351	−0.357	0.012	0.193	−0.046
		(0.006)	(0.054)	(0.054)	(0.006)	(0.053)	(0.053)
	Other Good	0.007	0.184	−0.191	−0.011	0.612	−1.031
		(0.005)	(0.029)	(0.030)	(0.005)	(0.030)	(0.029)
Aggregate	Gasoline	−0.459	−0.163	0.622	−0.457	−0.209	0.674
		(0.119)	(0.087)	(0.135)	(0.119)	(0.093)	(0.139)
	Labor	0.005	0.200	−0.204	0.034	−0.473	0.562
		(0.004)	(0.022)	(0.023)	(0.004)	(0.025)	(0.026)
	Other Good	0.020	0.157	−0.177	0.009	0.421	−0.631
		(0.004)	(0.017)	(0.018)	(0.004)	(0.019)	(0.020)

average over all one-adult households in that quintile (including those households excluded from the system estimation because they did not work or did not consume gasoline), and an analogous set of averages provide the representative two-adult households. For prices, these averages are weighted by the quantity of the good consumed or the quantity of labor supplied, so that for each good, the average price times the average quantity consumed equals the average spending on that good. Utility function parameters are given by the system estimation, except for the α_i parameters, which are set so that the resulting demand functions match the observed quantities for the representative households when evaluated at the observed prices. We then calculate incidence for each representative household and aggregate the one-adult and two-adult households in each quintile, weighting by the fraction of each household type in the quintile.

We consider three different assumptions about the revenue raised by the environmental tax: that it is discarded, that it is used to cut taxes on wage income, and that it is returned through a uniform lump-sum distribution. The first assumption implies that the net wage and lump-sum income for each household will remain constant: only the price of gasoline changes. The second assumption implies that net wages will rise because the recycled revenue will lead to a drop in marginal tax rates. We assume that this is an equal percentage-point cut in all brackets (equivalent to a cut in the Medicare payroll tax, for example). The third assumption implies that household lump-sum income will rise. We assume that this transfer is based on the number of adults in a household; thus, a two-adult household will receive twice the transfer that a one-adult household would get. Under the latter two assumptions, the government budget constraint is given by

$$G = \sum_i \sum_k \tau_i^k x_i^k - \sum_k T^k \qquad (4.16)$$

where x_i^k is the consumption of good i by household k, τ_i^k is the tax rate on good i for household k, T_i^k is the lump-sum transfer to household k, and G is government revenue (net of transfers). In each case, we calculate the demand for each good implied by a given income and vector of prices for each of the representative households, using the share equations (4.7). We then solve numerically for the tax cut or increase in the lump-sum transfer (depending on how the revenue is recycled) that will exactly offset the increased gas tax revenue and thus hold G constant.[28]

As discussed in the section on measures of tax incidence, we use four different incidence measures, which differ in the implicit assumptions they make about the system of demand elasticities. However, in each case, we still use the full system of demand elasticities to calculate the change in the labor tax rate or lump-sum transfer that is made possible by the increased gas tax revenue; thus, the change in the tax or transfer is the same across all three incidence measures. It is well-known that cross-price elasticities (and particularly the cross-price elasticity with leisure) play an essential role in determining the efficiency of a commodity tax in an economy with a pre-existing tax on labor income. In such an economy, a marginal increase in labor supply leaves the individual worker's utility unchanged (at the margin, the value of forgone leisure will exactly equal the after-tax wage), but produces an increase in tax revenue for the government, and thus an overall efficiency gain. Thus, taxing a good that is a substitute for leisure will produce an efficiency loss in addition to any effect in the market for the taxed good itself, because it will decrease labor supply (and thus labor tax revenue). Taxing a leisure complement will have the opposite effect, producing an additional efficiency gain.[29] Using the same tax and transfer changes holds this efficiency effect of cross-price elasticities constant across the different incidence measures, allowing us to focus on how cross-price elasticities affect distribution.

Comparing these incidence estimates across different income groups will demonstrate how regressive or progressive a particular tax shift is. Comparisons across the four different incidence measures will illustrate the importance (or lack thereof) of incorporating behavioral changes in incidence estimates.

Incidence Results

Table 4-3 presents the results of the incidence analysis (with standard errors computed using the same bootstrap approach used for the standard errors in Table 4-2). For each of the three assumptions about the use of the additional revenue from the increased gas tax, it shows the incidence on each income quintile as a percentage of total expenditure (on gasoline and other goods) for that quintile under each of the four different incidence measures. The table also provides the aggregate burden—the sum of the burden across all five quintiles, divided by the sum of expenditures—from the tax change, thus allowing a comparison of overall efficiency across the three uses for the additional revenue.

Finally, Table 4-3 provides an index of the overall progressivity or regressivity of each tax change. This index is given by

$$\sum_{i=1}^{5} t_i - \sum_{i=1}^{5} \left[\left(-t_i + 2\sum_{j=1}^{i} t_j \right) y_i \right] \tag{4.17}$$

where y_i is quintile i's share of total expenditure and t_i is the burden of the tax change on quintile i as a fraction of total expenditure for all quintiles. This index is closely related to the widely used Suits index (see Suits 1977); dividing this index by $\sum_{i=1}^{5} t_i$ (the aggregate burden of the tax change as a percentage of expenditure) gives the Suits index. The index will be positive for a progressive tax change, zero for a flat tax, and negative for a regressive tax change. A tax equal to 1% of total expenditure levied only on households at the top of the income distribution would have an index of 1%; the same tax levied on households at the bottom of the distribution would have an index of –1%. This index also aggregates easily; the index for a combination of two tax changes is simply the sum of the indices for each of the two changes taken separately.

We use this index rather than the closely related Suits index because the Suits index can produce misleading results for a tax reform such as those considered in this chapter, where one tax rate is raised and another lowered. The problem arises because the Suits index for a combination of two tax changes is equal to the average of the Suits indices for each of the changes, weighted by the total burden of the tax. For a reform that combines a tax increase and a tax decrease, one weight is positive and one weight is negative. When the increase in tax burden from the tax increase is similar in magnitude to the decrease in tax burden from the tax decrease—which tends to be the case for a revenue-neutral tax reform—this can greatly increase the magnitude of the Suits index. Indeed, if the increase and decrease exactly offset, so that the net change in the aggregate tax burden is zero, the Suits index is undefined. The index given by (4.17) avoids these problems.[30]

Table 4-3 shows that increasing the gas tax will generally be regressive, but that the use of the revenue greatly affects the degree of regressivity and can make the change progressive if the revenue is returned lump-sum. If the revenue is simply discarded, then the burden of the tax as a fraction of expenditure is lower for higher-income groups. The cost to either of the bottom two quintiles, as a fraction of expenditure, is roughly twice that for the top quintile. The progressivity index for this case is –0.31%. The equivalent Suits index for this case is –0.14 (note that there is no offsetting tax decrease in this case, so the Suits index produces valid results). As a comparison, the Suits index for the taxes considered by Suits (1977) ranged from –0.17 to 0.36, so the regressivity of the gas tax is similar to that of the most regressive tax (the payroll tax) considered in that study. This is not surprising; numerous studies have shown that gasoline has a larger budget share for lower-income groups, and thus that the gas tax tends to be regressive. An exception is the bottom quintile, which bears a slightly lower burden from the gas tax than the second quintile, because a substantial number of households in the bottom quintile do not consume gasoline at all—again, a result that matches those from prior studies.

If the revenue is used to cut the labor tax, all five quintiles are substantially better off than when the revenue is discarded. In addition to the obvious gain from not wasting the revenue, cutting the labor tax rate reduces the deadweight loss in the labor market. Consequently, this is the most efficient of the three options considered for how to use the gas tax revenue. This also reduces the regressivity of the tax; the progressivity index is –0.11%, indicating that the net effect is still regressive, but is less regressive than the gas tax increase alone. This is not surprising; labor provides a greater fraction of income for low-income households than for high-income households, so reducing the tax rate on labor will be somewhat progressive.

Using the revenue to provide a lump-sum transfer to households makes the overall effect much more progressive. The bottom quintile is actually better off as a result of the change, even though our estimates ignore the external

TABLE 4-3 Incidence of an Increased Gas Tax.

All Transfer or Tax Cut Calculations Based on Full Demand System

Additional gas tax revenue discarded:

	Quintile					Aggregate Tax Burden	Progressivity Index
	1	2	3	4	5		
Equivalent Variation Using By-quintile Elasticities	-2.78%	-3.01%	-2.88%	-2.49%	-1.60%	-2.25%	-0.31%
	(0.61)	(0.19)	(0.17)	(0.13)	(0.08)	(0.07)	(0.05)
Consumer Surplus Using By-quintile Elasticities	-3.06%	-3.04%	-3.00%	-2.53%	-1.67%	-2.33%	-0.32%
	(0.88)	(0.23)	(0.22)	(0.16)	(0.11)	(0.10)	(0.07)
Consumer Surplus Using Aggregate Elasticities	-3.04%	-3.27%	-3.00%	-2.53%	-1.52%	-2.29%	-0.38%
	(0.14)	(0.14)	(0.12)	(0.11)	(0.06)	(0.09)	(0.02)
No Demand Response	-3.55%	-3.82%	-3.50%	-2.95%	-1.77%	-2.67%	-0.44%
	(0.08)	(0.06)	(0.05)	(0.04)	(0.03)	(0.02)	(0.01)

Additional gas tax revenue used to cut labor tax rate:

	Quintile					Aggregate Tax Burden	Progressivity Index
	1	2	3	4	5		
Equivalent Variation Using By-quintile Elasticities	-0.78%	-0.59%	-0.44%	-0.24%	-0.13%	-0.30%	-0.11%
	(0.55)	(0.21)	(0.18)	(0.16)	(0.11)	(0.09)	(0.04)
Consumer Surplus Using By-quintile Elasticities	-1.14%	-0.65%	-0.61%	-0.32%	-0.21%	-0.41%	-0.13%
	(0.82)	(0.23)	(0.21)	(0.17)	(0.12)	(0.08)	(0.06)
Consumer Surplus Using Aggregate Elasticities	-1.10%	-0.87%	-0.60%	-0.31%	-0.06%	-0.37%	-0.18%
	(0.09)	(0.09)	(0.09)	(0.09)	(0.06)	(0.07)	(0.01)
No Demand Response	-1.59%	-1.40%	-1.09%	-0.71%	-0.31%	-0.74%	-0.24%
	(0.18)	(0.19)	(0.19)	(0.17)	(0.11)	(0.15)	(0.02)

TABLE 4-3 *(Cont.)*

Additional gas tax revenue used to provide lump–sum transfer to households:

	Quintile					Aggregate Tax Burden	Progressivity Index
	1	2	3	4	5		
Equivalent Variation Using By-quintile Elasticities	2.18%	0.05%	-0.60%	-0.81%	-0.70%	-0.44%	0.25%
	(0.54)	(0.24)	(0.18)	(0.13)	(0.08)	(0.08)	(0.05)
Consumer Surplus Using By-quintile Elasticities	1.86%	0.01%	-0.72%	-0.86%	-0.77%	-0.52%	0.23%
	(0.80)	(0.26)	(0.21)	(0.16)	(0.10)	(0.07)	(0.06)
Consumer Surplus Using Aggregate Elasticities	1.88%	-0.21%	-0.72%	-0.85%	-0.62%	-0.48%	0.18%
	(0.28)	(0.13)	(0.08)	(0.06)	(0.03)	(0.06)	(0.03)
No Demand Response	1.37%	-0.76%	-1.23%	-1.27%	-0.87%	-0.86%	0.11%
	(0.39)	(0.24)	(0.18)	(0.13)	(0.07)	(0.14)	(0.04)

Note: Each incidence measure is expressed as a percentage of total expenditures (on gasoline and other goods). Standard errors are reported in parentheses and are expressed in the same units as the estimates. Gas tax is increased to $1.39/gallon in all cases.

benefits of reduced gas consumption. In this case, the increased transfer that households in the bottom quintile receive more than offsets the higher price of gasoline. The effects nearly offset for the second quintile, which is insignificantly better off. And for the top three quintiles, the effect of the higher gas price dominates; those households are made worse off. In this case, the incidence as a fraction of expenditures is highest for the fourth quintile; the lower quintiles do relatively well because of the progressive lump-sum transfer, and the gas share for the top quintile is small enough that even though that quintile's relative benefit from the lump-sum transfer is smaller, it still does better than the fourth quintile. The progressivity index is 0.25%, indicating that the overall change is progressive; the lump-sum transfer is sufficiently progressive to overcome the regressivity of the gas tax.

This policy is less efficient than using the revenue to cut the labor tax rate; the difference in aggregate burden between the two policies is 0.14% of expenditures. This cost is relatively small in absolute terms, but it is substantial relative to the distributional effects. Using the revenue for a lump-sum transfer rather than for a cut in the labor tax makes the average household in one of the bottom two quintiles better off by $146 per year, but at an average cost of $222 to the households in the top two quintiles.

Now consider the effect of different assumptions about demand responses. Comparing the incidence measure that assumes no change in demand to any of the other measures shows that considering demand responses is important for measuring either the overall burden of the tax or the distribution of that burden. Omitting demand responses will lead one to substantially overstate the burden of the gas tax. Given the magnitude of the tax change involved, this is hardly surprising; households will consume significantly less gasoline in response to such a large tax increase, and that reduces the burden of the tax increase.

Omitting demand responses also makes the gas tax appear more regressive. Gas demand elasticities are relatively similar across most of the income distribution, but gas demand for the top quintile is substantially less elastic. Thus, the relative burden on the top quintile is larger under measures that include demand responses, making the tax more progressive. This effect can be substantial; a gas tax increase with the revenue used to provide a lump-sum transfer appears to be only slightly progressive (progressivity index = 0.11%) when demand responses are ignored. The result is more progressive under any of the other incidence measures, and using the equivalent variation gives a result that is more than twice as progressive (progressivity index = 0.25%) for this case. A similar pattern appears when the tax revenue is discarded or used to cut the labor tax.

Comparing the two consumer surplus measures shows the importance of allowing demand elasticities to vary by quintile. This distinction is unimportant for the estimated aggregate burden of the policy; under each of the three scenarios, the difference in estimated aggregate burden between the two consumer surplus measures is only 0.04% of expenditures. However, in estimating the progressivity or regressivity of the policy, this distinction is much more important; assuming that demand elasticities are constant across quintiles makes the gas tax look more regressive than if elasticities are allowed to vary across quintiles. Assuming equal elasticities produces higher estimates of the burden on the bottom two quintiles, and a lower estimate of the burden on the top quintile.

This is exactly what one would expect: if elasticities differ across quintiles, using the aggregate elasticity for all groups will produce lower estimates of burden for some quintiles and higher estimates for other quintiles, thus affecting the distribution of the burden but having little effect on the aggregate burden. Thus, it is clear that there is no need to use disaggregated elasticities when calculating the aggregate burden of the gasoline tax. But even when calculating the distribution of that burden, it is unclear that one would want to use disaggregated elasticities, because the standard errors are substantially smaller when using the aggregate elasticities. Again, this is exactly what one would expect; the estimates of the aggregate elasticities will be substantially more precise than the estimates of elasticities by quintile, and that difference will carry through into the precision of the incidence estimates.

Comparing the consumer surplus measure that allows elasticities to vary by quintile to the full equivalent variation shows that the three potential differences between these two incidence measures—cross-price effects, income effects, and the assumption of a constant-elasticity demand curve—together have relatively little effect; these two measures yield very similar results. Even for the large price changes considered here, these differences are sufficiently weak that taking them into account makes little difference in the incidence analysis.

The results in Table 4-3 were calculated using the same tax cut or lump-sum transfer for each case in order to isolate the effect of demand responses on incidence calculations. Thus, in each case the revenues from the gas tax (and the size of the tax cut or lump-sum transfer it can finance) were calculated using the full demand system, including cross-price effects.[31]

Conclusion

This chapter has analyzed the distributional effects of increasing the gasoline tax under a range of assumptions about how the revenue is recycled and for a range of different incidence measures. It shows that increasing the gasoline tax will generally be regressive, though it can become somewhat progressive if the additional revenue is used to provide a lump-sum transfer to households; the transfer is sufficiently progressive to outweigh the regressivity of the tax increase. Using the additional revenue to lower taxes on labor yields an efficiency gain and makes the policy more progressive, but not by enough to overcome the regressivity of the gas tax.

Our use of Consumer Expenditure Survey data and estimation of the almost-ideal demand system allows us to incorporate behavioral responses, including cross-price effects, in our incidence calculations. Similar data and techniques could be used to examine the incidence of other environmental taxes across income groups, regions, or urban versus rural households.

Using an incidence measure that incorporates demand responses results in significantly lower estimates of the tax burden on all groups, because gas consumption falls substantially in response to the increased tax. It also substantially lowers the estimated regressivity of the tax, primarily because of the effect on the top quintile, which has a substantially lower gas demand elasticity than do the other quintiles. Using an incidence measure that allows demand elasticities to vary across income quintiles has little effect on the measured aggregate burden of the tax, but it has a substantial effect on the distribution of that burden, yielding a less regressive estimate. Finally, our results suggest that ignoring cross-price effects introduces little error in incidence calculations, and that the same is true for the other differences between our equivalent variation and consumer surplus measures.

However, one must be careful in interpreting this result. All of our calculations used the same demand system estimates, and varied the incidence measure. It might be valuable for future research to compare these results to the alternative approach that would hold the incidence measure fixed and vary the assumptions made in the estimation.

Many debates on environmental taxation center on the burden a proposed tax would impose on low-income households. These debates often ignore the ways in which the additional environmental tax revenues can be used to reduce the overall tax burden on those families. Many European countries have already implemented environmental tax reforms that use environmental tax revenues to reduce labor taxes and thus generate efficiency gains in labor markets. Our results suggest that these labor tax cuts will also mitigate the regressive effect of environmental taxes. Devoting a portion of the revenue to a lump-sum transfer, or to another progressive policy such as an increase in the earned income tax credit, could even make the net effect of the policy progressive.

Appendix 4-1

Correcting for Selectivity Bias

This appendix explains corrections made for potential selection into work and into gasoline consumption. Results for the estimation discussed here are available from the authors by request.

The Work Decision

Since we do not observe wages for individuals who are not working, we follow Heckman (1979) to correct for the associated selectivity bias and obtain selectivity-corrected net wages for workers. Heckman's model is composed of a selection mechanism and a regression equation. Our selection mechanism specifies that z_h equals 1 if an individual works and 0 otherwise and:

$$z_h^* = \mathbf{w_h}\boldsymbol{\xi} + u_h \tag{4.A1}$$

where $\mathbf{w_h}$ includes variables that predict whether the individual works, $\boldsymbol{\xi}$ are parameters to be estimated, u_h is an error term,

$$z_h = 1 \text{ if } z_h^* > 0, \ z_h = 0 \text{ if } z_h^* \leq 0$$

$$\text{Prob}(z_h = 1) = \Phi(\mathbf{w_h}\boldsymbol{\xi}), \ \text{Prob}(z_h = 0) = 1 - \Phi(\mathbf{w_h}\boldsymbol{\xi})$$

and $\Phi(\cdot)$ is the standard normal cumulative distribution function.

The regression equation is used to predict an individual's net wage:

$$p_{lh} = \mathbf{x_h}\boldsymbol{\beta_h} + \varepsilon_h \tag{4.A2}$$

where net wage, p_{lh}, is observed only if the individual works $(z_h = 1)$, $\mathbf{x_h}$ include variables that predict net wage, $\boldsymbol{\beta_h}$ are parameters to be estimated, and ε_h is an error term. The Heckman model assumes that the error terms in (4.A1) and (4.A2) are jointly distributed bivariate normal $[0,0,1, \sigma_\varepsilon, \rho]$. We wish to correct net wage for the effect of selection into work and in so doing we must account for the correlation among error terms in (4.A1) and (4.A2).

Following Heckman (1979) we specify (4.A1) as a probit of the choice to work or not work. Heckman's "two-stage" model obtains predicted inverse Mills ratios from the probit:

$$R_{zh} = \frac{\varphi(\mathbf{w_h}\boldsymbol{\xi})}{\Phi(\mathbf{w_h}\boldsymbol{\xi})} \tag{4.A3}$$

where $\varphi(\cdot)$ is the standard normal density function. The "two-stage" model includes (4.A3) in (4.A2) and estimates the amended (4.A2) using OLS. Rather than estimate the model in two stages, we estimate (4.A1) and the amended (4.A2) jointly using full information maximum likelihood.

We estimate selection models separately for one-adult and for two-adult households, and within those samples, separately for men and for women. The one-adult probits include age, age squared, education, race, marital status, number of children, region, the log of gas price, the log of other good price, and state-specific quarterly unemployment rates. The two-adult probits contain the one-adult variables plus partner's earnings and partner's demographic information.

The wage equation contains the inverse Mills ratio in (4.A3), age, age squared, education, race, marital status, occupation, and region. We define the dependent variable as the log of net wage and obtain predicted net wages for workers to include in demand system estimation.

In principle, the model defined above would be identified even if the variables in (4.A1) and (4.A2) were the same. In that case, the model would be identified by its functional form and the normality assumption. Note, however, that we include number of children, the logs of gas price and the other good price, state-specific unemployment rates, and, for two-adult households, partner's earnings, in (4.A1), but not in (4.A2). Number of children affects the fixed cost of working and thus the participation decision. But we do not expect number of children to affect the wage, since we control for age, race, and gender; number of children is a standard exclusion restriction in the labor supply literature. Our demand system allows gas and other good prices to affect the continuous demand for leisure, and thus it is reasonable to assume that they also affect the discrete work choice. While high price regions may also be high wage regions, there is no reason to postulate that an individual facing a high gas price or other good price will have a higher wage, since we control for region in our wage equation. Unemployment rates proxy for job availability in a state and thus affect the likelihood of working, but it is not clear why they would affect wages. Partner's earnings proxy for an individual's nonwage income but should not directly affect an individual's wage; this is another standard exclusion restriction.

The Gasoline Decision

Since some households have zero expenditure on gasoline, another selection bias may arise. Heien and Wessels (1990) propose a procedure for dealing with households that do not consume a good. They estimate a probit of the dichotomous choice to consume or not consume a particular good and obtain inverse Mills ratios analogous to that defined in (4.A3). They include these inverse Mills ratios in all equations of the demand system.

Shonkwiler and Yen (1999), however, find that especially in the case of a large number of censored observations, the Heien and Wessels procedure is biased. They recommend an alternative unbiased approach that still allows for the demand system to be estimated over all households. However, we cannot use the approach suggested by Shonkwiler and Yen, because we need to use instrumental variables to estimate our system, and their approach would yield inconsistent estimates if used along with instrumental variables.

Like Heien and Wessels, we use the results of the probit on the dichoto-mous choice to consume or not consume gasoline to calculate an inverse Mills ratio (R_{gh}) for each household. We run separate probits for one-adult and two-adult households. Each probit includes the log of total goods expenditures, age, age squared, race, marital status, the number of children, region, an indi-cator for whether the household owns its house, and the logs of gas price and other good price. Demographic variables for both adults are included in the two-adult probit. Home ownership acts as a proxy for wealth and access to credit and thus increases the likelihood of owning an automobile and the like-lihood of consuming gasoline (see West 2004). We do not expect it to affect the continuous choice of gasoline and therefore use it as an exclusion restriction.

Also like Heien and Wessels, we include the inverse Mills ratio from the discrete gas choice (R_{gh}) in all equations of our demand system. We, however, estimate the system on *only households* that *consume gasoline* and in which *all adults work*, whereas Heien and Wessels estimated the system for *all* house-holds, not just those that consume positive amounts of all goods. In their case, it was particularly important to keep all households in the system estimation, because they estimated a demand system defined over some goods that the majority of households did not consume. We lose relatively few households that consume no gasoline; in the one-adult sample, 510 households buy no gasoline; in the two-adult sample 56 households buy none. And, by excluding households that do not consume gasoline from the estimation, we avoid the bias noted by Shonkwiler and Yen. In this sense our correction is exactly anal-ogous to that in Heckman (1979), in which the second stage regression is esti-mated only over uncensored households.

Acknowledgments

For their helpful comments and suggestions, we thank Don Fullerton, Larry Goulder, Michael Greenstone, Dan Hamermesh, Gilbert Metcalf, Ian Parry, Raymond Robertson, Ken Small, Steve Trejo, Pete Wilcoxen, participants in the November 2001 and May 2002 NBER Environmental Economics Conferences, three anonymous referees, and the editor. For their excellent research assist-ance, we thank Chris Lyddy, Trey Miller, and Jim Sallee.

Notes

*   This chapter is based on the following article: West, Sarah E., & Robertson C. Williams III. 2004. Estimates from a consumer demand system: implications for the incidence of environmental taxes. *Journal of Environmental Economics and*

Management, Elsevier, vol. 47(3). We gratefully acknowledge permission from *Journal of Environmental Economics and Management* to reprint it.

1   See Dahl (1986); Dahl and Sterner (1991a), and Espey (1996) for reviews of the gasoline demand literature. We focus on the cost side of environmental policies; we do not consider the distribution of benefits from reduced pollution (see Baumol and Oates 1988, Brooks and Sethi 1997 for general discussion of the distribution of benefits).

2   The recent survey on tax incidence analysis by Fullerton and Metcalf (2002) describes three different measures of the burden of a tax. Those three measures correspond to the incidence measures considered in this chapter, except that we consider two variants of the consumer surplus measure: one that allows elasticities to vary with income and one that does not, while Fullerton and Metcalf mention only one consumer surplus measure.

3   Parry and Small (2002) estimate the marginal external damage associated with gasoline consumption at 83 cents per gallon. The average before-tax price of gasoline in our sample is less than $1 per gallon.

4   Several other incidence studies consider the effects of different scenarios for the use of environmental tax revenue, including Brännlund and Nordström (2004), Cornwell and Creedy (1997), Metcalf (1999), and Symons et al. (1994). However, of these, only Metcalf examines the gasoline tax, and it does not include any behavioral responses to tax changes.

5   These scenarios are by no means exhaustive—there are more efficient uses of revenue than cutting wage taxes, or more progressive uses than a lump-sum transfer—but they provide a simple way of demonstrating how the use of the revenue affects the incidence of these policies and illustrate the equity-efficiency tradeoff involved. Environmental tax reforms—measures that use pollution tax revenue to reduce taxes on employment or investment—are now common in Europe (Denmark, Finland, Germany, Italy, the Netherlands, Norway, Sweden, and the United Kingdom have all implemented such reforms (Hoerner and Bosquet 2001)). Such reforms have also been proposed in state legislatures in the United States (Hoerner and Erickson 2000). Feldstein (2001) proposed a system of tradable gasoline vouchers to reduce gasoline consumption. Feldstein mentions a range of possible ways of allocating these vouchers, one of which (dividing vouchers equally across households) would be equivalent to a gasoline tax with the revenue returned lump sum.

6   The labor literature suggests that labor demand is much more elastic than labor supply (see Hamermesh (1993) for a survey of the labor demand literature), which implies that workers will bear nearly all the burden of labor taxes.

7   This error will be relatively small, given the low elasticity of demand for gasoline. Still, to the extent that gasoline suppliers bear part of the burden of the tax, our estimates will overstate the incidence on households that consume gasoline, and will understate the incidence on households that own firms that supply gasoline. If firm owners are concentrated in the top quintile, this would mean that our estimates would overstate the regressivity of the tax.

8   The theoretical problems with consumer surplus arise because these income effects can lead to inconsistencies in the relationship between utility changes and the dollar measure of consumer surplus. The equivalent and compensating variations avoid these potential problems. See Willig (1976) for more on the difference between

equivalent variation and consumer surplus, and for conditions under which the change in consumer surplus will be a close approximation to the equivalent variation.

9   One of the advantages of the Deaton-Muellbauer demand system used in this chapter is that it is sufficiently flexible to provide meaningful estimates of cross-price elasticities. The assumption of separability, which is imposed by many commonly used functional forms such as the linear expenditure system (LES), implies that cross-price elasticities are determined entirely by own-price and income elasticities. Thus, even when using an incidence measure such as equivalent variation that can account for cross-price elasticities, studies that use the LES (e.g., Cornwell and Creedy 1997) still do not incorporate cross-price elasticities in any meaningful way. Two other studies of the incidence of an environmental tax, Brännlund and Nordström (2004) and Symons et al. (1994), use the Deaton-Muellbauer demand system, but neither paper discusses the importance of such cross-price effects, nor does either paper provide a second set of estimates that exclude cross-price effects. Indeed, we are not aware of any studies—on environmental taxes or in the broader literature on tax incidence in general—that have looked specifically at the role of cross-price effects in determining the incidence of a commodity tax.

10   The percentage errors introduced by these three factors are proportional to the change in the marginal utility of income, the compensated cross-price change in demand, and the change in demand elasticity along the demand curve, respectively, induced by the price changes. Each of these goes to zero for sufficiently small changes in price.

11   Sevigny (1998) examines the incidence of the gas tax using a measure that differs in that it uses the quantity consumed after the imposition of the tax, rather than before, and thus understates the tax burden rather than overstating it.

12   The reader should be careful not to confuse the burden of a tax with the excess burden, which is the burden minus the government revenue. Ignoring demand responses will obviously lead to an underestimate of excess burden.

13   For discussion of the use of different price indices in the AIDS, see Moschini (1995) and Buse and Chan (2000).

14   We exclude those over 65 because their labor market behavior is significantly different from that of adults under 65. Doing separate estimation for the over-65 is not feasible; since so few of the over-65 households work (less than 5 percent) the by-quintile samples are very small (as small as 4 observations in one case). This exclusion influences our results in several respects. Those over 65 consume less gasoline than their under-65 counterparts, and so will bear less tax burden from the gasoline tax. They also are much less likely to work, so they will get less benefit from a cut in the labor tax rate. Because those over 65 have lower average incomes than the rest of the population, our results will overstate the regressivity of the gasoline tax and overstate the progressivity of the labor tax rate cut.

15   Estimates of marginal and average tax rates should be reasonably accurate given the level of detail in the TAXSIM model, which incorporates state and federal tax brackets, the earned income tax credit, the child tax credit, deductibility of state income taxes, and other important features of the tax code. Because of the other features of the tax code (such as credit phase-outs), the marginal tax rate is not necessarily equal to the statutory marginal rate. Therefore, marginal tax rates are calculated based on a $1,000 increase in earned income. For more detail on this and other features of the TAXSIM model, see Feenberg and Coutts (1993).

16   Use of estimates from regressions that make *no* correction for sample selection do not significantly affect incidence results.

17   The overall wage distribution in our sample closely follows the wage distribution in the 1997 Current Population Survey.

18   The equivalence scale also results in a more even distribution of household size across quintiles. If quintiles were based on household expenditure, without any adjustment for household size, 81% of the households in the bottom quintile would be one-adult households, as compared to only 17% in the top quintile. Under our equivalence scale, 64% of those in the bottom quintile are one-adult households, as compared to 36% in the top quintile. Results from sensitivity analysis with quintiles defined by unscaled total expenditures are available as supplementary material at http://www2.econ.iastate.edu/jeem/supplement.htm.

19   This equivalence scale is suggested by Williams et al. (1998), which also considers several alternatives and finds that the choice of equivalence scale has little effect on distributional estimates. We find a similar result: changing the relative weight given to adults and children in the equivalence scale has little effect on the ranking of households. Even if quintiles are based on total household income, without any adjustment for family size, the general pattern of results is unchanged (results for this case are available from the authors upon request).

20   Note that the probits used to correct for selectivity bias include at least one variable not included in demand system estimation.

21   Observations for workers in two occupation categories (farming, forestry, fishing, and groundskeeping is the first and the armed forces the second) are spread very thinly across states. For workers with these occupations, we instrument for net wage with the national mean net wages by occupation rather than the state level means. Gasoline prices may also be endogenous. In the absence of a good instrument for state-level gas prices, however, we do not control for this potential endogeneity (state-level gas tax rates turn out to be poor instruments for gas prices; they explain only about 1% in gas price variation in our sample).

22   Full estimation results are available as supplementary material at http://www2.econ.iastate.edu/jeem/supplement.htm. The first stage of 3SLS develops instrumented values for the endogenous variables. In our system the log of the net wage (already corrected for selectivity) is regressed on all exogenous variables in the system including the instrument, the mean of log net wage by occupation by state. This regression is used to obtain the predicted value of log net wage for each observation. A similar process is undertaken to instrument for the real income term. The second stage obtains a consistent estimate for the covariance matrix of the equation errors based on the residuals from a two-stage least squares estimation of each equation in the system. Last, a generalized least squares type estimation is performed using the covariance matrix estimated in the second stage with the instrumented values in place of the right-hand-side endogenous variables (see Greene 2003).

23   Equations for these elasticities are available as supplementary material at www.aere.org/journal/index.html.

24   These are short-run elasticities; households in our system do not respond to gas price increases by, for example, buying more fuel-efficient cars. To the extent that wealthier households may be more able than poor households to avoid gas taxes in the long run by switching vehicles, our use of short-run elasticities will result in incidence estimates that are biased towards greater progressivity.

25   Each bootstrap replication recomputed the division of households into quintiles and the calculation of the Heckman-corrected wage, as well as recomputing the regression coefficients. Thus, the standard error estimates incorporate not only variation in regression coefficients, but also variation in quintile means, corrected wages, etc. Standard errors based on the standard error estimates from the regressions would have been biased downwards because they omit those extra sources of variation.

26   Because the initial level of the gas tax differs across states, the average gas tax rate paid differs slightly across quintiles. Thus, because the gas tax is increased to the same level for all quintiles, the amount of the tax increase also varies slightly across quintiles. The results would be very similar, however, if we were to hold the tax increase, rather than the new tax rate, constant across quintiles; the difference between the highest tax rate faced by any quintile and the lowest is less than three cents.

27   The effects of an increase in gas prices on other goods is likely to be small: Metcalf (1999) uses Benchmark Input-Output Accounts to determine the effect of a 15% gas price increase on other goods' prices. He finds that other goods' prices increase by no more than 0.6%; most goods' prices increase by 0.2% or less. Our scenario increases gas prices by about 85%, implying that other goods' prices would rise by no more than 3.4%.

28   Note that this implies that the amount of revenue from the gas tax used to finance government spending (on road construction and maintenance, for example) remains constant; only the additional revenue raised by the increase in the gas tax is used to finance the lump-sum transfer or reduction in the labor tax rate.

29   The literature on second-best optimal environmental taxes has extensively investigated the efficiency implications of the cross-price elasticity between leisure and polluting goods. For a recent survey of this literature, see Goulder (2002). West and Williams (2004) focus specifically on the effect of the cross-price elasticity between gasoline and leisure on the efficiency of the gasoline tax.

30   Metcalf (1999) takes an alternative approach by calculating the effect of the tax reform on the Suits index for the entire tax system, rather than just calculating the Suits index for the tax change. This approach will yield similar conclusions to the index we use. For a tax system that is approximately flat and tax reform that is small relative to the entire tax system, the change in the Suits index for the entire tax system from a tax reform will be proportional to the index we use.

31   We also calculate an alternative set of results, with the size of the tax cut or lump-sum transfer calculated using the same assumptions about demand responses as in the incidence measures (i.e., the incidence measures that assume no demand response include a tax cut or transfer calculated assuming no demand response). Thus the size of the tax cut or transfer varies across the different incidence measures, unlike in Table 4-3. These results are available as supplementary material at http://www2.econ.iastate.edu/jeem/supplement.htm.

## CHAPTER 5

Fuel Tax Incidence in Costa Rica: Gasoline versus Diesel

ALLEN BLACKMAN, REBECCA OSAKWE, AND
FRANCISCO ALPIZAR

S purred by robust economic and population growth, Costa Rica's vehicle fleet grew at 3% per year between 1999 and 2007. By 2007, roughly 800,000 cars, trucks, and buses were registered in Costa Rica, one for every six citizens (Table 5-1). Seventy percent of these vehicles were in the greater metropolitan area (GMA) of San José, which is home to 60% of the country's population (Herrera and Rodríguez 2008).

As cars, trucks, and buses have proliferated in Costa Rica, so too have attendant negative externalities including air pollution, traffic congestion, and accidents. Costa Rica's vehicle fleet generates approximately three quarters of polluting emissions in the GMA (Herrera and Rodríguez 2005) where average annual levels of total suspended particulates, nitrogen oxides, and especially sulfur dioxide all exceed national or international norms (Table 5-2).[1] A contingent valuation survey conducted in the mid-1990s found that GMA residents viewed mobile source air pollution as their single most important environmental problem (Celis et al. 1996 cited in Johnstone et al. 2001).

Cars, trucks, and buses also contribute to untenable congestion on the GMA's poorly planned and maintained road network, particularly in the morning and evening rush hours when traffic is often completely deadlocked. Travel speeds in the GMA average less than five miles per hour, and a survey of car commuters found they are willing to pay half of the national average hourly wage for travel time reductions (Vega et al. 2004; Alpízar and Carlsson 2003).

Finally, traffic accidents are a major problem throughout Costa Rica. Despite safety campaigns, the number of accidents has grown by 5% per year in recent years (Table 5-1). In 2007, a fifth of disability pensions administered by the Costa Rican Social Security agency were for victims of traffic accidents (Ávalos 2007).

In the transportation policy literature, an often discussed means of addressing vehicle related air pollution, congestion, and traffic accidents is to impose a tax on motor fuel: higher taxes can spur cuts in driving, substitution

TABLE 5-1 Economic, Transport, and Fuel Statistics for Costa Rica, 1999–2007.

Year	Population[a] (millions)	GDP per capita[a] ('91 col.[e])	Vehicles[b] (no.)	Price Diesel[c] ('91 col./liter)	Price Regular Gasoline[c] ('91 col./liter)	Sales Regular Gasoline[c] (barrels)	Sales Diesel[c] (barrels)	Traffic Accidents[d] (no.)
1999	3.838	1,398	612,300	29	42	179,722	415,052	48,983
2000	3.810	1,423	641,302	39	56	229,768	394,918	50,358
2001	3.907	1,439	664,563	37	53	251,806	423,255	53,208
2002	3.998	1,480	689,763	34	49	259,584	430,905	58,380
2003	4.089	1,575	728,421	39	55	257,762	443,723	53,668
2004	4.179	1,642	705,975	43	60	265,521	460,323	52,362
2005	4.266	1,739	705,546	53	72	282,415	503,681	57,127
2006	4.354	1,892	729,487	57	82	299,301	587,228	68,627
2007	4.443	2,039	797,902	64	84	311,997	650,535	69,761
Avg. Annual % Change	1.85	4.86	2.94	14.25	9.58	7.43	5.95	4.82

Sources: [a] International Monetary Fund (IMF), World Economic Outlook Database; [b] *Ministerio de Obras Públicas* (MOPT), *Dirección de Planificación Sectorial;* [c] *Refinadora Costarricense de Petróleo* (RECOPE); [d] *Ministerio de Obras Públicas* (MOPT), *Dirección de Planificación Sectorial,* using data from *Consejo de Seguridad Vial;* [e] col. = colones.

TABLE 5-2 Mean Annual Measured Concentrations of Air Pollutants in San José, Costa Rica for Year 2000 and World Health Organization and European Union guidelines ($\mu g/m^3$)

	TSP^a	$PM10^b$	$SO2^c$	$NO2^d$
Measured	101	18	160	31
WHO[e] guideline	60	–	50	40
EU[f] limit	–	40	20	30

[a] Total suspended particulate matter; [b] Particulate matter smaller than 10 microns; [c] Sulfur dioxide; [d] Nitrogen dioxide; [e] World Health Organization; [f] European Union.
Source: Baldasano et al. 2003.

out of fuel-inefficient vehicles and, consequently, a reduction in polluting emissions, congestion, and traffic accidents (Sterner 2007; Timilsina and Dulal 2008). Although fuel taxes have the potential to generate deadweight welfare losses, welfare analyses for both industrialized and developing countries have found that given the large negative externalities associated with driving (and the potential for offsetting distorting labor taxes with fuel tax revenue) the net effect of fuel taxes on welfare is generally positive and substantial (see Chapter 4; Parry and Small 2005; and Parry and Timilsina 2008).

As discussed in the next section, notwithstanding this evidence, a common argument against raising fuel taxes in Costa Rica is that it would be regressive—poor households would bear an unfair burden. To our knowledge, this question has yet to be rigorously examined. Also, the literature on fuel tax incidence in other countries is equivocal. Studies of fuel tax incidence in industrialized countries, where vehicle ownership is widespread in all socio-economic classes, have generated mixed results (see Chapter 4; Parry et al. 2007; Poterba 1991; Santos and Catchesides 2005). Intuitively, one might expect fuel taxes to be less regressive in developing countries where vehicle ownership is concentrated in higher socioeconomic brackets (Sterner 2007). Some emerging research provides support for this hypothesis (see Chapters 8 and 14). Costa Rica is a particularly interesting case study of this issue because it is classified as an "upper middle income country" (World Bank 2009). If fuel taxes are not regressive in Costa Rica, they are unlikely to be regressive in poorer countries where vehicle ownership is even more concentrated in higher socioeconomic brackets.

We use data from a 2005 household income and expenditure survey to examine the distributional impacts of fuel taxes in Costa Rica. We find that the distributional impacts of a fuel tax are different for gasoline and diesel: a tax on gasoline is progressive and that on diesel is regressive, mainly because households in low and middle socioeconomic strata spend a significant share of their income on diesel bus travel. However, even in the short run, the impact of an increase in fuel taxes on these households would be modest: a 1% increase in diesel taxes would spur at most a 0.10% increase in spending on diesel, including via bus transportation.

The remainder of the chapter proceeds as follows. The next section presents background information on Costa Rica, specifically its fuel taxes and its public discourse about distributional effects of these taxes. The next section discusses distributional impacts of a fuel tax. The last section presents a summary and conclusion.

Fuel taxes

Costa Rica's 2001 Law of Tax Simplification and Efficiency (No. 8114) replaced a complicated system of fuel taxes and fees administered by a variety of regulatory agencies with a single tax administered by the Ministry of Finance. The tax is a fixed sum per liter for each type of fuel (regular gasoline, premium gasoline, and diesel) adjusted four times a year for inflation. In May 2009, the tax was 181 colones per liter for regular gasoline, 189 colones per liter for premium gasoline, and 107 colones per liter for diesel (*La Gaceta* 2009). Because the fuel tax is a fixed sum, its percent contribution to the total retail price of fuel depends on that pre-tax price of fuel. In recent years, this contribution has ranged from 28% to 52% for regular gasoline and has been within the range of the percent contribution of fuel taxes in other Latin American countries (Figures 5-1 and 5-2). The 2001 law that established Costa Rica's fuel tax mandated that revenues be allocated as follows: 66.4% to the Finance Ministry (*Ministerio de Hacienda*), 29.0% to the National Road Council

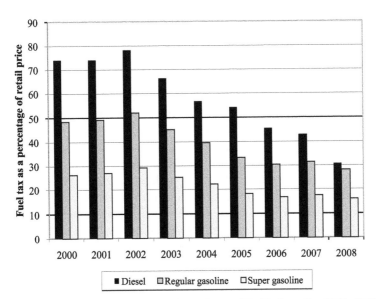

FIGURE 5-1 Tax as a Share of Average Annual Retail Price of Fuel in Costa Rica, 2000–2008.

Source: RECOPE.

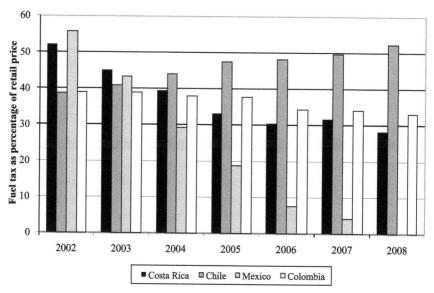

FIGURE 5-2 Tax as a Share of Retail Price of Fuel for Costa Rica, Chile, Colombia and Mexico, 2000–2008.

Sources: RECOPE, CEPAL.

(*Consejo Nacional de Vialidad*, CONAVI), 3.5% to the National Forestry Finance Fund (*Fondo Nacional de Financiamiento Forestal*, FONAFIFO), 1.0% to the University of Costa Rica (*Universidad de Costa Rica*, UCR), and 0.1% to the Ministry of Agriculture (*Ministerio de Agricultura y Ganadería*, MAG).

Distributional concerns

The notion that rising fuel prices unfairly burden the poor is a staple of Costa Rican political discourse and has been used to argue for reductions in fuel taxes. Concern has often focused on diesel, which is used to fuel thermoelectric generating plants, trucks, and buses.[2] In the past several years, faced with rising diesel prices, the Costa Rican Electricity Institute (*Instituto Costarricense de Electricidad*, ICE), along with associations of bus drivers and truck drivers have lobbied for a reduction in the diesel tax on the grounds that it would make electricity, bus transport, and food shipped by truck more affordable to low-income households. For example, in 2008, national legislation was introduced to exempt electric power plants from paying the diesel tax (January 2008, file 16759); exempt buses and taxis from the tax (October 2008, file 17132); and reduce all fuel taxes by 30% and disallow increases for a year (January 2008, file 16924). In addition, in July 2008, bus drivers threatened a national strike on the grounds that fuel prices had become unacceptably high (Cantero 2008).

Incidence Analysis

This section focuses on the distributional impacts of fuel taxes in Costa Rica. The first subsection discusses our methodology, followed by subsections presenting an analysis of household survey data that indicate which socioeconomic strata are most affected via direct spending on fuel. The last subsection presents results of a simple spreadsheet simulation of the magnitude of these effects.

Methods

The incidence of fuel taxes is typically assessed by examining the variation across economic strata of some measure of the change in households' welfare due to the tax (for example, the dollar amount of fuel taxes they paid or the change in their consumer surplus) normalized by pretax absolute level of welfare (for example, total income or expenditure). The literature on fuel tax incidence highlights four methodological issues. The first concerns the measure of total welfare used to sort households into economic strata and to normalize changes in welfare. Studies that use annual income to proxy for pre-tax welfare generally find that gasoline taxes are regressive (KPMG Peat Marwick 1990). However, these studies have been criticized on the grounds that "permanent income" (Friedman 1957), not annual income, drives households' consumption choices, and that socioeconomic strata defined by annual income include households whose income is much higher or lower than their permanent income (because they are headed by people who are particularly young and old or in the midst of a transient financial shock) and whose consumption choices, therefore, differ markedly from other members of these strata. Most recent analyses of fuel tax incidence use measures of permanent income, such as annual household expenditure, to proxy for pre-tax welfare (Poterba 1991; Walls and Hanson 1999; Metcalf 1999; Hassett et al. 2009). These studies tend to find that fuel taxes are less regressive than it would seem from studies that rely on annual income to measure welfare. Here, we use annual expenditure as a proxy for lifetime income.

The second methodological issue concerns indirect impacts of fuel taxes on households' welfare. The ideal analysis of fuel tax incidence would measure general equilibrium changes in consumer and producer prices of all goods and services in the economy due to a fuel tax increase, and then measure the effects of those price changes on household welfare in different economic strata. However, the informational requirements of such an analysis, which would include demand and supply elasticities for all goods and services and the distribution of ownership of firms across economic strata, are costly (see Chapter 4). Many studies of fuel tax incidence omit altogether consideration of indirect impacts of fuel taxes (Poterba 1991; Walls and Hanson 1999; Bureau 2011). Those that do not often make simplifying assumptions to facilitate modeling these effects—specifically, that supply is perfectly elastic so that impacts of fuel taxes on producer prices can be ignored and, in many cases, that demand for various goods and services is perfectly inelastic (Chapter 8; Metcalf 1999; Hassett et al. 2009). Despite the assumption that consumers do not respond to price changes, these studies often find that indirect impacts of fuel taxes are relatively small (Metcalf 1999). Here, we omit consideration of second order impacts.

A third methodological issue concerns variation across economic strata in own-price elasticity of fuel demand. Studies that use household panel data to estimate elasticities of fuel demand in the United States have found that demand is more elastic in lower economic strata, a feature that mitigates fuel tax regressivity (West 2004; Chapter 4). However, we lack panel data needed to estimate elasticities by socioeconomic strata and therefore assume an average elasticity for all strata.

A final methodological issue concerns recycling of fuel tax revenue. The incidence of fuel taxes will depend on whether and how tax revenue is used (Metcalf 1999; Wiese et al. 1995). Here, we abstract from this issue and focus on tax impacts absent recycling.

Spending on Fuel Taxes by Decile

To analyze the incidence of fuel taxes, we use data from the National Statistics and Census Institute (*Instituto Nacional de Estadística y Censos*, INEC) 2004–2005 Household Income and Expenditure Survey, which was derived from a random survey of 4,231 households. The analysis comprises four steps. First, we calculate the upper and lower bounds of total monthly expenditure deciles in the INEC data. Table 5-3 presents the results: the upper bounds of the deciles range from $178 to $52,796. Second, we identify five different types of monthly fuel expenditures included in the survey data:

1   direct expenditures on gasoline;
2   direct expenditures on diesel;
3   indirect expenditures on diesel via spending on bus transportation;
4   direct and indirect expenditures on diesel; and
5   all direct and indirect expenditures added together.

To calculate (3), we rely on the cost-based pricing model of the Public Service Regulatory Authority (*Autoridad Reguladora de Servicios Públicos*, ARESEP), the institution that sets bus fares in Costa Rica (ARESEP 2008). This model mandates that the cost of diesel accounts for 21% of bus fares. Accordingly we assume 21% of households' expenditures on bus transportation are devoted to diesel. Third, for each decile, and for each of the four categories of fuel spending, we calculate the average expenditure on fuel taxes as a percentage of total expenditure on all goods and services. Fourth, we compare these averages across deciles to determine whether the share of expenditures devoted to taxes

TABLE 5-3 Costa Rica 2004–2005 Monthly Expenditure Decile Brackets; Vehicle Ownership, by Decile; and Average Household Expenditure on Fuel Taxes as a Percentage of Total Expenditure, by Fuel Type and Decile.

Decile	Min. Total Expenditure (2004 US$)	Max. Total Expenditure (2004 US$)	Vehicle Ownership (%)	Gasoline Tax Expenditure (%)	Diesel Tax Expenditure (%)	Bus Diesel Tax Expenditure (%)	All Diesel Tax Expenditure (%)	All Fuels Tax Expenditure (%)
1	0	178	6	0.15	0.00	0.80	0.80	0.95
2	178	268	10	0.33	0.03	0.87	0.90	1.22
3	268	353	15	0.85	0.00	1.03	1.03	1.88
4	353	455	21	0.87	0.08	1.09	1.17	2.04
5	455	571	30	1.40	0.10	0.94	1.04	2.44
6	572	730	33	1.55	0.11	0.91	1.02	2.57
7	731	953	44	2.19	0.17	0.87	1.04	3.23
8	953	1,297	60	2.74	0.26	0.68	0.94	3.68
9	1,301	2,117	69	2.95	0.18	0.52	0.71	3.66
10	2,131	52,796	91	2.31	0.18	0.28	0.47	2.77
Suits index				0.076	0.103	−0.272	−0.185	0.022

for each category of fuel is greatest among lower deciles—in which case the tax is regressive—or higher deciles, in which case it is progressive. Finally, for each category of fuel taxes, we calculate the Suits index, a widely used measure of tax incidence that is the tax analog of the Gini coefficient used to measure income inequality (Suits 1977). The Suits index is bounded by −1 and +1. For a proportional tax, it is equal to zero, for a progressive tax it is positive, and for a regressive tax it is negative.

Table 5-3 and Figure 5-3 present the results of our analysis. They show that gasoline taxes are progressive. The percentage of total spending devoted to gasoline taxes is much higher in richer deciles than poorer ones, and the Suits index is positive (0.076).

Diesel taxes, however, are regressive. They are composed of taxes paid through direct spending on diesel and taxes paid through indirect spending on bus transportation. Each type of tax has a countervailing distributional impact. Taxes paid through direct spending on diesel are progressive. The average percentage of household total spending on such taxes is highest in the eighth decile, and the Suits index is positive (0.103). However, diesel taxes paid through indirect spending on bus transportation are regressive. Poorer households spend more on such taxes—a phenomenon that reflects their heavy reliance on Costa Rica's extensive, quasi-public system of bus transportation—and the Suits index is negative (−0.272). Moreover, in all but the richest deciles, indirect spending on diesel taxes via bus transportation is several times greater than direct spending. As a result, on net, diesel taxes are regressive, with a negative Suits index (−0.185).

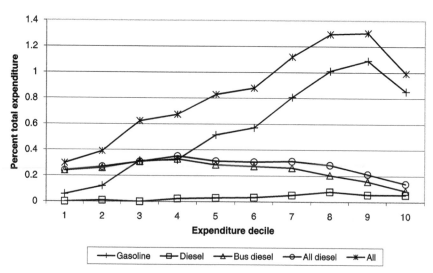

FIGURE 5-3 2004–2005 Average Household Expenditure on Fuel Taxes as a Percentage of Total Expenditure, by Fuel Type and Decile.

Finally, Figure 5-3 indicates that taxes on all types of fuel are progressive. The percentage of total expenditures devoted to taxes on all fuels is highest in the ninth decile. The progressivity of direct spending on fuel taxes offsets the regressivity of indirect spending on diesel taxes via bus transportation so that the overall effect of all fuel taxes is progressive, with a positive Suits index (0.22).

Hence, our analysis suggests that Costa Rica's tax on diesel is regressive, its tax on gasoline is progressive, and that when aggregated, taxes on both types of fuel are progressive.

Magnitude of Impacts

The foregoing analysis indicates which socioeconomic brackets would be most affected by a fuel price hike due to a tax increase, but it leaves open the question of the magnitude of those effects, which would depend on the pre-tax total expenditure on fuel in each decile, the size of the tax increase, and the own-price elasticity of demand for fuel. This section presents a simple spreadsheet simulation that estimates the increase in fuel expenditure due to a fuel tax by decile.

To derive an expression for the increase in fuel expenditure due to a fuel price hike, we begin with two simple definitions. Total expenditure on fuel, G, is defined as

$$G = PQ \tag{5.1}$$

where P is the price of fuel and Q is the quantity of fuel consumed. Own-price elasticity of demand for fuel, ε, is defined as

$$\varepsilon = \frac{dQ}{dP}\frac{P}{Q} \tag{5.2}$$

The total derivative of (5.1) is

$$dG = QdP + PdQ \tag{5.3}$$

Using (5.2) to substitute out dQ, (5.1) to substitute out Q, and simplifying yields

$$dG = GC \tag{5.4}$$

where

$$C = [(dP / P)(1 + \varepsilon)]$$

Hence, the increase in fuel expenditure due to a price hike is equal to the pre-tax expenditure on fuel times C, which is the percentage increase in the fuel price times one plus the own-price elasticity of demand for fuel. Equivalently, dividing each side of (5.4) by T, the total expenditure on all goods and services, generates an expression for the increase in the percent of total household expenditure on fuel due to a price hike (equal to the pre-tax percent expenditure on fuel times C).

$$dG/T = (G/T)C. \tag{5.5}$$

We use equations (5.4) and (5.5) to simulate the short-run impact of a 1% fuel price increase on four categories of fuel spending: gasoline, diesel, bus diesel, and the combination of all three. We use Hernández and Hernández's (1999) estimates of short-run, own-price elasticities of demand for gasoline (–0.33) and diesel (–0.20) in Costa Rica.[3] Note that long-run elasticities of demand for fuel are generally several times larger than short-run elasticities.[4] Given that we use short-run elasticities, our estimates of the magnitude of the fuel tax impacts can be thought of as an upper bound on the long-run impacts. Our simulation suggests that even short-run changes in average monthly household expenditure on fuel due to a 1% fuel price hike are modest (Figure 5-4).

The greatest change in expenditure on gasoline is one fifth of 1% in the ninth decile; the greatest change for non-bus diesel is one fiftieth of 1% in the eighth decile; the greatest change for diesel, including bus diesel, is just under one tenth of 1% in the fourth decile; and the greatest change for combined gasoline and diesel expenditures is just over one quarter of 1% in the eighth decile.

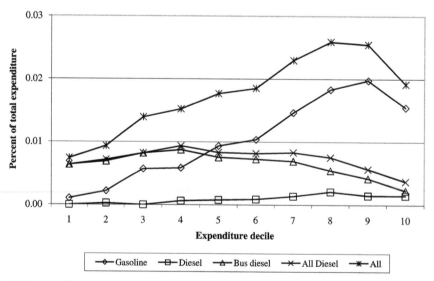

FIGURE 5-4 Change in Percent Total Household Expenditure Devoted to Fuel due to a 1% Increase in Fuel Tax, by Fuel Type and Expenditure Decile.

Conclusion

We have used 2005 household survey data to analyze the distributional impacts of fuel taxes in Costa Rica. We used the survey data to determine which socio-economic strata are most affected by fuel taxes and to conduct a simple spread-sheet simulation that sheds light on the magnitude of these effects.

We found that the distributional impacts of a fuel tax are different for gaso-line and diesel. The tax on gasoline is progressive. Households in the highest socioeconomic strata devote the greatest share of their spending to gasoline. The tax on diesel, by contrast, is regressive: households in lower and middle socio-economic strata devote the greatest share of their spending to diesel.

Simple spreadsheet simulations suggest that even in the short run when fuel demand is relatively inelastic, the impact of an increase in fuel taxes on household spending is modest. A 1% increase in the taxes on gasoline and diesel would spur at most a one-fifth of 1% increase in spending on gasoline (in the eighth expenditure decile) and a one-tenth of 1% increase in spending on diesel (in the fourth decile).

What are the policy implications of these results? In general, they suggest that in Costa Rica—and possibly in similar upper income developing coun-tries—increases in gasoline taxes, whether imposed to help mitigate vehicle negative externalities or for other reasons, would not exacerbate income inequality since wealthier strata would bear most of the burden of the increase. However, the same is not true of increases in diesel taxes, which have the greatest effect on lower and middle income strata, mostly because they would spur increased spending on bus travel.

One possible specific policy implication is that in Costa Rica and similar countries, policymakers can avoid adverse distributional consequences of fuel tax hikes by differentiating taxes on diesel and gasoline and reserving steep increases for the latter. This reasoning has not been lost on Costa Rican policy-makers. The 2001 Law of Tax Simplification and Efficiency set taxes on gaso-line 70% higher than taxes on diesel.

But while differentiating tax increases by type of fuel could mitigate distri-butional concerns, this policy may be problematic on other grounds. First, if consumers buy far more diesel than gasoline as in Costa Rica (Table 5-1), increasing gasoline taxes will generate less revenue than increasing diesel taxes. Second, if diesel vehicles are more plentiful than gasoline vehicles, increasing taxes on gasoline would probably do a worse job of mitigating negative vehicle externalities, particularly in light of the fact that compared to gasoline vehi-cles, diesel vehicles typically generate more fine particles that are particularly damaging to human health (Sterner 2002). Finally, taxing gasoline at a higher rate than diesel can create incentives for businesses and households to buy diesel vehicles. In Europe, for example, the share of such cars in the auto-mobile fleet is growing quickly and is strongly correlated with increases in the price of gasoline relative to diesel (Sterner 2007).

An alternative policy prescription, also supported by our findings, is to use revenue from fuel tax hikes to subsidize bus travel. In Costa Rica, and presumably countries like it, diesel taxes are regressive in large part because they raise spending on bus travel.

In the final analysis, in deciding whether to raise fuel taxes, policymakers in developing countries need to balance an array of distributional, political, fiscal and environmental goals. Our analysis demonstrates that distributional concerns need not trump competing goals.

Notes

1 Carbon monoxide levels also regularly exceed national limits (Alfaro and Ferrer 2001).
2 In addition, diesel is increasingly used in private vehicles because the relatively high price of gasoline has created incentives for households to switch from gasoline cars to diesel cars (Herrera and Rodriguez 2005).
3 To our knowledge, estimates of long-run elatisticies, that are more recent, and estimates that are broken down by income or expenditure category, are not available.
4 For example, Rogat and Sterner (1998) find that the average short-run elasticity of demand for gasoline in 13 Latin American countries is −0.17, while the long run elasticity is −0.85.

## CHAPTER 6

The Income Distribution Effects of Fuel Taxation in Mexico

Thomas Sterner and Ana Lozada

M exico is a prominent oil producer and exporter. The national oil company PEMEX is a state company that develops oil fields, explores and produces crude oil and gas, as well as refining and selling petroleum products domestically and internationally. On the national market, product prices (as well as all energy prices) are low and product taxation is low. The Mexican economy has for several decades been characterized by its high energy intensity and the dominance of the petroleum sector (see Sterner 1985, 1989a).

It is common practice in many developing countries with large petroleum resources to provide their domestic market with cheap petroleum products and other energy carriers (Sterner 1989b). This is typically popular with many customers and serves as a simple means for distributing part of the rent into the economy. It is also often (although not necessarily correctly) argued that this will speed up industrialization and modernization of the economy. It has been argued that it may in fact have the opposite effect as the debate on "Dutch Disease" has shown (Corden and Neary 1982). Either way the Mexican economy has grown—whether "because of" or "in spite of" the booming oil sector or other more or less controversial economic policies, such as joining the NAFTA free commerce area with the U.S. and Canada. For a decade and a half Mexico has been a member of the OECD, and the country has enjoyed a GDP growth rate as high as 7.5% in the period 1990–2007.

For a middle-income country, Mexican carbon emissions are fairly high, at 4 tons of CO_2 per capita (although clearly much lower than the U.S. level—around 20 t/capita). A large number of studies have shown that the most effective method of reducing carbon intensity in the economy is to raise the effective factor price. Unfortunately, we also know that the political economy of fuel taxation is fairly complex, as discussed in Hammar et al. (2004). Although economists know that fuel taxes are an efficient instrument with fairly low costs, there is a great deal of resistance. This has been shown on numerous occasions when U.S. or Mexican politicians have attempted to raise gasoline taxes, or—as in Mexico 2007—in fact succeeded in doing so.

Some of this resistance comes from affected parties who stand to lose from a carbon tax. The Mexican industry has evolved over many decades of heavily subsidized energy, and this has naturally affected both industrial structure and technology choices. Mexico specializes in industrial sectors that require a lot of (fossil) energy, and their technology choices are also energy intensive. This makes perfect economic sense in an environment of low fossil prices. Adapting to higher prices will naturally require considerable effort during a transition period. Also motorists and others have made their choices regarding transportation patterns under a regime of cheap fuel. Large, gas-guzzling cars imported secondhand from the U.S. are quite common, for instance. A fairly large and inefficient ownership of private vehicles is paired with a somewhat under-dimensioned public transport sector. In the short-run, this results in considerable resistance from large sectors of society to the readjustment that would inevitably follow higher prices. It is also politically difficult to motivate prices that are not lower than those in the U.S., another country with light taxation of motor fuels.

One cannot write about petroleum products in Mexico without mention of the very spectacular history of the oil industry. It was nationalized by President Lázaro Cárdenas in 1938 in a move so radical it would normally have led to U.S. intervention. That this did not happen was due partly to the outbreak of World War II, which changed priorities drastically for U.S. politics. Since then PEMEX—ranking 42nd on the *Fortune* list of 500 biggest companies—has been the cornerstone of Mexican energy policy, as well as an icon of radical nationalism. Thirty years ago, a giant offshore field, Cantarell, was discovered. It peaked at more than 2 million barrels of oil a day in 2004 but has since declined very sharply. This has marked the end of an era of easy oil for PEMEX, whose production has fallen to 2.7 million barrels of oil a day (MBD), down from 3.3 MBD at its peak in 2004. PEMEX earned $98 billion in revenues in 2008, but after paying $57 billion in taxes and royalties to the national treasury, its official profit was negative: it made a loss of over $8 billion.[1] The government relies on PEMEX for nearly 40% of its total tax revenue. When export prices fall and the company is politically committed to providing very cheap products on the domestic market, its losses accumulate, and PEMEX has insufficient funds for investment. Raising product prices on the domestic market would thus not only serve to reduce carbon emissions, it would also give revenue to PEMEX and the treasury through two mechanisms: first, the direct tax revenue on domestic sales, and second, by reducing domestic sales, more oil would be available for Mexican exports.

The potential gains are thus important (and there is actually some evidence that domestic product prices have been contracyclical—i.e., higher when international prices are low and vice versa, but the effect is weak). The rhetoric about the duty of PEMEX to supply almost endlessly cash to the treasury, employment to workers—with around 140,000 employees it has many more workers than companies number 1 and 2 on the *Fortune* list, Exxon Mobil and

Royal Dutch Shell, with about four times the revenue but 80,000 and 110,000 workers respectively—and cheap products is a strong force. The distributional issues are very important, and Mexico shares in many respects some characteristics of the U.S., where considerable concern has been expressed that fuel taxes will be regressive. We will in this chapter deal with this concern for the case of Mexico.

During 2010, the government and PEMEX[2] have actually been raising the price of fuels, consistently and frequently but in small steps of about 8 centavos or 1% per month. In the first price increase on January 1, for instance, the gasoline Magna rose from 7.80 to 7.88 pesos/liter. In June of 2010 six price increases totalling 6.2% brought the price of regular Magna to 8.28 pesos/ litre. Diesel and premium gasoline are both a little more expensive, but Magna has the biggest market share (67%).[3] Protests accompany almost every price rise, and it is quite common for the protests to focus on the lot of the poor. On December 19, 2009, upon the first (small) price increment for that year, the left-wing opposition parties PRI and PRD immediately accused the right wing PAN[4] government in parliament of breaking its promise not to raise fuel prices during the year. PAN, which tried to argue that subsidies to the petroleum sector had to be reduced and that they were regressive, was drowned in a chorus of voices saying this would hurt the ordinary Mexican citizen's economy.[5] In January 2010, Mexican Secretary of the Treasury Ernesto Cordero and Secretary of Energy Georgina Kessel were summoned to the Standing Committee of the Mexican Congress to explain and defend the price increases in petrol, diesel, gas, and electricity announced by the Executive, which was said to have caused hardship for the people through increased costs of transportation.

The Distributional Effect of Fuel Taxation in Mexico

As discussed in the introductory chapter, the notion that fuel taxes are regressive is widespread. Support for it exist in countries such as the U.S. even though the results appear to depend on a number of methodological issues. However, we have argued that this may not hold in poor countries, where low-income strata have little access to motorized transport. The question here is what the situation will be like in a middle-income country with some considerable similarities to—but also large differences from—the U.S. As we saw from the introduction, the debate is similar to that in many other countries where the government is regularly under fire if it tries to lower subsidies or apply taxes to this sector.

We might expect to find that the lower the average income of the country the higher the progressivity of its gasoline taxes, since really poor people hardly use any gasoline at all—at least not directly. We need, however, to recognize that poor people in poor countries also need transport; thus they indirectly consume fuels when they use public transport. Therefore, we have in this study included for Mexico the budget shares of both direct fuel consumption and public transport.

The data in Table 6-1 show clearly that the budget share for fuel[6] rises from 1.66% in the poorest decile I to a maximum of 4.88% in decile IX. The very richest in decile X, however, have a slightly lower budget share of 3.54%. Including the tenth decile one might thus actually say that there is some weak evidence of a Kuznets type curve, but with a turning point so high as to still imply in principle that a fuel tax would be quite strongly progressive (with the exception of that difference between the ninth and tenth deciles). Table 6-1 also shows that this correlates quite strongly with the budget shares for automobiles. Figure 6-1 shows that the sum of gasoline and vehicles is in fact progressive over all ten deciles. The conclusion thus far is quite clear that the direct income distribution effects of fuel taxation (or other policies that affect the cost of motoring) are strongly progressive. The reader should note that this would also apply if one looks exclusively at the fuel costs (with the exception of decile X).

Including Indirect Effects

As also shown in Table 6-1 (as well as in Figure 6-1), the budget shares for public transport exhibit the opposite pattern: The poor spend almost 12% of their income on public transport, whereas the rich spend negligible amounts.

TABLE 6-1 Percentage Share of Income Spent on Gasoline and Public Transport.[a,b]

Decile	Fuel ($)	%	Cars ($)	%	Public Transport ($)	%
I	29	1.66	11.23	0.64	206	11.71
II	65	2.05	24.14	0.76	339	10.66
III	91	2.16	27.62	0.66	375	8.91
IV	158	3.02	34.25	0.65	425	8.13
V	213	3.33	43.45	0.68	459	7.20
VI	293	3.77	79.14	1.02	514	6.60
VII	383	4.00	135.46	1.41	567	5.92
VIII	556	4.51	259.92	2.11	543	4.41
IX	839	4.88	415.82	2.42	471	2.74
X	1,442	3.54	1,772.76	4.35	291	0.71

Notes: All prices in Mexican pesos. [a] The first value of each item represents the average monthly expenditure of households ordered by income decile. The second value is the expenditure as a percentage of monthly income of households ordered by income decile. [b] Only includes homes located in cities with population over 100,000.

Source: own estimates based on the National Survey of Household Income Expenditure 2004 (ENIGH 2004) published by the National Institute of Statistics and Geography (INEGI).

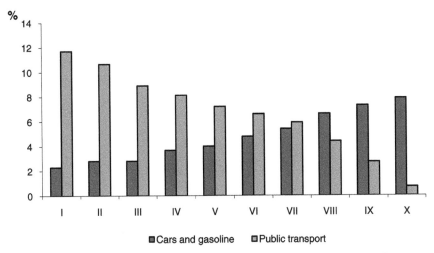

FIGURE 6-1 Expenditures on Gasoline Plus Cars and Public Transport as Percentage of Disposable Income.

The question thus arises: What would the total income distribution effects be in a model that had all indirect effects through a full input-output model to really trace the specific indirect effects of transport fuel use only? Food, for instance, is an important item, as we will see in Chapter 13 on Mali by Kpodar and Chapter 8 on India by Datta. The share of fuel costs in food is not so high, but because the budget share of food for the poor is very high that makes this item important. Still, Datta shows that fuel taxes continue to be progressive even when all indirect effects are considered. We believe therefore that we get a reasonable approximation if we include the effects of fuel prices on public transport prices.

To judge the importance of the indirect effects of fuel price increases, through the fares of public transport on the various income deciles, we need to know how much the cost of public transport would go up when fuel prices increase. Fares are partly regulated in Mexico, and no explicit formula relates fuel price changes to changes in public transport fares. We therefore make the most neutral long-run assumption that fuel prices are carried through in proportion to their importance in total costs. In other words, we assume either marginal cost pricing or at least that profit margins are constant. In this case the appropriate coefficient is simply the cost share of fuel in public transport.[7] To ascertain the cost share we carried out a small survey, which is summarized in Table 6-2.[8]

The cost share for full-size buses was not available but should be lower, given that fuel consumption per passenger is much lower for large buses. Ideally, we would need a passenger-weighted average figure for all the different kinds of public transport. In fact, it would, in principle, have been preferable to have different mixes for different deciles since the different income strata do

TABLE 6-2 Fuel Shares in Revenues for Different Types of Passenger Transport.

Type of Transport	Share of Fuel in Total Revenue (%)
Taxi	32
Combi	46
Microbus	42

Source: Own small-scale survey.

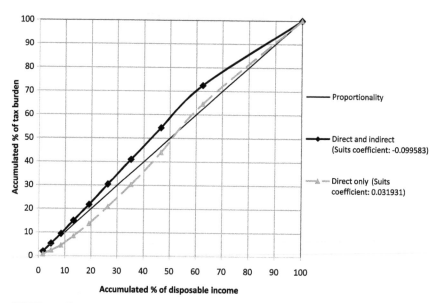

FIGURE 6-2 Concentration Curves and Suits Coefficients for Mexico.

Source: Authors' calculation based on the ENIGH 2004 survey from INEGI.

not use the same type of public transport. However, because we do not have this, and so as not to underestimate the effect on the poor who use microbuses frequently, we will assume an average figure of 42%. (This still seems quite a high number, and it is unclear whether total capital costs have been properly included. In Sweden the fuel share for taxis is under 5% and for buses lower—in spite of very expensive and heavily taxed fuel. However, in Mexico lower wages and older cars reduce the share of fuel—despite cheap fuel in Mexico.)

Using 42% as an estimate of the fuel content of public transport allows us to estimate the total budget shares including both direct and indirect (via public transport) use according to (6.1):

$$\eta_{Ti} = \eta_{Gi} + \sigma_{GP} * \eta_{Pi} \qquad (6.1)$$

where η_{Ti} is the total fuel household budget share for decile I. η_{Gi} and η_{Pi} are the corresponding budget shares for direct purchase of fuel and of public

TABLE 6-3 Share of Direct and Indirect Petrol Expenditures in Disposable Income.

Decile	% Fuel	% Public Transport	% Total Effect
I	1.66	11.71	6.58
II	2.05	10.66	6.53
III	2.16	8.91	5.90
IV	3.02	8.13	6.43
V	3.33	7.20	6.35
VI	3.77	6.60	6.54
VII	4.00	5.92	6.49
VIII	4.51	4.41	6.36
IX	4.88	2.74	6.03
X	3.54	0.71	3.84

transport. Finally σ_{GP} is the cost share of fuel in public transport tickets (in this case assumed to be 0.42). The actual numbers are easily produced by applying (6.1) to the data in Table 6-2, resulting in Table 6-3.

We see now that the progressivity found when looking exclusively at fuel is essentially gone. There is still in fact a very small progressive effect up to decile IX but then there is a strongly regressive effect for decile X. Basically, the distributive effect is now almost completely neutral; in fact the Suits coefficient shows a small regressive net effect (see Figure 6-2).

The result is clearer when we examine our obtained Suits coefficients. Generally, the fuel tax, with regard to both direct and indirect fuel use, is weakly regressive—the Suits coefficient is roughly –0.1. When only the direct fuel use is considered, the coefficient is weakly progressive (+0.032). One must remember that these results are obtained by using the conventional income approach, which in previous studies tends to give a higher degree of regressivity than the lifetime income/expenditure approach (e.g., Poterba 1991 or Hassett et al. 2009). It is thus likely that the results would be progressive if lifetime income were used.

In our study of Mexico, the tax burden from *indirect* fuel use falls more heavily on the poor than on the rich. Although the net effect is almost neutral, it could easily be modified. If the policymaker wants to ensure that the distributional effects of a fuel tax do not hurt the poor, then an increased fuel tax could be complemented with a tax relief or subsidy for public transportation. As shown by Bento et al. (see Chapter 3), this is easily achieved even in the U.S. case, where the policy tends to be more regressive to start with.

Conclusion and Discussion

Fuel taxes are a very efficient instrument for much needed climate policy. Resistance to these measures is based on various factors. In Mexico as in many other countries, lively and often acrimonious debate arises whenever the government tries to raise fuel prices. Whether this is due to populism, lobbying by the oil industry, or out of genuine conviction is not for us to say; certainly fierce opposition exists—in the case of Mexico, currently formulated by the left-wing PRD and PRI opposition. One might suspect that the PRI would follow similar policies were it to return to power in government, so one of the reasons the opposition attacks this particular policy might simply be populism. Either way, the critique and concern must be taken seriously. Earlier research has shown the importance of how the tax is presented and "marketed." In most countries it is probably desirable to present this as part of a tax reform rather than the increase of an already familiar tax—since the latter is bound to attract critique. Another may be that the tax is believed to be inflationary (even if there is little evidence, it may still be a common belief). It therefore seems important to present every tax-based instrument in a budget-balanced manner: Thus a fuel tax needs to be explicitly related to cuts in other taxes or raised spending.

Finally, fuel taxes are resisted because they are said to be regressive. We have studied this in the case of Mexico and not really found any support for this hypothesis. On the contrary: If only direct use of fuel is considered, the tax would be quite strongly progressive. If indirect use through public transport is included the tax would be roughly neutral or very weakly regressive. Had lifetime income been used, or had adaptation through individual elasticities for the different deciles been used, it is possible that we would have found progressivity even for the case of total impacts of fuel taxation. On the other hand, if the decisionmaker wants to increase the progressivity of a fuel tax, this is also very easy. Fuel taxation gives rise to new tax revenues which can either replace regressive taxes such as the VAT or be used to subsidize goods and services used by the poor (such as public transport). In the case of an oil exporter such as Mexico, constraints on domestic consumption also imply extra export revenues, which typically are higher—and much needed given the current state of PEMEX, which has record debts despite record revenues.

Acknowledgments

Valuable Research assistance by Emanuel Carlsson is much appreciated. Financial support from Sida to the Environmental Economics Program at the University of Gothenburg is gratefully acknowledged. Thanks also to Ashokankur Datta, Roland Kpodar, and one anonymous referee for valuable comments.

Notes

1   The situation is serious. In September, 2009 the head of PEMEX, Jesús Reyes Heroles, was fired and replaced with Juan José Suárez Coppel. However, as *Business Week* commented (Sept. 8, 2009), this superficial change in no way addresses underlying problems.

2   As a state company, PEMEX has strong connections with the ministries. Fuel pricing decisions and taxes may for purposes of simplification be thought of as government policies.

3   Asociación Mexicana de Empresarios Gasolineros (AMEGAS), June 2010.

4   PRI is the *Partido Revolucionario Institucional*; PRD: *Partido de la Revolución Democrática*; PAN: *Partido Acción Nacional.* Eight pesos is approximately 70 U.S. cents/liter.

5   See for example www.eluniversal.com.mx/notas/647305.html (accessed May 23, 2011).

6   Actually the data include gasoline, diesel, and LPG (e.g., propane). Since LPG is used mainly by low-income households, we would have had even found even stronger progressivity if we had had the data for auto fuels only.

7   If we assume public transport profitability is constant then the long-run (equilibrium) percentage increase in tariffs equals the percentage price increase in fuel multiplied by the share of fuel in the total costs for public transport.

8   Several dozen drivers were asked about revenues and several categories of cost.

CHAPTER 7

Is Fuel Taxation Progressive or Regressive in China?

JING CAO

The magnitude of the air pollution problem in China is alarming. While rapid economic development and urbanization over the last three decades have created a "China miracle" with average 8% to 10% GDP growth and lifted hundreds of millions of Chinese people out of poverty, the side effects are putting tremendous pressures on the environment. Escalating urban environmental threats to human health afflict many cities of China. According to the State Environmental Protection Administration of China (SEPA), roughly 40% of Chinese cities are facing either medium or serious air pollution, and 16 to 20 of the most polluted cities of the world are located in China. In addition, with the energy and carbon spike after year 2002, China has topped the list of carbon emitting countries, exceeding the United States. Figure 7-1 shows the rise in total carbon emissions and carbon per capita from fossil fuel combustion since economic reform thirty years ago. Though the trend is somewhat flattened out during 1995–2002, it increases dramatically after 2002.

Recently, China has made significant progress in terms of promoting energy saving in heavy manufacturing and controlling air pollution from coal combustion in urban areas. However, new challenges arise from the increasing number of automobiles (Figure 7-2), and vehicle emissions pose new challenges for urban air pollution control. For example, daily and hourly NO_x and ozone concentrations have exceeded national air quality standards; high concentrations of CO, HC, and SO_2 are found in the air along roadsides; potential threats of photochemical smog exist for many large cities. Vehicle emissions have become the main source of air pollution in some big cities, including Beijing.

To stabilize future carbon emissions and vehicle emissions, one class of economic-based instrument has proven very effective: a tax on transport fuels such as gasoline. In some cases, fuel taxes may be imposed first for non-environmental reasons, such as energy security; however, it has been shown that fuel taxes also have environmental effects as well. To put it simply, a higher gasoline price not only encourages people to drive less but also provides

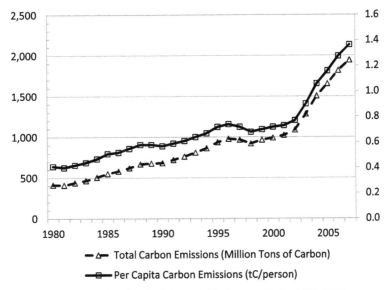

FIGURE 7-1 Total Fossil Fuel Carbon Emissions and Carbon per Capita, 1980–2007.

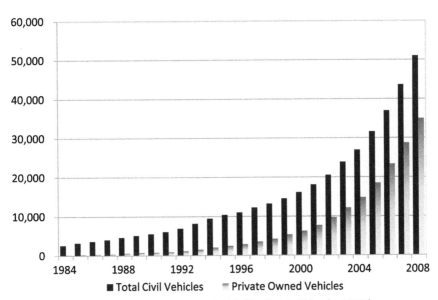

FIGURE 7-2 Trends in Civil and Private Motor Vehicle Numbers in China (in 1,000s).

Note: "Total Civil Vehicles" refers to vehicles that are registered and have received vehicle license tags according to the Work Standard for Motor Vehicles Registration formulated by the Transport Management Office under the Department of Public Ssecurity at the end of the reference period. Total civil vehicles can be divided into private vehicles and vehicles for public or business use (collective use, or company car).

Sources: Chinese Statistical Yearbook (2009), CEIC Data Company Ltd. database.

incentives for producers to build more efficient cars. The tax can change incentives from both the demand and supply sides. Fuel taxes impact people's behavior; the question is, how large are these impacts?

In addition, political barriers often arise during debate on the distribution of these impacts. A common argument against gasoline tax reform is the potential for regressive consequences, where the poor are hit hardest and bear an unfair tax burden. Studies of fuel tax incidence in developed countries offer mixed evidence (Parry et al. 2007; Poterba 1991; Santos and Catchesides 2005; and Chapter 4 of this book). The empirical literature on fuel tax incidence is quite plentiful, and this book provides more examples focusing on developing countries. Intuitively, one might think that in these countries, because only rich people can afford to buy automobiles, it would be likely that a fuel tax might be less regressive. Currently, no empirical study on fuel tax incidence in China exists. This chapter will fill the gap and shed some light on the policies of this tax.

Recent research has shown that even for developed countries, gasoline taxes may be mixed or even slightly progressive. As shown in various chapters of this book, it depends very much on methodological and other issues. For example, West (2004) concludes that the gasoline tax is more progressive between the low- and middle-income distributions, and less regressive between middle- and high-income distributions. In addition, the progressive or regressive results depend greatly on how the gas tax revenue is used. For example, West and Williams (see Chapter 4) suggest that one can easily make the gas tax highly progressive by using revenue derived from it to finance cuts in other more regressive taxes. Thus, the actual tax incidence depends fully on how gas tax reform is designed and how it interacts with other pre-existing taxes.

This chapter is organized as follows. We start with an overview of China's fuel tax reform, followed by estimates of price and income elasticities using a panel data set of major Asian countries. We explore distribution effects across different expenditure and income deciles, using a simple spreadsheet analysis combining our household survey data and estimated elasticities, which shed some light on the distributional effects of current fuel taxation reform in China, before we conclude.

Fuel Tax Reform in China: Background

Since January 1, 2009, China has launched a fuel tax reform on cars, buses, and trucks, initiated under the nation's top economic planner, the National Development and Reform Commission (NDRC), together with the Ministry of Finance and Ministry of Transport. This is a big breakthrough after 15 years of discussion and debates on fuel tax reform in China. In 2008, the remarkable fall in world oil prices offered a rare opportunity for China to push ahead with such a fuel taxation reform. In addition, it also provided some channels to revise the current fuel price mechanism, as well as resource tax reform. Though the fuel tax reform was kick-started by energy security concerns, it will in the long run give the market a larger role in pushing for better environment and energy efficiency.

The scheme of the current fuel tax removes six types of pre-existing fees on road and waterway maintenance and management while raising the fuel consumption tax from 0.2 yuan per liter to 1.0 yuan per liter for gasoline. Similarly diesel and jet fuel are now charged with 0.8 yuan per liter consumption tax. With the remarkable fall in international gasoline prices in 2008, an increase in gasoline tax was barely noticeable to consumers, and then it gradually increased as the world oil price increased. This will ultimately push Chinese consumers to lower their fuel demand.

However, fuel taxes in China are currently still low, and most of the increase in fuel price is simply offset by replacing the pre-existing road maintenance fees. In addition, the tax revenue is spent on compensating the losses in the Ministry of Transportation and other fuel users who previously were exempt from the road tolls, such as the airlines, army, and so on. Therefore, the tax rates are very modest at the current stage. So the question is whether China needs to deepen its fuel tax reform in the future. As in other countries, the fuel tax reform often faces resistance for political reasons of regressivity. Thus, we need to provide some empirical evidence to shed some light on the issue.

Fuel Demand Elasticity

Estimating fuel demand elasticity is very important in understanding the effectiveness of gasoline tax reform. There are some concerns that gasoline demand is not price elastic, and people may believe the fuel price change will only be temporary due to the fluctuations in global oil market. In this chapter, we conduct a comparative study of fuel demand elasticity for several Asian developing countries and Asian developed countries; we then compare these elasticity results with those of western countries.[1]

We use a standard estimation procedure to calculate demand elasticities using aggregate data. Like all products, we can assume that the demand for fuel is a function of income Y, and price P. A simple model that can capture this dynamic behavior is presented as follows:

$$Q_{it} = a \cdot P_{it}^{\alpha} \cdot Y_{it}^{\beta} \cdot Q_{it-1}^{\delta} e^{\mu_{it}} \qquad (7.1)$$

Q_{it} = Quantity of gasoline demand of country i in period t

P_{it} = Real price of gasoline country i in period t

Y_{it} = Real income in country i period t

Q_{it-1} = Quantity of gasoline demand country i in period $(t-1)$

This formulation is called the lagged endogenous model, and the lagged quantity of gasoline demand (Q_{t-1}) represents the inertia of economic behavior due to the slow turnover of the vehicle stock, etc. In the formulation, α, β, and δ are three parameters to be estimated. α represents the price elasticity, β represents the income elasticity, and δ reflects the strength in habit formation. The short-run price and income elasticities are α and β while the long-run elasticities can be found by dividing the short-run price and income elasticities by $(1 - \delta)$ (see for instance Dahl and Sterner 1991a and 1991b).

Though extensive studies have been conducted in the OECD, U.S., and other countries, the empirical studies of fuel demand elasticities in Asian countries—especially China—are still very limited. In this study, we construct a data set of aggregate fuel price, fuel use, and income data from the Chinese Statistical Yearbook; then we merge the Chinese data with another data set for India, Indonesia, Japan, Korea, the Philippines, Thailand, and Vietnam from the CEIC database "Macroeconomic, Industry, and Financial Time-Series Databases for Global Emerging and Developed Markets."[2] In order to make the data comparable across countries, we convert the per capita GDP data in all the countries using the World Bank Purchasing Power Parity (PPP) adjustment coefficients. Our data sample is from 1991 to 2007.

The results are summarized in Table 7-1. The short-run price elasticity for the whole Asian sample is –0.17, and short-run income elasticity is 0.25. The long-run elasticity is more than four times the short-run elasticities. Long-run price elasticity is –0.80, and long-run income elasticity is 1.14 for all the Asian countries. Then we divide the whole sample into two groups: Asian developing countries (China, India, Indonesia, the Philippines and Thailand) and Asian developed countries (Japan and South Korea). We can see that Asian developing countries have higher price and income elasticities than Asian developed countries in both the short and long run.

The large number of gasoline demand studies in the western world show that short-run elasticity is usually –0.2 to –0.3, and long-run elasticity is quite

TABLE 7-1 Gasoline Demand Elasticity Estimation Results for Asian Countries.

Sample Coverage	α	β	δ	$\alpha/(1-\delta)$	$\beta/(1-\delta)$
All Asian Countries	–0.1735 (–.33)	0.2490 (2.89)	0.7821 (9.74)	–0.7963	1.1427
Asian Developing Countries (China, India, Indonesia, Philippines, Thailand)	–0.2115 (–1.99)	0.3256 (2.03)	0.7465 (6.23)	–0.8343	1.2844
Asian Developed Countries (Japan, South Korea)	–0.1841 (–6.05)	0.2327 (4.39)	0.6895 (18.46)	–0.5929	0.7494

Note: The numbers in parentheses are t-statistic.

high, around −0.80 to −1 (Dahl and Sterner 1991a and 1991b); (Goodwin 1992); (Goodwin et al. 2004); (Hanly et al. 2002); (Graham and Glaister 2002 and 2004). We can see that our results in the subsample of Asian developing countries are in a similar range compared with the elasticity estimates in the western countries, though somewhat smaller. It appears that many people believe gasoline demand is not very price elastic given very little flexibility in changing demand; the reason behind this may only reflect the lower estimate of short-run price elasticities. However, the elasticities may be very large in the long run. Both our Asian country results and western studies suggest that in the long run the price elasticities are very high. Thus a gasoline tax can be quite effective in curbing carbon emissions and vehicle pollutions.

Incidence Analysis of Fuel Tax Reform in China

We focus on fuel tax incidence by presenting an empirical analysis using urban household survey data in China. Currently, due to China's rapidly rising vehicle emissions and road congestion, an intense policy debate on gasoline taxes versus road restriction policies is underway. Supporters of the road restriction policy often cite western studies that argue that a gasoline tax is likely to be regressive—poor households might bear an unfair tax burden either through direct use of fuel from private vehicles or indirect use of fuel in public transport. Such possible distribution concerns often stand in the way of reform such as raising the fuel tax rate. Meanwhile, local government has every reason to favor road restriction over gasoline taxes for political reasons. Motorists who have already been complaining about world oil price increases also don't like gasoline taxes. Therefore, the incidence of gasoline taxes is very important. If it were shown to be progressive—as some of the recent developing countries studies do, pointing to vehicle ownership mainly concentrated in higher socioeconomic brackets (Sterner 2007)—gasoline tax reform in China would receive more political support.

Expenditure on Fuel Use by Decile

To analyze fuel tax incidence in China, we use data from the annual Urban Household Survey (UHS) conducted by the National Bureau of Statistics. The UHS data set, covering all the provinces in China, uses a stratified multistage method to select samples. The survey asks the households to keep very detailed records of their various sources of income and expenditure every month. However, the survey was changed substantially in 2002, increasing the sample size from 21,000 to 56,000, and more detailed questions were included. These changes might cause an inconsistency for the variables before and after 2002. Figures 7-1 and 7-2 also suggest that year 2002 seems to be a structural break: the carbon emissions rise about 4.4% before 2002 but 11.4% after 2002;

the total civil vehicles increase by 11% before 2002 but 16% after 2002. So considering the data limitations and likely gap in 2002, we choose our data sample from 2002 to 2007.

The Chinese annual UHS survey includes number of vehicles, fuel cost for private cars, and very detailed public transport expenditure which can be further divided into: 1) transportation fee: by flight, by train, by long-distance bus, by local public transportation, by taxi, and others (in both value and quantities); 2) transport service fee, including vehicle maintenance fee, user fees, and so on. This data set also has very detailed questions on household income and expenditures, which may be useful to conduct some simple spreadsheet simulations of the effect of fuel taxes on expenditures as well. The limitation of this data set is that the fuel use in the survey includes all the gasoline, diesel, and other fuels aggregated; no data for subcategories is available. So we cannot distinguish between the different policy implications for gasoline versus diesel.

Using the micro level household survey, we divide the sample into both income deciles and expenditure deciles, and the latter are used as proxy for lifetime income. For the tax incidence analysis using income deciles, we focus on household level expenditure. For the tax incidence analysis using expenditure proxy for lifetime income approach, we adopt an equivalence scale adjustment similar to West and Williams (see Chapter 4), because, given the same total household income, households with fewer members clearly have a higher standard of living. We use the same parameter that West and Williams used to weight adults and children equally but allowing for economies of scale in consumption. More specifically, we divide total expenditure by (adults + children)$^{0.5}$; we then pool all the households and rank them based on equivalence-scale adjusted total expenditure on gasoline and other goods. In our sample, the mean household size ranges from 2.6 to 3.2 across all 10 income deciles with a standard deviation of roughly 0.8. In our analysis we also tried alternative OECD equivalence-scale functions, but the results are all similar.

Table 7-2 gives the mean disposable income, expenditure, vehicle ownership, and direct and indirect fuel use expenditure as a share of total expenditure for year 2007, which is based on annual expenditure decile brackets after our equivalence-scale adjustment. Figures 7-3(a and b) and 7-4(a and b) give both fuel use expenditure share for both income decile and expenditure decile for 2007. They show that the average percentage of total household expenditure devoted to fuel use expenditure is greatest for the highest decile for direct use, as can be seen from the concentration of automobile ownership in these households. For the indirect use from bus transportation, the average percentage of total household expenditure on fuel use is greatest not for the highest decile but the second highest expenditure decile instead. This suggests that for public transportation, both richest and poorest households will rely less on public transports, since rich people would use private vehicles instead, and the poorest households cannot even afford public transport or travel, and they use it even less when unemployed. The Suits index is positive

TABLE 7-2 Summary Statistics of Household Income (in 2007 yuan), Expenditure, Vehicle Ownership, and Direct and Indirect Vehicle Fuel Use Expenditure by Expenditure Deciles.

Decile	Mean Disposable Income per Capita	Mean Expenditure per Capita	Vehicle ownership (%)	Direct Fuel Use Expenditure (%)	Indirect Fuel Use Expenditure (%)	Total Direct + Indirect Use Expenditure (%)
1	10,082	6,105	14.20	0.25	0.31	0.56
2	13,922	8,671	14.98	0.31	0.38	0.69
3	15,625	10,443	19.08	0.33	0.40	0.73
4	18,378	12,176	23.28	0.34	0.47	0.80
5	19,618	13,898	26.52	0.44	0.45	0.90
6	22,771	15,892	28.57	0.47	0.49	0.96
7	25,750	18,245	27.61	0.57	0.51	1.08
8	29,546	21,759	32.81	0.69	0.53	1.22
9	35,699	27,357	41.73	1.08	0.55	1.63
10	53,572	52,994	59.68	1.88	0.43	2.31
Suits Index				0.35	0.014	0.24

Source: Author's own calculation based on data source from National Bureau of Statistics, China.

for direct expenditure on fuel use and very small for indirect expenditure via bus transportation—about 0.014% in 2007. In sum, the average percentage of total expenditure for combined direct and indirect fuel use expenditure is highest for the highest expenditure decile.

The distribution of budgetary share of private and public transport fuel costs as share of total household expenditure in Figures 7-3 and 7-4 are quite intuitive. In China, only very rich people can afford private cars; therefore, in Figure 7-3 the share of private fuel costs over total household expenditure is increasing with income, and the slope becomes much steeper in the tenth income and expenditure decile (the richest). On the other hand, as for indirect fuel costs related to public transport as a share of total household expenditure, the budget share tends to follow a slightly inverted U-shape, with lower shares in the poorest decile and richest decile: the poorest people cannot even afford bus fees, and the richest people tend to use private vehicles more frequently. The expenditure on public transportation is very low in our data, maybe because the Chinese government recently gave substantial subsidies to urban public transportation. For example, the bus fare is only 0.4 to 1 yuan per ride in many metropolitan cities such as Beijing— much lower than fares in western countries. In addition, we did not include long-distance travel by train and other transports, so the data is only limited to short-distance travel. We tried the same analysis from 2002 to 2007 and got the same progressive results as in Figures 7-3 and 7-4. Also note that our sample only focuses on urban households. Considering that the share

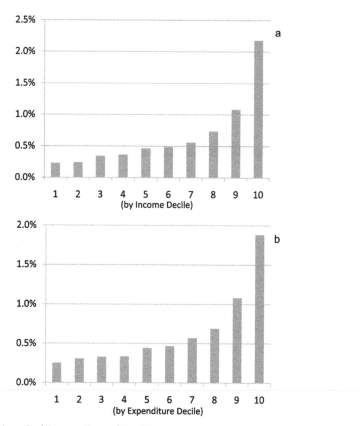

FIGURE 7-3 Private Fuel Costs as Share of Total Household Expenditures (2007).

of vehicle ownership in rural areas is even smaller, if we had considered the whole population our results would probably have shown even more progressivity, with vehicle ownership and fuel consumption concentrated even more on medium- and upper-income deciles. In addition, we should also note that the urban household sample might have a sample selection bias at the very high-end income decile (for instance, less than 5% decile). Because very rich people don't have incentives to fill out the questionnaire and record their monthly expenditures, it is hard to say whether they use more gasoline for luxury SUVs or commute less from jobs, or what fuel use shares compared to their overall expenditures would be. But the biases from omitting rural households and the richest people may offset each other, and the general results on progressivity should still hold.

Suits Index

In addition, we also calculate the Suits index to measure the average progressivity of the fuel tax across the entire income range. As Suits (1977) describes,

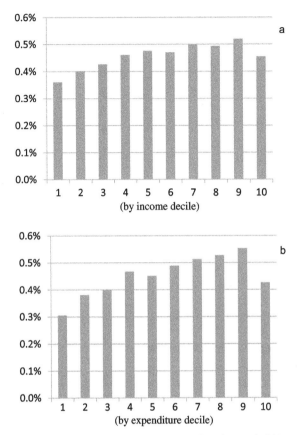

FIGURE 7-4 Public Transport Related Fuel Costs as Share of Total Household Expenditures (2007).

Note: Public transport here includes public bus, long distance bus, and taxi. We have interviewed bus drivers and taxi drivers in Beijing and Guangzhou to get a rough estimate of the fuel share at about 30%, and we use this coefficient to represent the whole country.

the Suits index is very similar to the Gini ratio that is widely used to measure inequality. The Suits index varies from +1 at the extreme of progressivity, where the entire tax burden is borne by members of the highest income bracket, through zero for a proportional tax, to –1 at the extreme of regressivity, where the entire tax burden is borne by members of the lowest income bracket. Poterba (1989) recommended using lifetime income for measuring distribution incidence of taxes.

Therefore, in this study, we adopt both the income and expenditure approaches to measure the Suits index for fuel cost expenditure in China. With limited data in China, we consider the expenditure approach a good proxy for the lifetime tax incidence. The results are given in Table 7-3. The direct tax burden is fuel expenditure on private vehicles, and indirect tax burden is the

TABLE 7-3 Suits Index of Fuel Tax in China.

	Direct Tax Burden		Combined Direct and Indirect Tax Burden	
	Expenditure Approach	*Income Approach*	*Expenditure Approach*	*Income Approach*
2002	0.18	0.26	0.09	0.17
2003	0.24	0.31	0.12	0.20
2004	0.28	0.35	0.15	0.22
2005	0.32	0.39	0.18	0.26
2006	0.35	0.43	0.20	0.29
2007	0.35	0.44	0.24	0.34

fuel cost implicitly included in the bus fares. Our results suggest that, not only is the Suits index quite large in China, its magnitude increased from 2002 to 2007. For both direct and combined direct and indirect tax burden, the Suits index derived using the expenditure approach will be smaller than the income approach. This result suggests that, unlike the regressive or mixed results in most western countries, China's fuel tax is indeed progressive.

Distribution Effects of the Fuel Tax Reform in China

Now let us try a very simple spreadsheet simulation to examine the impact of the fuel tax reform initiated since 2009 in China. If we assume all the households have the same price elasticity e, then we can derive a simple relationship between changes in expenditure in fuel and short-run or long-run price elasticity as in equation (7.2), where dP/P is the *ad valorem* fuel tax rate on the fuel price (see Chapter 6).

$$dG = G(dP/P)(1+e) \qquad (7.2)$$

Based on the price elasticity results in the previous section, the short-run price elasticity is –0.21, and the long-run price elasticity is –0.83. In China we use unit fuel tax on gasoline use—that is, 1 yuan/liter for gasoline and 0.8 yuan/liter for diesel. Since we cannot separate gasoline and diesel, here we ignore the small tax differences and assume the incremental price increase from the new fuel tax reform is 0.8 yuan/liter for all fuel (taking gasoline as representative).[3] In 2009, the gasoline price in China is roughly 6.5 yuan/liter, which included the tax already, so the pre-existing after-tax price is 5.7 yuan/liter, and the *ad valorem* tax rate is roughly 14% (= 0.8/5.7).

The results of impacts of changes on fuel use budget share are given in Figure 7-5. Here we rely on expenditure as a proxy for the lifetime income, and we include both direct and indirect use of fuel consumption. Our results shows that with a unit tax incrementally increased at 0.8 yuan/liter, the richest

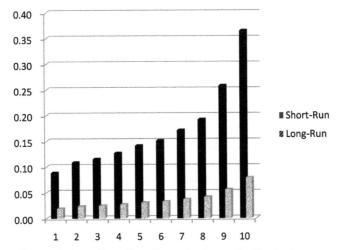

FIGURE 7-5 Changes in Expenditure Share Due to the Current Fuel Tax Regime.

decile experiences the biggest impact, and the fuel tax reform is progressive at all levels. The progressivity is smaller in the long-run case than in the short-run case, since people can adjust better in the long-run to the policy shock from the fuel tax reform, and thus the changes in terms of expenditure changes are smaller.

An alternative methodology is to use own-price elasticity of fuel demand for each decile, instead of assuming uniform price elasticities. In another paper (Cao and Zeng 2010), we found the households in the lower two deciles are very price elastic, but the medium- and upper-expenditure deciles are less elastic; therefore, using the individual elasticities by socioeconomic strata, our result of progressivity would be reinforced.

Conclusion

Fuel taxes have been proven to be very effective for combating vehicle pollution, global warming, and other externalities such as congestion and car accidents. However, political resistance often arises for a couple of reasons. First, people appear to believe that gasoline demand is not price elastic, given the impression there is little flexibility in demand. Politically local governments in China or other developing countries often favor a command-and-control type of traffic restriction policy over fuel taxes (for example, the road restriction policies implemented in Beijing and Mexico City). Second, the common argument against fuel taxes is that poor people might be hit harder and bear an unfair tax burden. Our studies suggest this argument does not hold for the Chinese case. Rather, we find the current fuel tax reform is very progressive, with the Suits index increasing in recent years. The progressivity finding suggests that it is desirable to clear up the political concerns on distribution effects of fuel tax reform, push forward a deepening reform of fuel tax, and gradually replace the temporary road restriction regimes in China.

Acknowledgments

The author would like to thank Thomas Sterner, Emanuel Carlsson, and Allen Blackman for their useful input and comments. The author is also grateful to Xiaolong Chen for assistance on the Chinese Urban Household Survey data, and Nan Zhong, Yang Xie, and Jieyin Zeng for their research assistance. This research was supported by funding from the Swedish International Development Cooperation Agency (Sida) via the Environment for Development (EfD) China Center and the Chinese National Science Foundation (Project No. 70803026).

Notes

1   We also compile time-series and provincial panel data to examine China's fuel demand elasticities. Our results show that both short-run price elasticities and income elasticities are slightly lower than our pooled Asian Developing countries results, but long-run price and income elasticities are very high. However due to having less variations in the price over the sample period, and especially limited time-series data at the aggregate level, the statistical inferences are not satisfactory.
2   See www.ceicdata.com (accessed May 23, 2011).
3   There is a pre-existing gasoline consumption tax at 0.2 yuan/L, so the incremental part is 0.8 yuan/L for gasoline.

## CHAPTER 8

Are Fuel Taxes in India Regressive?

ASHOKANKUR DATTA

C limate change is increasingly being accepted as a major problem by policy-makers around the world. India is the fifth largest emitter of CO_2 world-wide. It accounts for about 4% of world CO_2 emissions. Figure 8-1 shows the rise in total carbon emissions in India in the last century and the early years of this century.[1] It shows that the growth in emissions has increased rapidly in the last quarter of a century and shows no signs of tapering off.

Table 8-1 shows that fossil fuels are the single most important source of carbon emissions in India, accounting for over 95% of India's CO_2 emissions. Thus any meaningful reduction in emissions from India will require a reduction in fuel use. A rise in fuel prices via taxation or other means can be used to ensure such a reduction. However, so far the government has failed to appreciate the enormity of the problem of climate change and has done too little to deal with it. In the garb of socially beneficial intervention in the energy sector, it has often pursued policies of subsidization that have created negative incentives for emission abatement.

Although policymakers have often considered environmental taxes to be politically infeasible, Western Europe has long experimented with environmental taxes: directly in the form of carbon taxes in the 1990s and indirectly in the form of fuel taxes. Sterner (2007), reviewing several studies and calculating the hypothetical transport demand for OECD countries, concludes that fuel taxes are the single most powerful climate policy instrument implemented to date.

However, the harshest criticisms of fuel taxes are often on distributional grounds based on the popular perception that fuel taxes are regressive. The balance of academic evidence does not favor this view. In the early 1990s, many argued against fuel taxation on the grounds that it imposes a larger burden on poor people. Such claims were based on studies that used U.S. data on gasoline consumption (KPMG Peat Marwick 1990). However, in recent years, a large number of studies on this issue—Poterba (1991), West (2004), Santos and Catchesides (2005), Steininger et al. (2007) and others—have shown that it cannot be said without qualifications that a fuel (gasoline) tax is regressive.

Only a few papers on the distributional effects of fuel taxation are based on data from developing countries. Kpodar (see Chapter 13) on Mali, Yusuf and

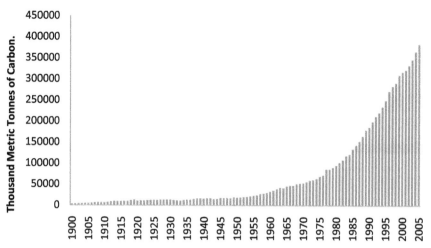

FIGURE 8-1 Total Fossil Fuel Emissions in India 1900–2005.

Source: http://cdiac.ornl.gov/ftp/trends/emissions/ind.dat (accessed May 24, 2011).

TABLE 8-1 Carbon Dioxide Emissions in India by Source.

Source	CO_2 Emissions (thousand metric tons)	Percent of Total Emissions
Solid Fuels	782,180	75.90
Liquid Fuels	192,950	18.72
Gaseous Fuels	45,045	4.37
Gas Flaring	3,344	0.32
Cement Manufacturing	47,339	4.59
Land Use Change	−40,307	−3.91

Source: Carbon Dioxide Emissions by Source 2005, Carbon Dioxide Information Analysis Center (CDIAC), World Resources Institute (WRI).

Resosudarmo (see Chapter 9) on Indonesia, the ESMAP report on Pakistan (World Bank 2001), and other chapters in this book show that there is no unanimous result on the regressivity issue.

This chapter studies the incidence of a tax on fuels in India. It specifically tries to check if fuel taxes or removal of fuel subsidies adversely affect the poor more than they do the rich. This study is important for two reasons: first, the energy sector in India is highly regulated by the government. The government provides subsidies to various fuels with an aim of helping the poorer sections of the society. It is important to see whether the poor really benefit from such subsidization schemes or if the removal of such subsidies might have a progressive effect. Second, India is one of the largest emitters of carbon dioxide worldwide. Emissions in India have been increasing at a rapid rate and are expected to do so in future. If fuel taxes are not regressive, they could be used to restrict fuel emissions without having serious adverse social implications. Besides, it is politically easier for governments to justify taxes that are progressive than those that are regressive. Hence, it is important to study the distributional effects of fuel taxes in India.

In comparison to other papers on the regressivity issue based on underdeveloped or developing countries, this chapter is different in two major ways: First, unlike other chapters in this book on developing nations, this chapter looks at the distributional effect of taxes on all major cooking, lighting, and transport fuels. Since solid fuels are a major source of emissions in India and a large part of energy produced in India is thermal power, we specifically look at the impact of a coal tax. It is also important to consider all fuels in a single framework, as major petroleum products like diesel and kerosene are produced jointly in refineries. Often fuels like kerosene and diesel are used as substitutes.[2] Thus taxing one of them while subsidizing the other can lead to problems of adulteration and black-marketing. Second, while calculating incidence, this chapter makes full inclusion of all indirect effects through an input-output structure. Most research (including most chapters in this book) restrict attention to the indirect effects in the public transport sector on the assumption that by doing so substantial part of the indirect effects is captured. This chapter tests the validity of such an assumption in a developing country like India. It also undertakes sensitivity analysis using elasticity estimates from secondary sources.

Fuel Pricing in India

With the objective of moving toward market-determined prices for petroleum products, the government of India abolished the Administered Price Mechanism (APM) in April of 2002. However, the Indian government continues intervention in the petroleum sector by absorbing state-owned oil company losses. Market-determined prices are considered to be politically infeasible.

Given that the government considers subsidization of cooking fuels to be an important social instrument in helping poorer households shift from biomass to modern fuel, the government in 2002 decided to continue providing subsidies for liquid petroleum gas (LPG) and kerosene *ex-ante* in the budget. The oil marketing companies (OMCs) were to adjust the retail selling prices of these products in line with international prices during this period. Subsidies were expected not to exceed 15% of the gas-import parity price and 33% of the kerosene-import parity price. The government had even thought of abolishing all budget subsidies within five years from 2002. However, in compliance with government directions, the OMCs did not adjust prices of kerosene distributed through the PDS and domestic LPG commensurately, resulting in losses on account of these two products.[3] In October 2003, the government decided that the OMCs would make good about a third of the losses on these two products from the surpluses generated by them on petrol and diesel while the balance losses would be shared equally by the upstream companies (ONGC/OIL/GAIL) and the OMCs.

In late 2003, international oil prices started rising rapidly and this burden sharing arrangement collapsed. This had two impacts.

First, the burden of subsidy on public distribution system (PDS) kerosene and domestic LPG increased sharply. The burden of subsidies in 2005–2006 was Rs.15,000 crores[4] on account of PDS kerosene and Rs. 11,000 crores on account of domestic LPG. Table 8-2 shows how implicit and explicit subsidies on PDS kerosene and LPG changed during the period 2002–2006.

Thus, since the abolition of APM, we have a decline in the explicit component of subsidy (that is allotted for in the budget) and an increase in the implicit component. In 2004–2005, the total subsidy on LPG and kerosene was Rs.20,772 crores, whereas the fiscal budget allotted for only Rs.2,930 crores. Thus "under-recoveries" constituted 85% of total combined subsidy on PDS kerosene and domestic LPG (see Table 8-3).

Second, the government started interfering in the pricing of petrol and diesel. It restricted the pass through of international prices to domestic consumers. As a result the margins available to OMCs during 2002–2004 on petrol and diesel thinned and then rapidly turned negative.

TABLE 8-2 Subsidy of PDS Kerosene and Domestic LPG.

Item	PDS Kerosene (Rs./Liter)			
	2002–03	2003–04	2004–05	2005–06 (est.)
Subsidy from fiscal budget	2.45	1.65	0.82	0.82
"Under-recoveries" to oil companies*	1.69	3.12	7.96	12.14
Total subsidy to consumer	4.14	4.77	8.78	12.96
	Domestic LPG (Rs./Cylinder)			
Subsidy from fiscal budget	67.75	45.18	22.58	22.58
"Under-recoveries" to oil companies*	62.27	89.54	124.89	147.74
Total subsidy to consumer	130.02	134.72	147.47	170.32

* On the gross before adjusting amount shared by upstream companies.
Source: Report of the Committee on Pricing and Taxation of Petroleum Products, 2006.

TABLE 8-3 Total Subsidy of PDS Kerosene and Domestic LPG (in Rs. Crores).

Item	2002–03	2003–04	2004–05	2005–06 (est.)
Fiscal Budget	4,496	6,292*	2,930	2,900
Oil Companies "Under-recovery"	5,430	9,274	17,842	23,704
Total	9,926	15,566	20,772	26,604

* Includes arrears of 2002–03 of 2213 crores.
Source: Report of the Committee on Pricing and Taxation of Petroleum Products, 2006.

In 2003–2004, oil companies made an under-recovery of Rs. 2,303 crores on petrol and diesel.

Thus, five years since the dismantling of APM, India has not made much headway towards market pricing of petroleum products and gas.

However, contrary to popular belief, the fuel sector is not a story of one-way flow of subsidies. While on the one hand subsidies are in place, the same commodities are subjected to various taxes. Both the central government and the state government impose taxes which pull up retail prices. While on one hand oil PSUs (public sector undertakings) are advised not to revise prices in conformity with crude rates, the government imposes excise duties and a plethora of other taxes on these items. The result is that the Indian retail prices for petroleum and diesel are the highest in Southeast Asia. For cooking fuels, budget subsidies are coupled with various central and state level taxes. Table 8-4 shows the component of taxes in retail prices.

However, subsidies outweigh the taxes for cooking and lighting fuels.[5] The situation for transport fuels is not clear. Subsidies for transport fuels mainly take the form of under-recoveries, and official statistics don't give any information on under-recoveries per unit of petrol or diesel. Rough calculations reveal that taxes outweigh implicit subsidies for transport fuels.[6]

Thus we see that even after abolition of the APM, there has been substantial government intervention in the fuel sector, especially the petroleum products sector.[7] In such an environment, it is important to know the distributional impact of additional taxes or removal of existing subsidies. This is especially important in an underdeveloped country like India, where the funds provided as fuel subsidies can be alternately used in socially productive investment in health and education, which can directly benefit the poor.

Data and Methods

The objective of this chapter is to use information from an India-wide consumption survey and an input-output transaction matrix to answer the question: Would fuel taxes or any other policy that is equivalent in the sense of raising fuel prices in India be regressive, imposing a higher percentage burden on the poor compared to the rich?

TABLE 8-4 Component of Taxes in Retail Price.

Product	Central Taxes	State Taxes	Total Taxes
Petrol	38%	17%	55%
Diesel	23%	11%	44%
Domestic LPG	0%	11%	11%
PDS Kerosene	0%	4%	4%

Source: Report of the Committee on Pricing and Taxation of Petroleum Products, 2006.

In a partial equilibrium framework, I first examine the direct effects and then allow for indirect effects. Too little information is available for a credible complete general equilibrium analysis.

I start with the simplest of the measures: a measure of direct tax burden, ignoring indirect consumption of fuel through consumption of commodities that use fuel as an input. The following assumptions are made:

1   The production function of the taxed commodity shows a fixed coefficient technology. Thus the supply curve of the taxed commodity is perfectly elastic. Consumers bear the entire burden of the tax.[8]

2   The taxed commodity is not an intermediate input and so does not change the price of any other commodity in the economy. This assumption will be relaxed later.

3   Hicksian (compensated) demand for fuel is inelastic. Sensitivity checks will later be performed to check if relaxation of this assumption changes the result.

Under the above three assumptions, we can comment on the progressivity or regressivity of tax just by looking at the budget share of the taxed commodity for different levels of monthly per capita expenditure (MPCE). If a tax on a particular fuel is regressive (or progressive), the budget share will fall (or rise) as we move from lower to higher deciles.[9]

Data on consumer expenditure on fuel and other commodities is obtained from the consumption schedule of the 61st round of the National Sample Survey (NSS) conducted by the National Sample Survey Organization (NSSO) of the Government of India during the period of July 2004 to June 2005. This quinquennial round has a sample size of 124,584 households. The rural sample consists of 79,258 households and corresponding figures for the urban sample are 45,326 households. The national sample survey uses a stratified two-stage sampling design, first sampling clusters (which are villages in rural areas and urban blocks in urban areas) and then selecting 10 (or 12 as in the case of the 55th round) households within each cluster (called FSUs, or first-stage sampling units). I measure the incidence of a fuel tax across expenditure classes since, due to consumption smoothing, expenditure is a better measure of long-term economic welfare than income. In any case, the NSS does not report income. I use consumption figures that are based on 30-day recall for both non-durables and durables. The survey elicits consumption expenditures (and consumption quantity for consumption items) for the household for the month preceding the date of survey. For items on which quantity information is collected, prices can easily be obtained by dividing consumption expenditure by quantity. As consumption quantity of cooking and lighting fuels is reported, their prices can be calculated. However, for transportation fuels like petrol and diesel only expenditure values are stated. Thus price information on these two items is not obtained from the consumption survey.

The measure discussed above takes only direct consumption of fuel into account. However, fuel is an important input into the production of various commodities. When households consume such commodities, they indirectly consume fuel. A tax on fuel increases the price of such fuel-using commodities, which in turn imposes an additional burden on consumers. The second measure that I consider takes such indirect consumption into account. (Assumption 2 above is relaxed.) This measure of incidence requires the following additional assumptions:

1   A closed economy.
2   Unchanged value added per unit output. Primary factor markets unaffected.

Given these assumptions, I calculate the fuel tax–induced price changes in all sectors. Due to the assumption of inelastic compensated demand curves, we can calculate the compensating variation as a percentage of total expenditure. If this percentage increases (or decreases) with MPCE deciles, the tax imposed is progressive (or regressive).[10]

To calculate the economy-wide price changes I at first use the input-output coefficient matrix for 2003–2004, published by the Central Statistical Organization (CSO). The original matrix for 2003–2004 has disaggregated information on 130 sectors. In order to make it compatible with the NSSO data, the 2003–2004 CSO matrix has been aggregated to make an input-output matrix with information on 46 broad sectors. This obviously introduces an element of error, but compatibility between NSSO data and CSO data demands such aggregation. CSO's input-output matrix has fossil fuel sector information at a very aggregative level. The four fuel sectors are: Crude petroleum, Natural gas, Petroleum products, and Coal and lignite. In order to study the distributional effects of petroleum and gas products separately we use the input-output table (2003–2004) constructed by the National Council for Applied Economic Research (NCAER) in a report titled "Study of Macroeconomic Impact of High Oil Prices" (2006). This matrix has 27 sectors, of which eight are fuel-related sectors. They are: Petroleum, Oil and lubricants—crude, Motor gasoline, Diesel, Aviation turbine fuel (ATF), Liquefied petroleum gas, Kerosene, Other Petroleum Products, and Gas and Water Supply. However, here the coal sector is merged with all the other minerals to form a sector called "Mining and Quarrying." However, the CSO 2003–2004 matrix reveals that coal constitutes about 65 percent of total mineral production in India. Thus this sector can be used to calculate incidence of a coal tax.

This chapter also calculates the Suits index for each of the fuels. We calculate the index separately for direct, indirect, and combined effects. Formulated by Daniel B. Suits (1977) to measure the progressivity of taxes, the Suits index is inspired by and related to the Gini ratio. At the extreme of progressivity (+1) the entire tax burden is borne by members of the highest economic bracket, zero is a proportional tax, and –1 represents extreme regressivity, with the entire tax burden borne by members of the lowest economic bracket. This index allows us to compare the progressivity of different taxes.

Results

In a low-income country, we may expect transport fuel taxes to be progressive because poor people don't own cars. This is especially true for India, where 34% of the population is below the Purchasing Power Parity (PPP) of $1 a day, and 27.5% of the population lives below the abysmally low national poverty line.[11] In India, most rural households use biomass fuels instead of fossil fuels for cooking purposes. However, kerosene—an important petroleum product—is widely used as lighting fuel. Thus it cannot be said for sure whether cooking and lighting fuels will have distributional impacts different from that of transport fuels.

Figure 8-2 shows distribution of household MPCE in India as a whole. As expected, the per capita expenditure distribution for India is highly positively skewed, with a median MPCE of Rs.550.[12] Figure 8-2 gives us a clear idea as to where the MPCE deciles stand in respect to absolute levels of consumption.

In the context of India, it should be kept in mind that not only is a vast majority of Indians poor, Indian society is also highly unequal. Table 8-5 gives us an idea about the levels of inequality in India. The deciles are deciles of the per capita expenditure distribution. The third column notes the average total expenditure for that decile, while the fourth shows the percentage of total economy-wide consumption that can be attributed to that per capita expenditure decile.

Hence it is important to ensure that taxation policy is not regressive, accentuating the existing inequality in the society.

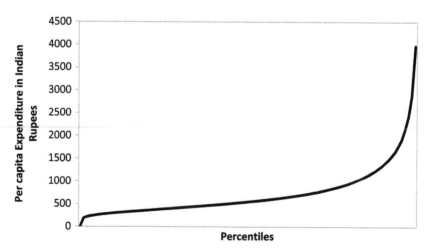

FIGURE 8-2 Distribution of Monthly Per Capita Expenditure.

TABLE 8-5 Distribution of Consumption Expenditure Across per Capita Expenditure Deciles.

Per Capita Expenditure Deciles	Average Per Capita Expenditure (in rupees)	Average Total Expenditure (in rupees)	Percentage of Total Consumption Expenditure
1	245	1,423	4.39
2	331	1,834	5.66
3	391	2,071	6.39
4	449	2,288	7.06
5	515	2,528	7.80
6	592	2,785	8.59
7	697	3,152	9.73
8	857	3,670	11.33
9	1,152	4,524	13.96
10	2,506	8,134	25.10

Direct Effects

We now look at the incidence results. Figure 8-3 shows the direct budget share of fuel[13] as a whole. The combined budget share of all major fuel products (coke/coal, petrol, diesel, kerosene, gas, and so on) is seen to be higher for higher consumption deciles. The budget share of fuels stays constant for the first three deciles, but it increases thereafter, indicating that an overall fuel tax would be strongly progressive. There is a difference of around 4% between the budget shares of the highest and lowest deciles. The Suits index is calculated to be 0.19, which indicates progressivity.

It will be interesting to see what is going on behind these figures. To determine that, we calculate the incidence results separately for transport fuels and cooking and lighting fuels. Figure 8-4 shows that the budget share of all

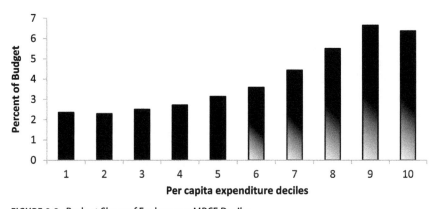

FIGURE 8-3 Budget Share of Fuels across MPCE Deciles.

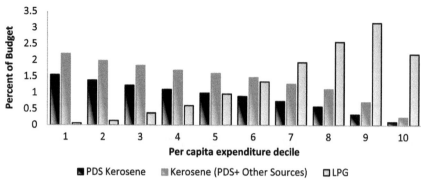

FIGURE 8-4 Budget Share of All Cooking and Lighting Fuels.

cooking fuels is constant for the first four deciles, but increases thereafter. It falls substantially for the highest decile. If we consider kerosene and LPG, then kerosene's budget share decreases with consumption while the budget share of LPG increases with consumption. The budget share for gas falls substantially for the last decile. This is expected because the urban non-poor are the major users of gas as a cooking fuel. The budget share curve for coke and coal has an inverted U shape (but its budget shares are negligible and hence not included in Figure 8-4).

The Suits index for taxes on coal and coke, kerosene, and LPG are –0.24, –0.32, and 0.26, indicating that taxes on cooking fuels other than LPG are regressive. If we only consider a tax on kerosene sold through the PDS, the Suits index falls to –0.38.

In order to understand these results, we need to understand the pattern of fuel use in India. Since such patterns are very different for rural and urban sector, it is important to look at the two sectors separately. In addition to the information on household expenditure on different fuels, the NSS reports the main cooking and lighting fuel for each household. Figure 8-4 shows primary cooking fuel usage across deciles and sectors. The deciles denote deciles from the country-wide distribution of per capita expenditure. For each decile households are divided into urban and rural sectors. For each sector in every decile, I calculate the percentage of households using a fuel as a primary cooking fuel. For example, 90% of rural households, who belong to the lowest decile of the all-India MPCE distribution, use biomass fuels as their primary cooking fuel. Figure 8-5 shows the percentage of households using kerosene as their main lighting fuel across sectors and deciles. For example, 68% of rural households in the lowest decile use kerosene as their main lighting fuel.

In rural India, very few households use gas. Only 0.19% of rural households, belonging to the poorest decile, use gas as their main cooking fuel. The figure increases to 43% for the top quartile. Thus, only people in the upper end of expenditure distribution use gas. The percentage of rural households using gas as their main cooking fuel increases monotonically with the level of

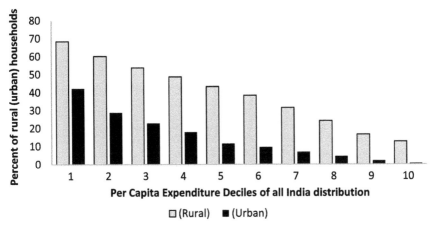

FIGURE 8-5 Percentage of Households Using Kerosene as Main Lighting Fuel.

expenditure. In urban India, LPG is a popular cooking fuel, especially among the middle and high expenditure groups. The percentage of urban households who use LPG as their main cooking fuel rises across expenditure deciles. From 4% for the first decile, the percentage of households using gas as the primary cooking fuel increases to 80% for the top decile. When budget shares of cooking gas across deciles is calculated separately for the rural and urban sectors (creating separate deciles for the rural and urban distribution), budget shares increase with deciles for the rural sector, while the budget share curve is an inverted U shaped for the urban sector.[14] However, the urban sector is richer and smaller than the rural sector. Households that are at the middle of the urban expenditure distribution are in the top end of the overall distribution. This explains the fall in budget shares of cooking gas as deciles rise (Figure 8-4). Thus a removal of existing subsidies on cooking gas is progressive.

Kerosene is the popular lighting fuel in rural areas, especially for the poor. As consumption increases, people move towards electricity, subject to its availability in villages. For cooking purposes traditional biomass fuels—firewood, dung, and agricultural residue—are generally used. With an increase in income people start shifting toward more convenient fuels like kerosene and gas. The budget shares of kerosene fall with income, indicating the expected regressivity from the use of kerosene as a lighting fuel outweighs the expected progressivity from its use as a cooking fuel.[15] This is expected as kerosene is not a very popular cooking fuel in rural areas, as is evident from Figure 8-6. In the urban sector electricity is used for lighting, almost universally. Only 8% of urban Indian households do not state electricity to be their main lighting fuel. These urban households use kerosene. The majority of these households come from the lower deciles. However, around 10% of urban households use kerosene as their main cooking fuel, and they are distributed across the urban deciles. In both sectors, budget share of kerosene falls with per capita expenditure. Even

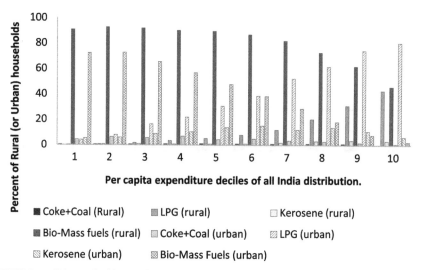

FIGURE 8-6 Primary Cooking Fuel Usage, Percentage of Households.

when kerosene sold through the PDS is considered separately, the budget shares decile with per capita expenditure in both sectors are taken together. Thus removal of existing subsidies on kerosene will be regressive in nature.

In Table 8-6, we see the per capita monthly consumption of subsidized kerosene and LPG per household for urban and rural deciles. Assuming that the subsidy is the same on all units of kerosene sold through PDS, the table reflects the bias in the distribution of existing subsidies on cooking fuels. When seen by decile group, per capita purchases of PDS kerosene steadily increase with expenditure decile in rural areas. Thus the kerosene subsidy received by an individual from the lowest rural decile is less than the kerosene subsidy received by an individual from the highest rural decile. The rural subsidy therefore has a bias in favor of the rich. In the urban sector, per capita purchases of PDS kerosene peak in the middle decile groups and then slowly decline until they fall off sharply in the top deciles. The third and fourth columns in Table 8-6 reflect the distribution of LPG subsidy. As might be expected, the per capita consumption of LPG increases with expenditure decile in both the sectors.

The literature on the distributional effects of a fuel tax concentrates on transport fuels like gasoline. In Figure 8-7 we consider the budget shares of petrol (gasoline). Diesel and other transport fuels are rarely used for personal transport in India. Diesel is mainly used for freight and for public transport. It is only in cities like Delhi that the public transport fleet uses cleaner fuels like compressed natural gas. Kerosene is used to adulterate diesel (NCAER 2005), yet we cannot make incidence calculations taking that into account as we don't know how such adulterated kerosene is distributed across deciles.

As is evident from Figure 8-7, the budget shares of transport fuels are strictly increasing with per capita expenditure[16] as only the very rich can

TABLE 8-6 Monthly per Capita Household Consumption of Subsidized Kerosene (Liters)/LPG (kg) per Household

	Kerosene		LPG	
Decile	*Rural*	*Urban*	*Rural*	*Urban*
1	0.38	0.43	0.00	0.19
2	0.48	0.54	0.01	0.57
3	0.52	0.55	0.02	0.87
4	0.53	0.50	0.04	1.26
5	0.56	0.47	0.07	1.53
6	0.60	0.44	0.11	1.97
7	0.63	0.28	0.17	2.23
8	0.65	0.32	0.28	2.54
9	0.67	0.24	0.55	2.84
10	0.60	0.12	1.32	3.37

Source: Calculated by the author from 2004–2005 NSSO data.

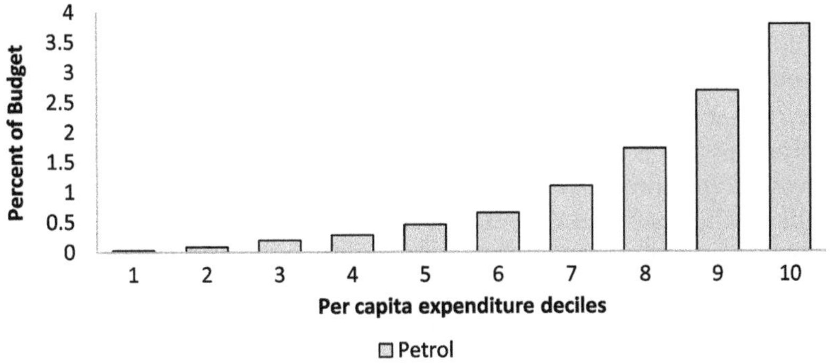

FIGURE 8-7 Direct Budget Share of Transport Fuels.

afford private transport and hence don't require public transport fuels. A large majority of Indian households (more than 80% according to 61st round NSS data) do not buy either petrol or diesel.

The Suits indexes for taxes on petrol and diesel are 0.45 and 0.56, indicating that taxes on transport fuels are very highly progressive. Table 8-7 summarizes the results of direct incidence of taxes on various fuels.[17] Of all the fuel taxes that I have considered, a tax on diesel is most progressive, followed by a gasoline tax, a tax on LPG, and a tax on coke and coal. A tax on kerosene is the most regressive. The magnitude of regressivity increases when we consider a tax that is imposed only on subsidized kerosene sold through the PDS. If we consider a tax on all fuels where the taxes on fuels are proportional to their existing prices, we get a situation of progressivity. The Suits index of such a tax is 0.19.

TABLE 8-7 Suits Index of Fuel Taxes (Direct Incidence).

	Suits Index	*Incidence*
Coke and Coal	−0.24	Regressive
Liquefied Petroleum Gas	0.26	Progressive
Kerosene	−0.32	Regressive
PDS Kerosene	−0.38	Regressive
Petrol	0.45	Progressive
Diesel	0.56	Progressive
All Fuels	0.19	Progressive

Combined Effects: Inclusion of Indirect Consumption of Fuel

Till now, I have ignored indirect consumption of fuel. Transport and cooking and lighting fuels are consumed not only by households but by industries as an intermediate input into the production of goods and services (electricity, transport services, agricultural products, and so on). This intermediate consumption of fuel is not negligible for many fuels and hence should not be ignored while calculating incidence of fuel taxes. Coal and diesel are mostly consumed indirectly, so it is important to take this into account. According to the 2003–2004 input-output data, inter-industry consumption of coal and petroleum products constitutes about 98% and 80% of the sum of total home production and imports respectively. According to basic statistics published by India's Ministry of Oil and Natural Gas, diesel sale in India is five times the sale of motor gasoline (petrol). This shows why we must take indirect effects into account when we calculate incidence. We now report combined budget shares that include both direct and indirect consumption of fuel. Such budget shares equal the compensating variation as a percentage of total expenditure, when fuel taxes rise by a unit amount.[18] An increase in such budget shares with increase in MPCE indicates progressivity. Figure 8-8 reports the combined budget shares that are calculated using CSO's 2003–2004 input-output table.

Figure 8-8 shows that inclusion of indirect consumption does not change the conclusion that taxation of coal and petroleum products[19] would be progressive. Inclusion of indirect consumption actually *reverses* some regressivity results that we had obtained earlier. Earlier the budget share of coal was highest for the middle deciles and was lower at the two ends (Figure 8-3). However with the inclusion of indirect consumption, we have a situation where the combined budget shares increase as expenditure rise. Hence it can be said a coal tax is progressive. The Suits index of a coal tax when all indirect effects are taken into account is 0.07 compared to −0.24 when only direct consumption was compared.

This result is quite intuitive. Coal is an important input in the production of energy and for the manufacturing sector. The rich spend a much bigger

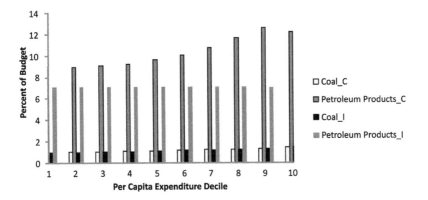

FIGURE 8-8 Combined and Indirect Budget Shares of Coal and Petroleum.

Note: _C = combined; _I = indirect.

proportion of their total expenditure on energy and consumer goods. This in turn changes the earlier result.

The budget shares for petroleum products are almost unchanged for the first few deciles. They start increasing thereafter. Thus at an all-India level, taxes on petroleum products are progressive. The Suits index of a petroleum products tax with all the indirect effects is 0.067. However, petroleum products include petrol, diesel, kerosene, LPG and various other industrial fuels. Thus this cannot be compared directly to the earlier direct results. If we only study the indirect effects, we see in Figure 8-8 that the bars don't show any clear pattern. However, the Suits index calculated separately for the indirect effects shows a value of –0.008, which is lower than the Suits index of the combined effects. This shows that the popular opinion that indirect effects of a petroleum products tax impose a larger burden on the poor is not true. It is neutral. Table 8-8 summarizes the results.

Many studies on gasoline taxation that look into indirect effects restrict their attention to the impact of gasoline tax on public transport, rather than calculating price changes for all sectors in an input-output framework. They argue that public transport is the most important sector that uses gasoline as an intermediate input, and so it suffices to capture the effect of a gasoline tax on

TABLE 8-8 Suits Index (Incorporating the Indirect Effects).

	Suits Index	*Incidence*
Coal (all effects)	0.070	Progressive
Petroleum Products (all effects)	0.067	Progressive
Coal (indirect effects)	0.076	Progressive
Petroleum Products (indirect effects)	–0.008	Neutral

transport. In the Figure 8-9 we show the percentage of indirect effects of a petroleum products tax that can be attributed to price changes in public transport.

It is evident that although public transport is affected by a petroleum products tax, it is not an overwhelmingly substantial part of the indirect effects. This might be because of that fact that the petroleum products sector includes products other than transport fuel. It includes gas, kerosene, and other industrial fuels which are often used as intermediate inputs in the industrial sector. It is interesting to note that the percentage of indirect effect that can be attributed to transport rises with deciles. This shows that unlike developed nations, in India the poor don't use public transport as much as the rich. Figure 8-10 shows the budget share of public transport across deciles. Values of budget share have been arrived at after adjusting for transport margins in the consumption values of other commodities.[20]

The budget shares rise with expenditure till the eighth decile and declines slightly thereafter. Thus, in India, including the indirect effect of public transport usage strengthens the progressivity result rather than weakening it.

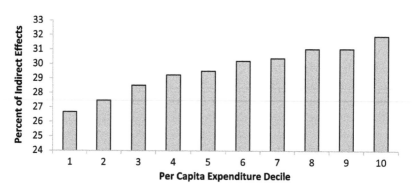

FIGURE 8-9   Percentage of Indirect Effects of a Petroleum Product Tax Attributed to the Transport Sector.

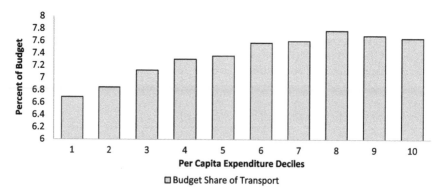

FIGURE 8-10   Budget Share of Public Transport (Adjusting for Transport Margins in the Consumption of Other Commodities).

It is important to understand why a tax on coal or petroleum products is progressive even when indirect effects of such a tax are taken into consideration. I provide some intuition for why the inclusion of indirect effects maintains the progressivity result. We know that the poor have low budget shares for petroleum products (except kerosene) and slightly high budget shares for coal, compared to the rich. In spite of that, the fact that fuel prices affect prices of other commodities and the possibility that the poor might have high budget shares for such commodities may change the direction of the incidence results. For example, the poor might be affected adversely if food prices are highly sensitive to fuel prices. The fact that the budget share of food for the poor is high might depress the progressivity result obtained earlier. Table 8-9 gives the difference in the budget shares of the last and the first decile for some important sectors. It shows that the "poor" (bottom decile) have a higher budget share for food items, forestry and logging, coal and lignite, edible oil, and toiletries, as well as services of the trade sector compared to the "rich" (top decile). These sectors have the potential to depress the progressivity obtained earlier by comparing direct budget shares, but only if the products

TABLE 8-9 Budget Share for the Poorest and the Richest Decile (After Adjusting for Trade and Transport Margins).

Sectors	Decile 1	Decile 10	Difference
Major Food Crops and Their Products	28.94	5.28	–23.66
Other Crops	9.01	4.30	–4.71
Milk and Milk Products	2.85	5.21	2.36
Other Animal Products	3.44	1.79	–1.65
Forestry and Logging	5.25	0.30	–4.95
Coal and Lignite	0.03	0.01	–0.02
Sugar	1.84	0.73	–1.11
Edible Oil	4.86	1.75	–3.11
Textiles	1.13	4.19	3.06
Miscellaneous Manufacturing	0.63	2.67	2.04
Petroleum Products	1.89	5.39	3.49
Health	2.60	9.10	6.50
Toiletries	4.76	2.69	–2.07
Electrical, Electronic Machines, and Appliances	0.12	1.54	1.41
Transport Equipment	0.11	3.47	3.36
Electricity	1.35	3.82	2.47
Other Services and Communication	3.10	9.99	6.89
Trade	13.43	7.84	–5.60
Hotels and Restaurants	0.19	2.08	1.89
Education	0.43	4.22	3.79

of these sectors are highly sensitive to fuel prices. On the other hand, textiles, milk products, petroleum products, health, electricity, transport services, other services, education, hotels and restaurants, and education have lower budget shares for the poor, compared to the rich.

Figure 8-11 shows the price changes in all sectors in response to an increase in the unit taxes on coal and petroleum products by one unit. Since the initial prices are normalized to one, it can be treated as the percentage change in price in response to an increase in a coal or petroleum tax by one unit. Food items, forestry and logging, edible oil, and toiletries (all important for the poor) are not very responsive to a coal tax. On the other hand, sectors like electricity, coal tar products, cement, metal products, and various machinery sectors (important for the rich) are highly responsive to a coal tax. Of all the sectors, electricity is most responsive to a coal tax; electricity consumption can have an important role in determining the incidence of a coal tax.

As none of these sectors are sectors for which the rich have higher budget share than the poor, the indirect effects lead the results towards progressivity. Direct consumption of coal is extremely rare (very few households use coal as a cooking and heating fuel), so the indirect effects matter more in the determination of combined incidence. The strength of the indirect effect ensures that the combined budget share of coal is rising in expenditure.

Now we look at the impact of a petroleum products tax on the prices across the economy. As a whole, prices in the economy are more responsive to petroleum products tax than they are to a coal tax. Thus indirect effects are expected to play a major role in determining the combined effects. Sectors like electricity, transport services, chemicals, and cement are highly responsive to a petroleum tax. Even food prices are affected to a larger extent than they were by coal taxes. Thus electricity and transport services consumption can have an important role in determining the incidence of a petroleum products tax. Because most of the tax-sensitive sectors are ones for which the rich have larger direct budget shares than the poor, the indirect effects reinforce the direct effects, making the combined effect of a petroleum products tax progressive.

The petroleum products sector of CSO's input-output table includes transport fuels like gasoline (petrol) and diesel, cooking fuels like LPG, lighting fuels like kerosene, and various other industrial fuels like fuel oil, bitumen, and petroleum wax. Thus, the indirect effects calculated using the matrix aggregate all the effects. However, the indirect effects of different fuels can conceivably run in different directions. For example, furnace oil is generally used in the high-end manufacturing sector. Kerosene, on the other hand, might be used as an intermediate input in very small-scale eateries in the informal sector and can thus have regressive indirect effects. But such differences get lost in the aggregation. To deal with this, I combine information from NCAER's report titled "Study of Macroeconomic Impact of High Oil Prices" (2006) to create a new matrix with disaggregated information on the energy sector. I then use this matrix to calculate incidence results at a disaggregated

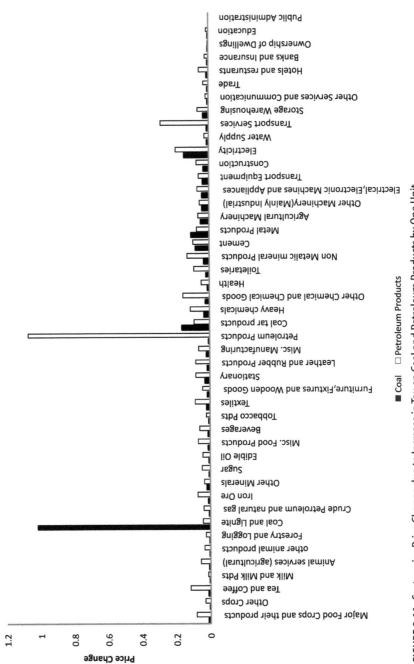

FIGURE 8-11 Sectorwise Price Change due to Increase in Tax on Coal and Petroleum Products by One Unit.

level. This matrix has disaggregated information on petroleum products. (As mentioned above, coal is included in the mining and quarrying sector.) Using this matrix, I see the impact of a tax on major petroleum products like petrol, diesel, LPG, kerosene, and ATF. None of the petroleum products other than diesel and ATF have indirect effects that are non-negligible. For these fuels the combined effects mimic the direct effects. Even after indirect effects have been considered, all of them (except kerosene) continue to be progressive. Kerosene remains regressive. Aviation turbine fuel has no direct consumption by households, so the indirect effect coincides with the direct effect. As expected they are progressive. However, the budget shares for all the deciles are negligible. Thus Figure 8-12 shows only the results for diesel only. Diesel is rarely used by households directly but has significant indirect uses.

Figure 8-12 illustrates the substantial indirect effect of a diesel tax, as the combined budget shares are much greater than the simple direct effects. The combined budget share for diesel rises with expenditure marginally, thus indicating very low progressivity. The Suits index of the combined tax burden is 0.02—lower than the direct effects Suits index of 0.56. The Suits index for the indirect burden of diesel tax is 0.01. Thus it shows that indirect effects do have a dampening effect on diesel tax progressivity.

Since indirect effects play such a crucial role in the overall determination of incidence of a diesel tax, it is important to identify the sectors that play an important role in the determination of the indirect effects. In Table 8-10 we calculate for each sector the percentage of indirect effects of a diesel tax that can be attributed to it and show the percentages for each of the deciles. It can be noted that the importance of each sector varies by decile. For example, food agriculture is a very important component of the indirect effects of diesel tax for the lowest decile: Around 31% of the indirect burden can be attributed to

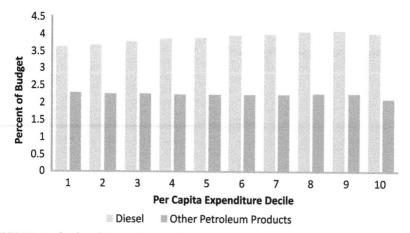

FIGURE 8-12 Combined Budget Share of Petroleum Products (Calculated Using Information from NCAER Matrix).

TABLE 8-10 Share of Major Sectors in the Indirect Effects of a Diesel Tax.

Decile	Food Agriculture	Non-food Agriculture (Livestock, Forestry)	Consumer Durables	Consumer Non-Durables	Road Transport	Electricity	Services
1	31.05	4.78	0.46	13.76	32.60	5.44	8.79
2	27.95	5.19	0.62	14.48	33.09	6.74	8.92
3	25.40	5.37	0.78	14.60	34.14	7.90	8.73
4	23.41	5.45	0.90	15.01	34.86	8.42	8.78
5	21.76	5.52	0.99	14.99	34.97	9.43	8.92
6	20.07	5.36	1.10	14.72	35.69	10.20	9.15
7	18.18	5.10	1.24	15.20	35.39	11.04	9.58
8	15.80	4.64	1.49	14.89	35.76	12.31	10.15
9	13.40	4.08	1.91	14.43	35.15	14.11	11.21
10	8.42	2.69	7.03	12.97	33.15	14.25	14.33

Note: Calculations using information from NCAER (2005). Only the sectors that play an important role in determining the indirect burden have been included.

it. It should be noted here that food agriculture prices are not very sensitive to a fuel price hike. However, the fact that food controls an overwhelmingly large share of a poor person's budget ensures that food agriculture has an important role in determining indirect effects. The share of food agriculture declines with deciles until it is just 8% of the indirect effects for the richest decile.

Many studies on gasoline taxation focus on the indirect effect on a gasoline tax on transport, ignoring the effect on other sectors. The argument is that public transport is the most important sector that uses gasoline as an intermediate input, so by focusing on transport we can capture a large part of the indirect effects. Table 8-10 shows that the road transport share of indirect effects is very high indeed, but other sectors like food are equally important. For none of the deciles does the share of road transport in indirect effects exceed fifty percent. Thus this study shows that, for developing and underdeveloped countries like India, one should not restrict attention to public transport while calculating indirect effects.

Conclusion

Fuel taxes have often faced criticism on the grounds of regressivity. Given that taxation is the outcome of a political process, such criticisms makes it difficult for the political class to use fuel taxes as a means of dealing with high levels of fuel emissions. Such criticisms are western in origin. Since the developing world in general and South Asia in particular has a very different consumption pattern, these criticisms might not hold true for a country like India. As this chapter demonstrated, the regressivity criticism does not apply to a low-income country like India. When direct consumption alone is considered, taxes on transport fuels (petrol and diesel) are highly progressive for both urban and rural sectors. Similarly, all cooking fuels—with the exception of kerosene and coal—show definite signs of progressivity. Critics might argue that, to the extent that the poor consume public transport and other goods and services that use fuel as an intermediate input, they are also affected by a fuel price rise even when they don't consume fuel directly. Yet as this study has shown, for most fuels the progressivity results remain unchanged even when indirect consumption is included in the analysis. A tax on coal turns progressive once indirect effects are taken into account. The combined effects of a petroleum products tax are also progressive. When components of the petroleum products sector are studied separately, all fuels other than kerosene are found to be progressive. For most fuels other than diesel and coal, the direct effects are found to be close to the combined effects. In other words, the indirect use of kerosene, cooking gas, and petrol are negligible. Sensitivity checks using elasticity estimates confirm the earlier results for kerosene, cooking gas and transport fuels.

An environmental fuel tax seeks to reduce emissions by reducing fuel consumption. Hence such taxes do not make sense for fuels that have low Marshallian elasticity of demand. Besides, given the objective of emissions reduction, taxes should be imposed on fuels that have high emissions potential. Transport fuels—where each liter of transport fuel burned emits around 2.3 kg of CO_2—satisfy these criteria, so they are an appropriate case for a fuel tax for environmental purposes. Although transport fuels demand is inelastic in the short run, it responds to price changes in the long run and has a long-run elasticity of −0.84. (Sterner 2007). However, Ramanathan and Geetha (1998) report a lower long-run elasticity of −0.42 for India. Still, transport fuel demand is sensitive to price changes.

The results presented here show that a tax on transport fuel is progressive even when indirect consumption is considered. Thus a tax imposed on transport fuels achieves the desired objective of emissions reduction without having any adverse distributional effects, thus making a strong case for transport fuel taxation.

Coal is the largest emitter of carbon dioxide in India. The results here show that a tax on carbon is regressive when only direct consumption is considered. Once the indirect effects are incorporated, a coal tax turns out to be progressive. Given that there is no regressivity, there is a strong case for taxing coal. Such a tax would encourage innovation and diffusion of unconventional carbon-free energy like solar and wind energy.

The issue of taxing cooking and lighting fuel is more complex. It is difficult to make an unqualified recommendation for a tax. Contrary to popular perception, studies by Gundimeda and Köhlin (2008) show that elasticities of cooking and lighting fuels are not low for all sections of the society. According to their study, the elasticity of gas is close to unity for almost all sections of the society, ranging from –0.92 for the urban rich to –1.05 for the urban poor. However, gas is a cleaner fuel compared to its counterparts, so the case for a gas tax (or equivalently, the case for a removal of the existing gas subsidy) is not strong in spite of the fact that such a tax is progressive. The case for a gas tax becomes reasonable only when the government can couple it with incentives for using electricity for cooking purposes. Electricity is an efficient cooking fuel. But because a large percent of electricity in India is coal based, a movement away from gas to electricity might accentuate emissions problems. Thus the case for a gas tax is not strong. A pure carbon tax (a tax on the carbon content of fuels) would be much better since it would have a higher incidence on coal than on gas.

In India, kerosene is an important cooking and lighting fuel. While urban households use kerosene as a cooking fuel, rural households use it for light. The demand for kerosene is responsive to prices, especially in the rural sector. It ranges from –0.7 for the rural rich to –0.5 for the middle expenditure group. Kerosene is a poor lighting source and more expensive than electricity (Barnes, Plas and Floor 1997). The results from this chapter show that because 35% of rural households light their homes with kerosene, a removal of its subsidy would be regressive. Besides its regressivity, a tax on kerosene has another concern: Any tax on kerosene causes the poor to substitute fuelwood, which has strong adverse health and regional climate implications, as well as leading to deforestation. According to Gundimeda and Köhlin (2008), a 1% increase in the price of kerosene increases fuelwood use by 0.7% for the rural poor and 0.4% for the urban poor. Thus, any tax proposal should be preceded by compensatory proposals for the poor—perhaps in the form of targeted electricity and LPG subsidies and coupled with a rural electrification program. Baland et al. (2007) point out that the targeted gas subsidy might also help in forest conservation. The need for gas subsidy can be reduced by subsidizing biogas whenever viable.

When it comes to CO_2 emissions, a relatively small wealthy class of the population in India is hiding behind a huge proportion of poor people. It is India's poor who keep per capita CO_2 emissions really low. Thus it is natural that a policy designed to tackle GHG emissions should impose a larger burden on the rich. The evidence from this study shows that an environmental fuel tax would do just that. The progressivity result is robust even with inclusion of indirect fuel consumption. It should be a bit surprising when people speaking for the Indian underclass in the polity come down heavily against any fuel price hike proposal on the grounds that it imposes a higher burden on the poor than on the rich. While this is true for kerosene, it is not true for any other fuel.

Perhaps such criticisms arise out of ideological biases against governmental intervention in the form of taxes, or perhaps out of the political expediency in not taking on an elite who benefit from subsidization.

Acknowledgments

I would like to thank my supervisor, Dr. E. Somanathan, and Dr. Thomas Sterner for their suggestions on this research. I am grateful to Mr. Ramesh Kolli, who has helped me understand data published by the CSO. I would also like to thank all seminar participants at Indian Statistical Institute, New Delhi; University of Gothenburg; and Gujarat Vidyapeeth, Ahmedabad, for many useful comments on this research. Errors, if any, are my own.

Appendix 8-1: Figures and Diagrams

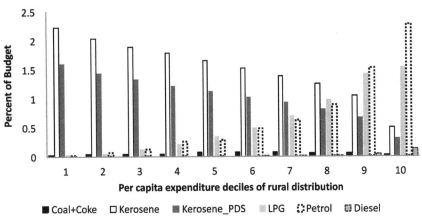

FIGURE 8-A1 Direct Budget Shares of Fuels across Rural Deciles.

Note: The rural sample is divided into ten deciles. Budget shares are calculated for each decile.

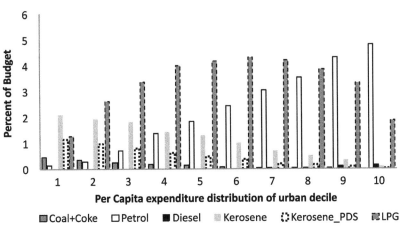

FIGURE 8-A2 Direct Budget Shares of Fuels across Urban Deciles.

Note: The urban sample is divided into ten deciles. Budget shares are calculated for each decile.

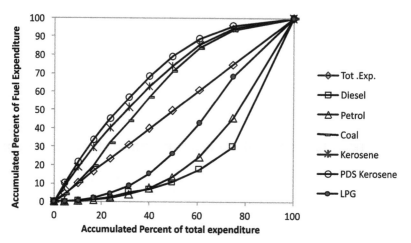

FIGURE 8-A3 Concentration Curves for Suits Index Calculation (Only Direct Effects Considered).

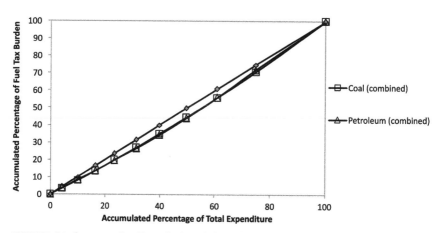

FIGURE 8-A4 Concentration Curve for Suits Index Calculation, Combined Burden (Indirect Effects Calculated Using CSO's 2003–2004 I-O Matrix).

Appendix 8-2: Tax-Subsidy Situation in the Indian Petroleum Sector

States in India impose taxes on petroleum products at different rates. To illustrate the relative tax-subsidy scenario we use information from the National Capital Territory of Delhi. According to Graczyk (2006), the tax situation in Delhi is as shown in Table 8-A1.

According to data published by the Petroleum Planning and Analysis Cell of India's Ministry of Petroleum and Natural Gas, subsidy (from fiscal budget and under-recoveries to oil companies) to consumers is Rs.12.92 per liter of PDS kerosene and Rs.175.04 per cylinder of domestic LPG. Thus for cooking fuels, subsidies outweigh the taxes.

TABLE 8-A1 Petroleum Retail Price Buildup in New Delhi 2005–2006.

	Price Without Tax	Customs Duty	Excise Duty	Sales Tax	Total Tax	Retail Price
Kerosene (liter)	8.7	0	0	0.35	0.35	9.05
Diesel (liter)	18.42	1.94	4.93	3.16	10.03	28.45
Petrol (liter)	17.04	1.6	14.74	6.75	23.09	40.13
LPG (14.2 kg cylinder)	215.49	0	16.06	18.17	34.23	249.72

Source: Graczyk (2006). *Note:* Taxes and prices are in Indian Rupees.

According to statistics published by the Ministry of Petroleum and Natural Gas, in 2005–2006 the total consumption of petrol (Mogas) and diesel (HSDO) was 8,647,000 tonnes (12,010 million liters, assuming a density of 0.72 kg/liter) and 40,191,000 tones (48,000 million liters, assuming a density of 0.85 kg/liter) respectively. Assuming that the tax structure of New Delhi (given in Table 8-A1) prevails across the country, this translates into tax revenue of Rs.7,643,240 million. According to data published by the Petroleum Planning and Analysis Cell, the total under-recovery on petrol and diesel was Rs.153,700 million. Thus the tax revenues outweigh the implicit subsidies. However, under-recoveries have grown very fast in the last few years.

Appendix 8-3: Calculation of Direct and Combined Incidence

Direct Incidence: For a given product, its budget share corresponds to the price elasticity of total spending, assuming volume of demand constant.

$$\theta_{ij} = \frac{\partial \log M_j}{\partial \log P_i} \qquad (8.1)$$

where M_j is money income/total expenditure of jth household, P_i is the price of the ith good, and θ_{ij} is the budget share of ith commodity in j^{th} household's budget.

$$Tax\ Burden\ due\ to\ tax\ on\ X = \frac{(P+t)X}{E}$$

$$= \left(1 + \frac{t}{P}\right) * Initial\ Budget\ Share\ of\ X \qquad (8.2)$$

where P is the initial price of X, t is the unit tax, X is the amount purchased and E is the expenditure.

Thus, under the three assumptions from the start of the chapter we can test whether a tax on a particular fuel is regressive simply by comparing the budget share of that fuel across different expenditure deciles. The budget share of a decile is the average (weighted mean) of the budget shares of all households in that decile.

Combined incidence calculated using input-output tables: Using the absorption (Use) matrix at producer prices and the make matrix, we create a (130 × 130) input-output matrix at producer prices. Each cell (i,j) denotes the total amount (inclusive of taxes paid) of input j used in the production of commodity j. At the end of each column is the gross value added and indirect taxes on each commodity. The sum for each column is thus the sum of value of intermediate input at producer prices, gross value added, and the indirect taxes, which in turn is the total value of output of that commodity at producer prices. By dividing each column by the output for that column, we get the coefficient matrix. Under the assumption that all prices are normalized to one, the cell (i,j) in this coefficient matrix denotes the amount of input i used to produce one unit of input j.

Let $A = [a_{ij}]_{n \times n}$ be the $(n \times n)$ input-output coefficient matrix, where n is the number of commodities considered. a_{ij} is the quantity of ith commodity output used to produce one unit of commodity j.

Let the price formation equation be

$$P_j = \sum_{i=1}^{n} a_{ij}P_i + VA_j + t_j, \quad i = 1, 2, \ldots, n \tag{8.3}$$

or,

$$(I - A^T)_{n \times n} P_{n \times 1} = VA_{n \times 1} + t_{n \times 1} \tag{8.4}$$

where I is a $n \times n$ identity matrix, t is a $n \times 1$ column matrix of tax rates, and P and VA are column vectors showing prices and value added of the n sectors.

Let us assume that the tax of the ith commodity changes by dt_i. Then,

$$(I - A^T)_{n \times n} dP_{n \times 1} = dt_i e_i \tag{8.5}$$

where e_i is a column vector with 1 in the ith place and 0 in every other place.

Taking inverse (assuming inverse exists) we have,

$$dP_{n \times 1} = (I - A^T)^{-1} dt_i e_i \tag{8.6}$$

The tax burden of the kth household is

$$TB_k = \left[\frac{X_k (I - A^T)^{-1} e_i}{Y_k} \right] dt_i \tag{8.7}$$

where X_k is the $(1 \times n)$ vector of quantities purchased by household k and Y_k is consumption expenditure of household k. Denote the term within parentheses by S_k. This can be interpreted as the share of commodity i in household k's expenditure taking all indirect effects into account.

Since the terms outside the parentheses are the same for all households, we only need to calculate the term within to comment on distribution of tax burden. Information on X_k and Y_k is obtained from NSSO data; information about other matrices is obtained from CSO's input-output table. In the NSSO survey, the consumption values of any commodity are at market prices. However, the CSO matrix used is at producer prices. The difference between the two arises because of the trade and transport margins. Thus when a consumer reports to buy Rs. X worth of commodity i in the NSSO survey, he is actually buying the output of three CSO sectors (commodities): sector I, trade sector, and transport sector. Thus a part of the consumer expenditure on sector I should be transferred to trade and transport sector. The CSO provides information on the trade and transport margins for all sectors. We assume that the overall margins for the economy apply to all consumption deciles. After adjusting for trade and transport margins, we obtain X_k required to calculate tax burdens using (8.7).

The CSO input-output transaction matrix is actually the cost share matrix $\{C_{ij}\}_{n \times n}$ where $C_{ij} = a_{ij} \times (P_i/P_j)$. We chose physical units in such a way that initially (before tax) $P_1 = P_2 = \ldots\ldots = P_n = 1$. Given this assumption $\{C_{ij}\}_{n \times n}$ is the same as $\{A_{ij}\}_{n \times n}$ and the tax burden can be easily calculated. We can calculate it for each household corresponding to tax changes in coal, natural gas, kerosene and petroleum products.

Notes

1   Marland et al. 2008.
2   According to a study titled "Comprehensive Study to Assess the Demand and Requirement of SKO" carried out by the National Council of Applied Economic Research (NCAER) and commissioned by the petroleum and natural gas ministry, 36% of the total amount of kerosene distributed in the country through PDS is diverted. The study further found that 18% of diverted kerosene finds its way to households through the open market. Assuming that the whole of the rest is used to adulterate diesel, around 18% of PDS kerosene supplied is used to adulterate diesel.
3   In the oil sector, under-recoveries and losses are often used interchangeably. This is not correct, as they are two distinct concepts. Refining of crude oil is a process industry where crude oil constitutes around 90% of the total cost. Since value added is relatively small, determination of individual product-wise prices becomes problematic. The OMCs are currently sourcing their products from the refineries on import parity basis, which then becomes their cost price. The difference between the cost price and the realized price represents the under-recoveries of the OMCs. The under-recoveries as computed above are different from the actual profits and losses of the oil companies as per their published results. The latter take into account other income streams like dividend income, pipeline income, inventory changes, and profits from freely priced products and refining margins in the case of integrated companies.
4   Conversion rates: 1 crore = 100 lakhs = 10 million.
5   For details, see Appendix 8-2.
6   For details, see Appendix 8-2.

7   The pricing of coal was fully deregulated after the Colliery Control Order, 2000, the central government has no power to fix the prices of coal.

8   If the supply curve is not perfectly elastic, then a part of the tax burden is shifted to the producers. Calculation of the burden would then require information on the demand and supply elasticities of different industries and the distribution of ownership of firms in those industries. This information is not available. Besides, a number of studies suggest that in the short to medium run, the burden of a carbon tax will be mostly passed forward into higher consumer prices (Bovenberg and Goulder 2001 and Metcalf et al. 2008).

9   Budget share of fuel x for a decile is the ratio of the "average expenditure on x in the decile" and "average household expenditure on all goods in the decile."

10 For technical details, see Appendix 8-3.

11 NSSO, 61st round. Eleventh Five Year Plan 2007–2012, Government of India. Information obtained from UNDP website: http://data.undp.org.in/CountryOffice_FactSheet08.pdf (accessed May 24, 2011).

12 Based on new statistical calculations of PPP exchange rates published in 2005 by the International Comparison Program (ICP) of the IMF, the PPP-adjusted exchange for India is Rs14.7/PPP-adjusted US$.

13 "Direct budget share" means directly consumed fuel (excluding indirect consumption through consumption of commodities using fuel as input) as a percentage of total expenditure. Consumption reported in the household consumption survey is treated as direct consumption. As noted earlier, budget share of a decile is the ratio of the average (weighted) fuel expenditure and average (weighted) total expenditure for that decile.

14 Figures 8-A1 and 8-A2 in Appendix 8-1 show incidence separately for rural and urban sectors. While calculating budget shares separately for urban and rural sectors, deciles are constructed separately from the rural and urban distribution of MPCE.

15 Figures 8-A1 and 8-A2 in Appendix 8-1 show budget shares separately for urban and rural sectors.

16 The curve for diesel is almost flat for first few deciles and showing some upward slope for top consumption deciles. It is very close to zero showing that a negligible amount of Indian households use diesel vehicles for private transport. Hence, it is not shown in the Figure 8-4.

17 For details, see Figure 8-A3 of Appendix 8-1.

18 For details on how information from the consumption survey and information from I-O tables are combined, see Appendix 8-3.

19 In addition to diesel, petrol, kerosene, and LPG, petroleum products also include lubricating oil and other industrial fuels. Coke is no longer a part of the coal sector. According to CSO classification, it is a part of the coal tar products sector. While reporting direct budget shares, we had provided the combined share of coke and coal.

20 For details on adjusting for trade and transport margins, see Appendix 8-3.

## CHAPTER 9

Is Reducing Subsidies on Vehicle Fuel Equitable? A Lesson from Indonesian Reform Experience

ARIEF ANSHORY YUSUF AND BUDY P. RESOSUDARMO

Today, fuel consumption is still subsidized in many parts of the world. It is estimated that fossil fuel related consumption subsidies amounted to US$ 557 billion in 2008. A recent assessment projected that phasing out these subsidies by 2020 would result in a reduction in primary energy demand at the global level of 5.8% and a fall in energy-related carbon dioxide emissions of 6.9%, compared with a baseline in which subsidy rates remain unchanged (IEA/OPEC/OECD/World Bank 2010, p. 4).

There are many reasons to call for reductions in fuel subsidies: Fuel subsidies create distortion by disregarding the economic value of the fuel, creating excess consumption, and discouraging energy substitution. They are also regarded as a major cause of environmental problems, not only from the pollution created by excessive fossil fuel combustion by industry and vehicles, but also due to excessive traffic and the inconvenience it causes. Fuel subsidies also discourage development of a more traffic-free public transport infrastructure. In most big Indonesian cities, this is already a major public concern.

In addition to the efficiency-related problem stated above, fuel subsidies are often deemed inequitable, as vehicle owners benefit greatly from the subsidy. Fuel pricing reform has been widely advocated as a means of promoting efficiency as well as equity.

However, in the Indonesian context, the biggest concern is the fiscal burden of the subsidy. In 2009, the Indonesian government spent more than US$ 6 billion to subsidize fuel consumption. This was almost 8% of total government spending. In a country with much more pressing issues such as poverty, education, and health, the allocation of such a large share of public spending on subsidizing fuel consumption seems to be inconsistent with rational resource allocation principles. However, the current situation has improved compared to the past few years. For example, in 2000 the fuel subsidy amounted to 40.9 trillion rupiah, or almost a third of total central government spending (see Table 9-1). Since the government always has a political constraint against reducing this subsidy, spending has been heavily limited by the fluctuation in the world oil prices.

When the world oil price started to increase rapidly from 2004 onward, the government saw no option but to reform its fuel pricing policy radically. In October 2005, the government made a big adjustment in fuel prices, increasing retail fuel prices for gasoline, kerosene, and diesel. The price of gasoline was increased by 87.5%, diesel by 104.7%, and—surprisingly—kerosene by 185.7%.

Over the past few years, reduction of the fuel subsidy has been one of the Indonesian government's main agendas. Gradual reform in fuel pricing policy through adjustments in fuel prices has been undertaken since 1999 (see Figure 9-1).

However, strong opposition from the people and parliament has slowed the reform. Most arguments against the reform are based on the concern that an increase in fuel prices would translate into an increase in other prices that would reduce purchasing power and exacerbate poverty. The fear is that the rise in fuel prices would create a chain reaction, affecting other costs such as transportation and other important commodities, thereby hurting the economy and those most vulnerable. In addition, some concerns are also related to the reform's effect on household income. Since economic activity is dependent on fuel, the increase in price of fuels would translate into an economic slowdown, reducing employment or real factor returns. It is common to see arguments that the impact through this channel will hurt the most vulnerable in society. These two possible channels through which a reduction in fuel subsidy will have an impact on households suggest that the economic consequences of reform as discussed above are inherently general equilibrium issues.

TABLE 9-1 Fuel Subsidy, Government Budget, and Oil Prices, 1999–2010.

	1999	2000	2001	2002	2003	2004	2005	2006	2007	2008	2009	2010
Fuel Subsidy (Rp Trillion)	40.9	53.8	68.4	31.2	30.0	59.2	89.2	62.7	83.8	139.1	54.3	59.0
Government Spending (Rp Trillion)	201.9	188.4	260.5	322.2	376.5	430.0	411.6	470.2	504.6	693.4	696.1	699.7
Percent	20.3	28.6	26.3	9.7	8.0	13.8	21.7	13.3	16.6	20.1	7.8	8.4
World crude oil price ($/barrel)	17.1	27.1	22.7	23.5	27.1	34.6	49.9	60.3	69.7	97.0	61.0	60.0

Source: Ministry of Finance (2010), and IEA (2010)

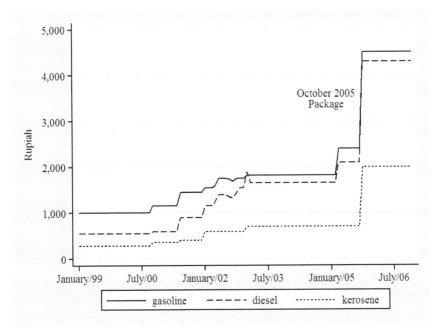

FIGURE 9-1 Prices of Gasoline, Diesel, and Kerosene around the Period of the Reform, 1999–2006.

Source: PERTAMINA (2008).

The main objective of this chapter is to analyze the distributional impact of fuel pricing reform using the Indonesian experience as implemented in October 2005 with its focus on vehicle fuel prices. Questions addressed are whether a reform of reducing the vehicle fuel subsidy in Indonesia constitutes a progressive reform and through which channel such a distributive impact can be explained. INDONESIA-E3, a computable general equilibrium (CGE) model with highly disaggregated household, is used for the analysis.

The chapter is organized as follows: the next section discusses relevant previous work. The subsequent section briefly describes the model, followed by discussions of the scenarios and results before a brief conclusion.

Previous Studies

World Bank (2006) is the only available study that assesses the distributional impact of the October 2005 package. Other studies exist but do not explicitly analyze the October 2005 reform. These are Clements et al. (2003), Sugema et al. (2005), and Ikhsan et al. (2005), all published before the October 2005 package was implemented.

World Bank (2006) looks at the impact of the increase in various fuel prices on household expenditure using data from the 2004 National Socioeconomic Survey (SUSENAS). The result suggests that in the absence of any compensatory measures, the October 2005 package would have led to a 5.6% increase in poverty incidence. Compensation in the form of an unconditional cash transfer to poor and near-poor households would have, on average, more than offset the negative impact of the fuel price increase.

Ikhsan et al. (2005) analyzes the distributional impact of the March 2005 fuel price adjustment that increased the price of kerosene to industry[1] by 22.22%, gasoline by 32.60%, diesel for transportation by 27.27%, diesel for industry by 33.33%, and diesel oil and fuel oil by 39.39%. Ikhsan et al. (2005) used a combination of a CGE model[2] and a simulation using household survey data. That result suggested that poverty rises by 0.24% without compensation, whereas poverty falls by 2.6% with fully effective compensation, and by 1.89% with compensation that is 75% effective (Ikhsan et al. 2005, Table 9). The policies simulated slightly reduce inequality.

The March 2005 fuel price adjustment is also analyzed by Sugema et al. (2005), where the poverty impact analysis is carried out using a SUSENAS-based micro-simulation, and the macro impact is analyzed using an ORANI-based CGE model. The result suggests poverty would rise by 1.95%. This poverty impact seems relatively high considering the petroleum price only rises by 29%.

Clements et al. (2003) examine the scenario of increasing the price of petroleum products by 25% using a CGE model. The study suggests aggregate real household consumption falls from 2.1% to 2.7% following a 25% increase in the price of petroleum products. Urban and high income households suffer the most, indicating the progressivity of the reform.

From a methodological point of view, the weaknesses of these studies lie in either the incompleteness or inaccuracy of the distributional analysis. In a market economy, the effect of a policy shock on a household's welfare works through both the market of commodities (through changing commodity prices) and the market of factors of production (through changing factor prices or employment). The change in real expenditure of various households then depends on both expenditure and factor ownership patterns of each of the respective households. Taking into account one without the other provides only a one-sided story. This incompleteness could be solved by an economy-wide framework, by using a model that has highly disaggregated households to maintain accuracy in the distributional story. It is very important to acknowledge that commodity prices, household expenditure, factor prices, and household income are all endogenous and solved simultaneously in the model. Household heterogeneity is also inherent and should be integrated in the model.

The World Bank's (2006) assessment, then, is a partial equilibrium story which overlooks the factor-income effect. The Ikhsan et al. (2005) study, despite combining a CGE model and a micro-simulation, is a top-down approach where commodity prices are exogenous in the micro-simulation, and commodity prices from the CGE model are determined by a single representative household. There is no connection between the economy-wide model and SUSENAS micro-simulation in the Sugema et al. (2005) study. Finally, the Clements et al. (2003) study is only based on 10 representative households, preventing accuracy in distributional analysis.

The innovation used in the model for this chapter is that of including highly disaggregated households within one integrated economy-wide framework. Because the households are classified by centile of real expenditure per capita, the model is not only able to capture expenditure and factor ownership patterns of households but also to assess the distributional impact more accurately. The cumulative density function (CDF) of real expenditure per capita before and after the shock can be pictured. Therefore, an objective answer to the question of whether a policy shock is progressive or regressive is readily available. This is the first time this model has been used for Indonesia. Warr (2006), for example, used this approach for Laos in assessing the poverty impact of large-scale irrigation investment.

The Computable General Equilibrium (CGE) Model

The analysis uses the INDONESIA-E3 (Economy-Equity-Environment) model, a CGE model with a strong focus on distributional analysis. The structure of the model is built based on the ORANI-G model (Horridge 2000) with two significant modifications. The first is allowing for substitution among energy commodities, and also between primary factors (capital, labor, and land) and energy. In this respect, this model has 38 industries and 43 commodities with detailed energy sectors. Energy commodities include coal, natural gas, gasoline, automotive diesel oil, industrial diesel oil, kerosene, LPG, and other fuels. Second, the model has been adapted to have multiple households not only on the expenditure or demand side of the model but also on the income side of the households. The household demand system follows the Linear Expenditure Demand (LES) system, and its parameters are estimated econometrically.[3]

The integration of highly disaggregated households into models adequate for accurate distributional analysis is made possible by constructing an Indonesian Social Accounting Matrix (SAM) 2003, which serves as the core database for the CGE model. The SAM is specially constructed and consists of 181 industries, 181 commodities, and 200 households (100 urban and 100 rural households grouped by centile of real expenditure per capita). The SAM (with a size of 768 x 768 accounts) constitutes the most disaggregated SAM for Indonesia yet to date at both the sectoral and household levels. The data used for constructing the SAM include the Indonesian Input-Output Table and—most important—household level survey data from SUSENAS. A detailed description of the SAM can be found in Yusuf (2006).

Scenario and Simulation Strategy

The scenario to be simulated is increasing the price of gasoline by 87.5% and diesel by 104.7%, as implemented by the Indonesian government in October 2005. To exactly represent the rate of the price increase as announced by the government on October 1, 2005, in the simulation the price of fuels is set exogenously and the subsidy rate is set to be determined endogenously in the model.

The simulation is carried out with the following three assumptions: (1) Capital is specific or immobile across sectors representing the short-run scenario to be reflected in the analysis; (2) All kinds of labor are mobile across sectors, but wage rigidity allows for aggregate employment to change; and (3) To isolate only the impact of the fuel subsidy reduction, the government revenue from the reduction of the subsidy is saved by endogenizing the government budget surplus.

Results and Discussion

This section looks first at macroeconomic and industry results before examining distributional results.

Macroeconomic and Industry Results

Table 9-2 shows the impact of the simulation on selected macroeconomic variables and income from each factor of production. GDP declines by 1.72% relative to the baseline, mostly due to the decline of aggregate employment (−3.32%). Aggregate real consumption expenditure falls by 2.61% relative to baseline. It should be noted, however, that this contractionary effect greatly reflects the short-run scenario represented by the assumption of wage rigidity as well as the amount of extra revenue from the reduction of subsidy which is kept as a budget surplus by the government.

The simulated impact on income from the ownership of factors of production reveals a distinct picture. Capital income falls much more proportionately than labor income, and among labor income categories, the fall in urban labor income is relatively larger than that of rural labor income. The fall in formal labor income is relatively larger than that of informal labor income. A plausible explanation is related to how the decline in output varies across sectors, as will be explained later. As we will see, this also has important implications for the distributional impact across various household groups through the income channel.

As shown in Table 9-3, industry output falls as a result of the increase in the price of vehicle fuels. The increase in fuel prices lowers fuel demand, resulting in an immediate reduction in the output of the refinery industry.

TABLE 9–2 Simulated Macroeconomic and Factor Market Results (% Relative to Baseline).

Macroeconomic		*Income from Factor Ownership*	
Real Gross Domestic Product	−1.72	Capital	−6.76
Real Household expenditure	−2.61	Agriculture, rural, formal labor	−1.20
Export (volume index)	−1.72	Agriculture, urban, formal labor	−0.87
Import (volume index)	−2.38	Agriculture, rural, informal labor	−1.05
Aggregate Employment	−3.32	Agriculture, urban, informal labor	−1.09
Consumer Price Index	1.12	Production, rural, formal labor	−2.20
		Production, urban, formal, labor	−3.93
		Production, rural, informal, labor	−2.06
		Production, urban, informal, labor	−2.58
		Clerical, rural, formal labor	−2.10
		Clerical, urban, formal, labor	−3.06
		Clerical, rural, informal, labor	−2.62
		Clerical, urban, informal, labor	−2.75
		Managerial, rural, formal labor	−1.49
		Managerial, urban, formal, labor	−2.26
		Managerial, rural, informal, labor	−2.56
		Managerial, urban, informal, labor	−2.99

Source: Author's calculation.

The final (new equilibrium) reduction in the output of petroleum refineries is 4.57% (relative to baseline without the reform). Other industries which experience big reductions are those closely related to the petroleum refinery sector. These are road transportation (−3.53%), other transportation (−5.18%), utility sectors (electricity by −1.98%, and water and gas by −2.98%), and some manufacturing industries (automotive by −3.10% and rubber and its products by −3.34%). After simulating an increase in the price of petroleum products, Clements et al. (2003) report that industries that experience a large drop in output are similar to the types of industries just mentioned.

As shown by the data on the input intensity of each sector, the industries which experience relatively larger contractions are energy intensive sectors which, in general, happen to be capital intensive or skilled-labor intensive. This explains why returns to capital have declined more proportionately than returns to other inputs such as unskilled labor. This has an important implication on the distributional impact as it affects the functional distribution of income by the tendency to hurt households endowed with capital more proportionately.

TABLE 9-3 Input Share and Simulated Change in Output (% of Baseline).

	Share in Total Input					% Change in Output
	Labor	Capital	Land	Energy	Other	
Paddy	51.4	16.56	14.5	0.00	17.54	−0.57
Other food crops	57.36	17.35	15.2	0.01	10.09	−1.04
Estate crops	52.73	11.21	8.88	0.29	26.89	−1.57
Livestock	42.39	8.08	3.41	0.03	46.08	−1.67
Wood and forests	36.37	21.78	21.34	0.42	20.09	−0.96
Fish	38.96	8.77	27.19	1.98	23.09	−1.08
Coal	8.09	72.01		13.17	6.73	−0.06
Crude oil	5.8	80.74		6.41	7.06	−0.14
Natural gas	5.81	80.97		6.44	6.77	−0.19
Other mining	26.38	47.48		2.16	23.98	−0.52
Rice	6.17	8.32		0.02	85.49	−0.52
Other food (manufactured)	15.23	18.59		0.84	65.34	−1.91
Clothing	14.58	19.1		0.83	65.48	−2.14
Wood products	18.24	25.36		1.05	55.35	−1.1
Pulp and paper	13.92	22.79		1.51	61.79	−2.32
Chemical product	11.88	14.7		3.56	69.86	−2.79
Petroleum refinery	7.54	57.83		8.04	26.6	−4.57
LNG	1.66	51.77		40.05	6.51	−0.83
Rubber and products	15.8	14.82		1.82	67.56	−3.34
Plastic and products	7.81	20.26		0.82	71.11	−1.97
Nonferrous metal	20.4	34.71		6.82	38.07	−1.09
Other metal	9.9	14.06		1.79	74.26	−1.53
Machineries	9.35	13.31		0.63	76.71	−3.16
Automotive industries	15.67	29.73		0.8	53.8	−3.1
Other manufacturing	14.06	26.56		1.4	57.98	−2.36
Electricity	5.92	50.14		19.33	24.62	−1.98
Water and gas	17.27	26.37		13.42	42.93	−2.98
Construction	23.09	9.55		4.76	62.6	−0.15
Trade	35.27	26.83		1.48	36.42	−1.95
Hotel and restaurants	36.93	10.99		0.04	52.04	−2.13
Road transportation	21.4	22.11		8.31	48.17	−3.53
Other transportation	12.48	18.17		10.33	59.01	−5.18
Banking and finance	18.9	53.47		0.25	27.38	−1.63
General government	53.98	5.62		2.14	38.26	−0.08
Education	43.72	8.54		1.1	46.65	−1.58
Health	54.5	9.02		0.19	36.29	−1.5
Entertainments	17.24	18.11		0.1	64.55	−2.26
Other services	25.06	34.83		0.31	39.79	−2.36

Source: Author's calculation.

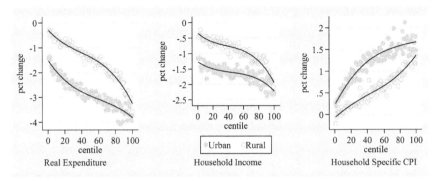

FIGURE 9-2 Incidence Curve and its Decomposition.

Source: Author's calculation.

Distributional Results

Figure 9-2 illustrates the impact of the reduction in vehicle fuel subsidy on household real expenditure, income, and household-specific consumer price index (CPI) for urban and rural households as well as across percentile of real expenditure per capita. The percentage change in real expenditure is used to calculate the Gini coefficient.

The change in the household nominal income and household-specific CPI indicates how the expenditure and factor income patterns of each household contribute to the distributional results. Household-specific CPI is a consumption-weighted average of the price increase of every commodity consumed by the respective household. It reflects the contribution of its household expenditure pattern and behavior. Still, the change in household income reflects changes in all sources of household income, comprising income from labor by skill types, capital, land, and transfers.

As clearly indicated by Figure 9-2, we can conclude that the reduction in vehicle fuel subsidy is progressive. The fuel subsidy on gasoline and diesel has been indeed inequitable, so that cutting the subsidy on these vehicle fuels would be a progressive reform. The progressivity of the reform is confirmed by the decline in the Gini coefficient (see Table 9-4). The calculation of the

TABLE 9-4 Simulated Distributional Impact (Gini Coefficient)

	Urban	*Rural*	*Urban + Rural*
Ex-ante Gini coefficient	0.347	0.277	0.350
Ex-post Gini coefficient	0.344	0.272	0.345
Change	−0.003	−0.005	−0.005

Source: Author's calculation.

Gini coefficient also suggests that when combining urban and rural house-holds, the reform is also progressive, reducing overall inequality.

The declining pattern of the fall in real expenditure over centiles of expenditure as shown in Figure 9-2 illustrates the progressivity. This happens in both urban and rural areas. This progressivity is driven both by household consumption (by the pattern of the change in household CPI) and income pattern (by the pattern of the change in household income). High-income households spend more intensely on vehicle fuel and related services such as transportation than poorer households; hence they tend to experience a greater increase in consumer prices. The fall in income in high-income house-holds is also higher compared to poorer households, reflecting the adjustment in the factor market which does not favor the factor endowment of high-income households.

Conclusion

Although the negative effect of distortionary fuel subsidies has been well known, they are still being implemented by governments in many parts of the world. The pressure to reform such policy has been discouraged by the concern over the distributional impact of such reform. The worry is that the increase in the price of fuels will reduce overall purchasing power of the most vulnerable households, and its contractionary impact on the economy will reduce employment and income. This line of argument is inherently a general equilibrium issue.

To contribute to this debate, using a CGE model with disaggregated house-holds that allows for a rich and accurate distributional story, we simulate the impact of increasing the price of gasoline and diesel, part of the October 2005 reform package of the Indonesian government.

The result suggests that the reform is strongly progressive. Higher income households experience a greater decline in welfare than lower income house-holds, reducing overall inequality. The novel feature of a general equilibrium analysis helps explain how this can be the case. The progressivity comes from both price and income channel. Higher income households tend to spend more intensively on vehicle fuel and commodities closely associated with it. They are also more dependent on income from factors of production, which are more intensively used in the production sectors that are heavily affected by the reform.

Notes

1   Kerosene for domestic household use was not increased.
2   It used The INDOCEEM model, an Indonesian CGE model based on ORANI-G, developed initially by Monash University and the Indonesian Ministry of Energy.
3   For a more detailed description of the model, please refer to Yusuf (2008).

## CHAPTER 10

Distributional Consequences of Transport Fuel Taxes in Ethiopia

ALEMU MEKONNEN, RAHEL DERIBE, AND
LIYOUSEW GEBREMEDHIN

Many studies show that Earth's climate is changing mainly as a result of the increasing concentration of greenhouse gases in the atmosphere that are emitted from burning of fossil fuels (Keeling and Whorf 2004). Thus, minimizing the consumption of fossil fuels, undertaking energy conservation measures, and renewable energy development activities have gained international focus, and many countries are engaged in the implementation of such activities.

Taxes on environmentally damaging activities are one of the instruments used to address environmental problems. In addition to generating revenue for governments, environmental taxes also contribute to emission abatement. In some cases, such as taxes on imported fossil fuels in countries like Ethiopia, they also discourage consumption and help save scarce foreign exchange that could be used to import other necessities for the poor.

Ethiopia, being one of the least developed countries in the world with a predominantly rain-fed agrarian economy, is most vulnerable to climate change and with least capacity to respond (ILRI 2007). Ethiopia already suffers from extreme weather events, manifested in the form of frequent drought (1965, 1974, 1983, 1984, 1987, 1990, 1991, 1999, 2000, and 2002) and recent flooding (1997 and 2006). At the national scale the link between drought and crop production is widely known (FAO 2005).

Aggregation of the 1994 CO_2, CH_4 and N_2O emissions resulted in a total of about 48,000 gigagrams CO_2-equivalents excluding CO_2 emissions/removals from the Land Use Change and Forestry (LUCF) sector (which was about 0.9 tonnes of CO_2-equivalent per capita per year). A comparison of GHG emissions between the years 1990 and 1995 shows that total (gross) CO_2 emissions (i.e., emissions from the energy and industrial process sectors) increased by about 24% while emissions of CH_4 and NO_x increased by 1% and 119%, respectively (MWR 2001). Almost 90% of the country's total CO_2 emissions came from fossil fuel combustion in the energy sector, and the transport (road) subsector is the main emitter of CO_2 within this sector. Petroleum is wholly imported and takes over 75% of total export earnings (MWR 2001).

In Ethiopian cities—especially Addis Ababa—the number of motor vehicles is growing at an increasing rate. In just three years the total number of vehicles on Addis Ababa roads increased by 22.4% (EMTCTA 2009). These vehicles use gasoline and diesel. Furthermore, about 53% of vehicles in Addis Ababa are more than 20 years old (EMTCTA 2009). Diesel and gasoline combustion in motor vehicles release carbon dioxide (CO_2), nitrous oxides (NO_x), sulfur dioxide (SO_2), volatile organic compounds (VOCs), and lead and particulate matter (PM). Many studies show a strong association between these air pollutants and respiratory symptoms and ear and throat infections (Randem et al. 2004, WHO 2005).

Because Ethiopia's economy has shown significant and continuous growth, especially in recent years, the country's energy needs are growing fast. For example, during the last three years, petroleum imports have grown from 1,406,899 metric tonnes in 2005/06[1] to 1,881,272 metric tonnes in 2007/08[2] (see Appendix 10-1, Figure 10-A2). The increasing cost of fossil fuels affects the whole economy as well as the environment, so fiscal policies and strategies to address this issue must be put in place.

For instance, a new biofuel energy strategy was formulated in September 2007 (MoME 2007). The need for this strategy was based on two important points. First, the imbalance between fossil fuel demand and supply and unstable prices have threatened the sustainability of Ethiopia's economic growth and hurt its trade balance. Biofuels can help the country reduce fuel imports and supply the international market, thereby strengthening its trade balance. Second, fossil fuel use is a cause for atmospheric air pollution and global warming.

Excise tax, value added tax (VAT), road funds, municipal taxes, and a price stabilization fund are among the instruments applied to consumers of gasoline and diesel oil in Ethiopia to raise government revenue. Such instruments also have the effect of discouraging environmental damage caused by fossil fuel consumption. The recently revised fuel tax regime is one of the efforts. Table 10-1 shows the current tax system (per liter of fuel type) compared with earlier fossil fuel taxes; all were increased. The increase in VAT on all fuels (except kerosene and aviation fuel, or JET A1), as well as the increase in payments to the price stabilization fund, are especially noteworthy. Note also that the excise tax, which has also increased, is applied only to gasoline and JET A1.

Income distribution and fairness are important issues in Ethiopia, home to a large population of very poor people. This chapter examines the distributional consequences of transport fuel taxes in Ethiopia and derives fiscal and environmental policy implications for Ethiopia and other similar developing countries.

TABLE 10-1 2009 Reform of Fossil Fuel Taxes in Ethiopia (Birr per Liter).

	Gasoline-Ethanol mix		Gasoline		Kerosene		ADO		JET A1	
Type of Tax	*Before*	*After*	*Before*	*After*	*Before*	*After*	*Before*	*After*	*Before*	*After*
Excise tax	1.79	1.79	1.58	1.99					1.43	1.61
VAT	1.17	1.17	1.03	1.30			0.77	1.02		
Road fund	0.095	0.095	0.095	0.095			0.08	0.08		
Municipal tax	0.02	0.02	0.02	0.02			0.02	0.02		
Stabilization fund	1.22	1.79	0.33*	1.00	0.20*	0.58	0.78*	1.27	0.91*	1.07
Total	4.29	4.86	3.06	4.41	0.20	0.58	1.66	2.39	2.34	2.68
Price as of March, 2009	7.47	8.37	7.96	7.32	4.12	5.5	5.44	7.13	5.8	7.7

Notes: "Before" refers to August 2006; "After" refers to March 2009. All values marked with * were effective as of February, 2009. ADO = Automotive diesel oil; Birr is the Ethiopian currency; in March 2009 the market exchange rate was about 11 Birr = 1 USD.
Source: Ethiopian Petroleum Enterprise, 2009.

Methodology and Data

Tax burden can be measured using either income or expenditure to estimate incidence. Poterba (1991) used expenditure and included U.S. households that own vehicles as well as those that do not own vehicles in his analysis. On the other hand, Walls and Hansen (1999) used annual income and a measure of lifetime income to analyze the distributional effects of vehicle pollution control policies. Many researchers believe that taxes should be compared with a household's long-term income rather than its annual income (see Chapter 6). Measuring tax burden relative to permanent income provides an estimate of household's ability to bear a tax over a lifetime. Annual income could significantly underestimate the long-term ability of some households to pay a tax. In this chapter, to examine whether fuel taxes are regressive or not, we use total expenditure rather than annual income, which is a way of approximating lifetime income.

Suits coefficients based on expenditure approach are calculated in order to examine whether fuel tax is regressive or progressive (Suits 1977). Direct fuel expenditure of households (from owners of motorized vehicles) and consumers who pay indirectly (from use of public transport) are taken into account. We also report tax burden from private (direct) consumption only, public (indirect) consumption only, and direct consumption and indirect consumption combined.

We make the following assumptions in the calculations of shares and indices. Following Chapter 6 on Mexico, which calculated the fuel share of public transport expenditure to be 42%, we assumed the fuel share of public transport expenditure for Ethiopia to be 50%. Ethiopia is poorer than Mexico, and salaries are generally lower; hence the fuel share of public transport costs should be somewhat higher. According to 2006 data from the Ethiopian Transport Authority, the total shares of gasoline and diesel from all types of fuels in Ethiopia are 50% and 50% respectively. Moreover, considering that private transport in Ethiopia largely uses gasoline, we assumed that 90% of private transport vehicles used gasoline and 10% used diesel. On the other hand, considering that public transport in Ethiopia largely uses diesel, a share of 70% diesel and 30% gasoline in public transport is assumed. In addition, during the study period the share of taxes in fuel prices was 38% for gasoline and 14% for diesel (Ethiopian Petroleum Enterprise, 2009).

Data

We used Household Income, Consumption and Expenditure (HICE) survey data collected by the Ethiopian Central Statistical Agency (CSA) in 2004–2005.[3] The survey covered over 21,000 households from all rural and urban areas of the country except all zones[4] of Gambella region,[5] the nonsedentary population of three zones of Afar, and six zones of Somali regions. From rural areas 794 enumeration areas (EAs) and 9,500 households were covered by the survey; from urban areas 768 EAs and 12,100 households were covered. In addition, time-series data on Addis Ababa retail price build-up and quantity of petroleum products imported were also gathered from the Ethiopian Petroleum Enterprise (Appendix 10-1).

Empirical Analysis

To analyze the distributional effect of fuel taxes, data from the household survey is allocated in expenditure deciles and the budget share spent on transportation calculated. It is here argued that the well-being of individuals often depends on the income of the entire household rather than on one individual's income, so income deciles are arranged according to household income.

Table 10-2 reveals total expenditure on all goods and services by decile group in 2004/2005. The results show that close to one third of the total expenditure was spent by the richest 10%, decile 10. On the other hand, households in the first two deciles—the very poorest 20% in terms of expenditure—used only about 7% of the total expenditure.

Distributional Consequences of Fuel Taxation using Expenditure Shares

We might expect higher progressivity especially for gasoline in a country like Ethiopia, where the average income is very low because really poor people hardly use any gasoline at all—at least not directly. However, we need to

TABLE 10-2 Mean Household Expenditure (Birr per year) by Expenditure Decile.

Expenditure Decile*	Mean Household Expenditure per Year	As % of total
1	2,906	3
2	4,245	4
3	5,133	5
4	5,939	6
5	6,768	7
6	7,713	8
7	8,944	9
8	10,636	11
9	13,691	14
10	29,522	31

* The first decile is the lowest expenditure group, the tenth the highest. In 2004/05, ~8.5 birr = US$ 1. According to the International Monetary Fund (2008), gross domestic product/capita (GDP) at purchasing power parity (PPP) per capita was US$ 897. Nominal gross domestic product (GDP/capita) was US$ 88.

Source: Authors' computation from CSA data.

recognize that poor people in poor countries also need public transport and indirectly consume fuels when they use it. Hence, in this study, we have included the budget share of both direct transport fuel consumption and that of public transport (indirect transport fuel consumption).

In Figure 10-1, public transport expenditure shares are presented for each expenditure decile. In this study, public transport includes air, taxi, bus, and train fares. Figure 10-1 illustrates that as household expenditure increases, share of public transport tends to increase in Ethiopia up to the 9th decile group. Nearly half of the total expenditure on public transport was spent by the two highest expenditure groups in 2004/05. On the other hand, about 1.3% of the total expenditure on public transport was spent by the poorest 10% of the population. The budget share of public transport does not fall with increase in expenditure; in fact, the share increases with rise in expenditure (except deciles 3 and 10). This suggests that public transport does not show signs of being an inferior good in Ethiopia.

Table 10-3 depicts mean household expenditure on private transport fuel (birr/year). Private transport fuel refers to fuel consumption for different kinds of privately owned vehicles (diesel and gasoline). About 97% of the expenditure on private transport fuel in Ethiopia is spent by the richest 10% of the sample: private vehicles are used almost exclusively by the richest 10% of the population. Their budget share is 0.8% while all other budget shares are virtually 0%.

Figure 10-2 shows combined shares of household fuel expenditure. As total expenditure increases, the share of household transport fuel expenditure tends to increase. It is particularly noteworthy that the richest 10% of the population spent more than 40% of the total expenditure. This is at least partly

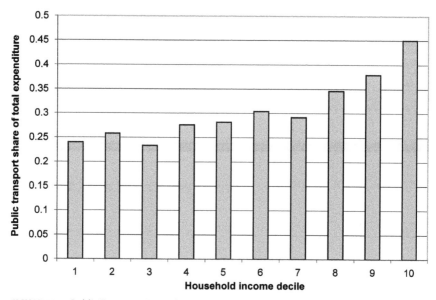

FIGURE 10-1 Public Transport Expenditures as (%) of Total Budget by Decile.

Note: The first decile is the lowest expenditure group, the tenth the highest.
Source: Authors' computation from CSA data.

TABLE 10-3 Mean Household Expenditure on Private Transport Fuel (Birr per Year) by Expenditure Decile.

Decile	Private(Direct) Transport Fuel			Cumulative (%)
	Mean Household Expenditure (in Birr)	Private (Direct) Transport Fuel as Share (%) of Total Expenditure	Cumulative (%)	
1	0.00	0.00	0	0.00
2	0.00	0.00	0.00	0.00
3	0.29	0.01	0.12	0.65
4	0.03	0.00	0.13	0.71
5	0.12	0.00	0.18	0.90
6	0.29	0.00	0.30	1.33
7	0.87	0.01	0.65	2.42
8	3.51	0.03	2.07	6.15
9	3.28	0.02	3.40	8.86
10	238.24	0.81	100	100.00

Notes: The first decile is the lowest expenditure group, the tenth the highest. In 2004/05, ~8.5 birr = US$ 1. According to International Monetary Fund (2008), gross domestic product (GDP) at purchasing power parity (PPP) per capita was US$ 897. Nominal gross domestic product (GDP) per capita was US$ 88.
Source: Authors' computation from CSA data.

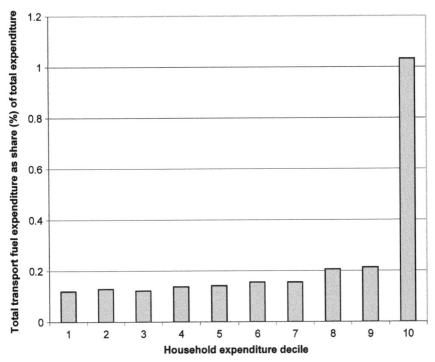

FIGURE 10-2 Household Total Transport Fuel Expenditure as Share (%) of Total Expenditure.

Note: The first decile is the lowest expenditure group, the tenth the highest.

Source: Authors' computation from CSA data.

due to the fact that 85% of Ethiopians are living in rural areas with limited use of modern modes of transport. People with a low level of expenditure prefer to walk or use unmotorized means of transport rather than spending their money on motorized transport.

Table 10-4 presents fuel tax paid related to public transport fuel, private transport fuel, and share (%) of each category of total expenditure. In all scenarios, the burden is the largest in the last decile group (highest expenditure category). This suggests that higher taxes are likely to have more impact on the categories with the highest expenditure.

Based on the analysis of the tax burden for private transport fuel, public transport fuel, and all combined presented so far, we observe that transport fuel taxes are progressive in Ethiopia.

Suits Coefficients and Concentration Curves

Table 10-5 and Figure 10-3 depict the calculated suits coefficients and concentration curves for taxes paid for public, private, and total (public and private combined) transport fuel. The obtained Suits coefficients for the tax burden

TABLE 10-4 Tax Paid for Public Transport Fuel and Private Transport Fuel.

Fuel tax paid related to public fuel transport	Fuel tax paid related to public fuel transport as % of budget	Fuel tax paid related to private fuel transport	Fuel tax paid related to private fuel transport as % of budget	Fuel tax paid related to total transport fuel	Fuel tax paid related to total transport fuel expenditure as % of budget
5.99	0.21	0.00	0.00	5.99	0.21
10.72	0.25	0.00045	0.00	10.72	0.25
12.67	0.25	0.1043	0.00	12.78	0.25
17.73	0.30	0.0108	0.00	17.74	0.30
22.65	0.33	0.0419	0.00	22.69	0.34
27.65	0.36	0.104	0.00	27.76	0.3
39.62	0.44	0.309	0.00	39.93	0.45
58.03	0.55	1.249	0.01	59.28	0.5
85.81	0.63	1.168	0.01	86.98	0.64
217.35	0.74	84.81	0.29	302.16	1.02

TABLE 10-5 Suits Coefficients Calculated for Different Types of Fuel Taxes Paid.

Public Transport Fuel Tax Paid	Private Transport Fuel Tax Paid	Total Transport Fuel Tax Paid
0.146	0.668	0.245

Source: Authors' computation from CSA data.

calculated are all positive. For public transport fuel expenditure the coefficient is 0.146, while the coefficient only for private transport fuel expenditure is 0.668. We can, therefore, conclude that fuel taxes in Ethiopia are progressive. In relative terms, fuel taxes are less progressive for public transport than for private transport. These results are consistent with those presented above in terms of expenditure deciles.

The results can be explained by the fact that about 39% of the population is below the poverty line (MoFED 2007), and most people cannot afford to use their own private transport. Most of the poor also cannot afford to pay even for public transport and therefore use other means of transport such as walking and unmotorized transport.

Conclusion and Policy Implications

Some taxes are progressive while others are regressive or neutral. However, the literature suggests that regressivity or progressivity of taxes depends on socioeconomic conditions as well as the nature of the goods and services

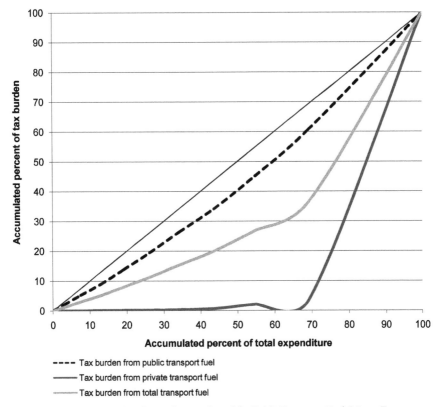

FIGURE 10-3 Concentration Curves (Lorenz Curve) for Public Transport Fuel, Private Transport Fuel, and Total Transport Fuel Tax Paid

considered (Aasness and Larsen 2002). This chapter shows that fuel taxes in Ethiopia are progressive, not regressive. The degree of progressivity varies a bit depending on the items considered. This may be explained by the fact that about 39% of the people live below the poverty line and cannot afford to use modern transport in general. While some of the poor could use public transport, it is not much of an option for the very poor who may find it cheaper to walk or use other unmotorized means of transport. The results indicate that taxes on transport fuels would not only generate government revenue, they can also achieve the objective of abating emissions and saving foreign exchange from the associated decrease in fuel consumed. Increased fuel taxes are likely to affect the rich disproportionately, and hence could be pro-poor.

Acknowledgments

The authors would like to thank Thomas Sterner, Emanuel Carlsson, and two anonymous reviewers for their useful inputs and comments.

Appendix 10-1

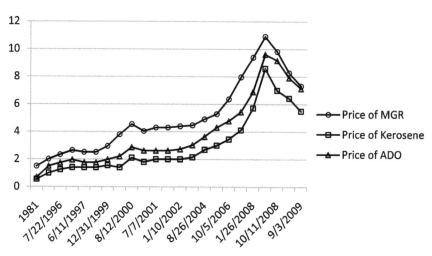

FIGURE 10-A1   Addis Ababa Retail Price of MGR (Gasoline), Kerosene, and Diesel (ADO) 1981–2009.

Source: Authors' computation from Ethiopian Petroleum Enterprise (2009) data.

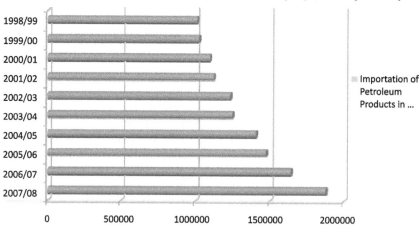

FIGURE 10-A2   Importation of Petroleum Products in Ethiopia (1998–2008).

Source: Authors' computation from Ethiopian Petroleum Enterprise (2009) data.

Notes

1   2005/06 refers to 1998 in Ethiopian calendar.
2   2007/08 refers to 2000 in Ethiopian calendar.
3   2004/05 refers to 1997 in Ethiopian calendar.
4   Zone refers to the second largest administrative unit in Ethiopia (after region).
5   Based on figures for the year 2000 from CSA, Gambella has an estimated total population of 259,000, or about 0.04% of the country's total population. The total population of Afar region at about the same time is a little over 1 million and that of Somali region over 3.5 million. However, as noted, only the nonsedentary populations of Afar and Somali regions were omitted from the survey.

## CHAPTER 11

Political Petrol Pricing: The Distributional Impact of Ghana's Fuel Subsidies

Wisdom Akpalu and Elizabeth Robinson

As we started writing this chapter, on June 5, 2009, petrol and diesel prices in Ghana jumped 30% at the pump overnight—a consequence of the recently elected new government removing the implicit fuel subsidy.[1] Soon after, public transport prices increased far out of proportion to the increased fuel costs. Food price increases were predicted to follow. The initial reaction on the radio and in the newspapers was one of outrage. One commentator asked despairingly what had happened to the pro-poor policies that the government had promised. The opposition parties used the price increase as an excuse to chide the government for harming the poor. Over the course of the week, the responses became more nuanced, as increasingly people started to talk about the distributional impact of the fuel subsidy and its subsequent removal. However, even several months after the increase in fuel prices, headlines bemoaning the price increase continued to crop up in the newspapers.

It is against this backdrop that we take a look at the impact of fuel subsidies for transportation on income distribution in Ghana. We look at the literature on taxation policies in low-income countries in the following section, focusing particularly on sub-Saharan Africa. There is a general consensus from the literature that the indirect effects of fuel taxation and subsidies are particularly important in low-income countries, where most of the population rely on public transportation, and where food takes a high proportion of households' income. We then take a historical perspective on fuel pricing policies in Ghana and the role of the National Petroleum Authority (NPA), which was established in part to reduce the politicization of fuel taxes by governments. Further, we consider the extent to which fuel subsidies in Ghana are progressive or regressive, separating direct spending on gasoline products from indirect spending on transportation. Finally, we look at some of the policy implications of fuel subsidies in Ghana and in developing countries in general.

Literature

Interest in taxation policies in low-income countries has re-emerged recently, motivated in large part by the large number of tax reforms that have been introduced in many of these countries (Gemmell and Morrissey 2005). Yet although the literature addressing fuel taxation and subsidies is growing, papers that specifically address the distributional impact of taxes, particularly in sub-Saharan Africa, are still scarce. Gemmell and Morrissey (2005) provide a useful summary of the earlier evidence of overall tax incidence from the literature from the 1960s and 1970s. For six African countries (Côte d'Ivoire, Ghana, Guinea, Madagascar, Tanzania, and Uganda), they find that taxes on kerosene/paraffin were regressive for all but Côte d'Ivoire and Tanzania (where the evidence is inconclusive); taxes on gasoline, however, were progressive for all six countries. Based on these data they suggest that uniform taxation of fuels is likely to be problematic because gasoline and kerosene/paraffin have very different consumption patterns—kerosene being disproportionately consumed by the poor, and gasoline being disproportionately consumed by the rich. Hope and Singh (1995) find that typically overall reducing or eliminating subsidies on fuel is unlikely to harm the poor because overall, subsidies are regressive.

From the 1980s and 1990s, a small body of literature addresses fuel tax policies in sub-Saharan Africa, often in the context of structural adjustment programs. Longhurst et al. (1988) focus on structural adjustment and its impact on the poor in Sierra Leone, where the removal of petrol subsidies in the 1980s was part of a structural adjustment package. The removal of the petrol subsidy resulted in petrol prices doubling over a period of eight months and transport prices consequently increasing. However, the authors do not analyze in detail the impact on income distribution. Bolnick and Haughton (1998) address excise taxes (that is, taxes levied on particular goods and services), specifically those on tobacco, petroleum, and alcohol products, in sub-Saharan Africa. Excise taxes were shown to be an important source of government revenue in the region in the 1990s. The authors found a wide variation in tax rates on gasoline, with a quarter of countries setting rates of at least 110% but almost half setting rates of 40% or less. Tax rates on diesel fuel were significantly lower, possibly reflecting the importance of diesel for agriculture. Although the authors found overall fuel taxation to be typically progressive in African countries, they recognized the importance of understanding the pass-through effects of higher fuel prices on goods and services such as transport and electricity, but it is difficult to find information to verify this assertion. The authors therefore suggest focusing on the indirect effects of fuel taxes on the most important components of household income, particularly public transportation.

A number of more recent papers have revisited the specific issue of fuel subsidies in Africa, often using data from more recent household budget surveys. For example, Leigh and El Said (2006) look at the implicit fuel subsidies that have been in place in Gabon since 2002, a consequence of domestic fuel prices remaining virtually unchanged while international prices have surged. They find that higher income households have benefited from these implicit subsidies significantly more than lower income households, with the top 10% of individuals receiving around one third of the subsidy and the bottom 30% receiving just 13%. Abouleinein et al. (2009) use a social accounting matrix (SAM) to focus on the recent phasing out of fuel subsidies in Egypt. The authors find that, overall, reducing fuel subsidies affects higher income households most, but lower income groups are most affected by the reduction in the subsidy on natural gas. In Egypt, indirect effects are felt mainly in the electricity sector (electricity prices are predicted to increase 6.9% for every 10% increase in the price of natural gas) and the food sector (which consumes 83.2% of fuel oil). Van Heerden et al. (2006) used a computable general equilibrium (CGE) model to analyze data on South Africa to compute the marginal excess burden by income class and found that a combination of increased energy taxes and reduced food taxes may reduce emissions, bridge the income gap between rich and poor, and increase economic growth.

Ghana has been the focus of a number of papers looking at fuel subsidies and income distribution. Overall taxes on petroleum products in Ghana were found to be progressive by Younger (1993), as reported in Bolnick and Haughton (1998). A breakdown of the data revealed taxes on kerosene to be regressive but on gasoline to be progressive, with the tax on direct spending on gasoline to be highly progressive, and the indirect effect on transportation costs to be approximately neutral. A similar finding is reported from Côte d'Ivoire (Yitzhaki and Thirsk 1990, reported in Bolnick and Haughton 1998), using household survey data from the 1985–1986 Côte d'Ivoire *Living Standards Survey*. Osei and Quartey (2005) focus on tax reforms in Ghana. They provide an assessment of the literature specifically looking at the incidence of taxes in the country. Although there appears to be a general consensus in the literature that petroleum taxes are progressive in Ghana, it is difficult to take into account all indirect effects—consumption of goods and services that use petroleum products as inputs—and all spillovers, such as when groups with market power use petrol price increases to increase their margins more than is justified by the price increase. So the taxes may not be as progressive as the data suggest (Addison and Osei 2001; Coady and Newhouse 2006).

Fuel Consumption and Pricing in Ghana

In response to variations in fuel prices, governments can choose whether to allow the market to set the price; whether to use a strict pricing formula; or whether to implement ad hoc price changes (Baig et al. 2007). Over the past decade and more, domestic petroleum prices have fluctuated significantly, often in response to changing international prices (Figure 11-1). But changing fuel pricing policies have also been a feature in Ghana. Fuel pricing in Ghana has fluctuated between periods where a pricing formula is used to periods when the government strongly influences the price. For example, in 2003 Ghana introduced a pricing formula, but it was rapidly abandoned when the government stopped passing on continued increases in world prices (Coady et al. 2006).

Explicit fuel subsidies in Ghana have variously been estimated to be 0.2% in 2003 rising to 0.9% in 2005 according to Baig et al. (2007), and 2.2% of GDP according to Coady et al. (2006). A direct consequence of not passing through fuel price increases in Ghana has been the large debt that has been accrued: in 2003, petrol subsidies amounted to about US$ 200 million, equivalent to the donor receipts by the country for that year (Sowa 2006).

The National Petroleum Authority (NPA) was established by the Ghanaian government in 2005 in an effort to remove governmental influence over pricing (Coady et al. 2006). The key roles for the NPA were to ensure that the pricing mechanism was implemented correctly and so remove from the government the responsibility for determining petrol prices (Coady et al. 2006). The NPA typically reviews petroleum prices every two weeks. By setting up the NPA,

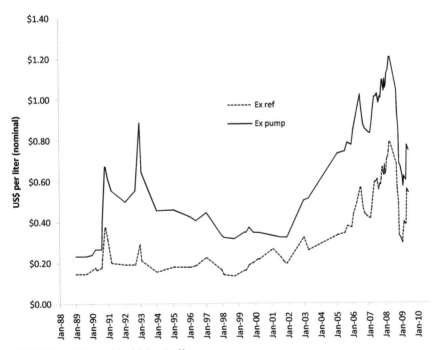

FIGURE 11-1 Historical Fuel Prices in Ghana.

Note: RFO (residual fuel oil) ex-depot price plus margins = RFO ex-pump price; RFO margins are negotiated between the oil marketing companies (OMCs) and the consumer; ex-pump prices are those at the pump in US$ per liter.

Source: NPA Ghana website www.npa.gov.gh.

Ghana was attempting to move from what had become an ad hoc petroleum pricing system, influenced by political rather than economic considerations, to a more automatic system.

Before the NPA was established, nominal petrol prices in Ghana had been constant for all of 2004, despite changes in the international price, demonstrating clear government influence over the price, as can be seen in Figure 11-2. Soon after the NPA was set up in 2005, petroleum prices were raised 50%. To make this fuel price increase more acceptable to the population, Ghana prepared a poverty and social impact assessment (PSIA) for fuel that was used to explain and justify the increase. The report suggested that price subsidies benefited the better-off more than the poor, and it demonstrated that Ghana's fuel prices were low relative to other countries in the region (Bacon and Kojima 2006b). After this jump in prices, the nominal price at the pump was virtually stable for a year, reflecting in part relatively stable international prices, and the decision by the NPA not to mirror the small international price adjustments. Between 2006 and early 2008, the NPA made more frequent small adjustments in price (see Figure 11-2), reflecting in part changes in international prices and the depreciation of the cedi.

Although the NPA is ostensibly independent of the government and was set up explicitly to separate petrol pricing from politics, it takes into consideration the political implications of its actions. Specifically, the government, and therefore the NPA, are likely to be concerned about the "ripple effects"—that is, the indirect effects—from increases in petrol prices, most importantly the impact on transport costs and food prices. Moreover, the strength of the transportation unions, such as the Ghana Road Transportation Association, ensure that the NPA cannot, in practice, make decisions totally independent from the government. Consequently, and similar to many other low-income countries, despite the NPA, Ghana has often not fully passed through increases in international fuel prices to consumers (Baig et al. 2007). This is consistent with the finding of Sterner (2002) that economic interests shape (gasoline) policies,

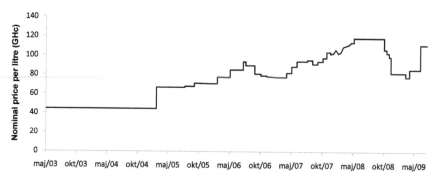

FIGURE 11-2 Nominal Price of Premium Petrol at the Pump in Ghana.

GHc = Ghanaian cedi. *Source:* NPA website (www.npa.gov.gh/?s=RFO)

and governments' propensity to raise taxes is negatively influenced by the extent of the dependence on motoring among the electorate or population.

Petrol pricing in Ghana cannot be separated from politics. Pre-election periods are often times when administrations may be tempted to lower petrol prices and opposition parties may be tempted to promise further cuts. In 2008, after holding nominal petrol prices constant, and after it failed to win the first round of the 2008 general elections, the New Patriotic Party (NPP), which was in power in Ghana at the time, reduced petrol prices at the pump from GHc 118.53 per liter to GHc 82.00 (*The Chronicle* 2009). At this stage the NPA's pricing mechanism had clearly been overridden. The National Democratic Congress (NDC), then in opposition, pledged to reduce the price further if it won, again marginalizing the role and purpose of the NPA.

Post-election periods are often a time for governments to introduce tough and unpopular policies, despite earlier promises (Baig et al. 2007). The NPA, established soon after a new government was elected in early 2005, soon increased nominal petrol prices by 50% (Figure 11-2). Later, six months after the NDC gained power in early 2009, nominal prices at the pump were increased 30% (Figure 11-2). Since being elected, in almost all of his comments on the fuel price increases new President John Atta Mills justified removal of the subsidy by citing the government's large budget deficit, inherited from the previous administration of President John Kufuor. Little official comment has been made on the rationale for subsidies with respect to income distribution issues—with or without an inherited budget deficit.

Indeed, overall the current government appears unwilling to discuss the general principle of subsidies, and fuel subsidies in particular. Many on the government's advisory board most likely do recognize the problems of persistent fuel subsidies. The government's public stance has tended to be that it is the pro-poor party, and removing the subsidies harms the poor. Naturally, people will be worse off in the short run, all other things equal, when a fuel subsidy is removed. But over the years that the fuel subsidy has been in place, the poor could have been better helped if the fuel subsidy had been directed towards goods and services most heavily used by the poor.

The government's decision to remove the subsidy does not appear to have been an easy one for officials.[2] The fuel price increases, anticipated for several months, were postponed again in May 2009 when the government paid GHc 7 million directly to the NPA to avoid anticipated increases in the last two weeks of that month. Similarly, a payment of GHc 21.7million was made during the first two weeks in June, as part of the recovery financial support (Robinsson et al. 2009). In March 2010 the NPA was once again in the headlines for absorbing a further fuel price increase (*Cedi Post* 2010). And in the last quarter of 2010 total fuel subsidies came to US$70million before petrol prices were raised again in early January 2011 (*The Business Analyst* 2011).

Impact of Ghana's Fuel Subsidies on Income Distribution

Although most academic papers suggest that fuel subsidies are largely regressive, in Ghana there is a strong feeling that removing them damages the poor. In part this is because the immediate impact is indeed rising costs of fuel, public transportation, and food, which affects everyone negatively. But there are also concerns about the disproportionate importance of kerosene for poorer households; the greater use of public transport by poorer households; and the high cost of food, which has accounted for approximately half of household budgets over the past decade. (The food component of the Consumer Price Index in Ghana in 2009 constituted 44.9% of the basket, and food expenditure constituted about 50.5% of total expenditure, according to recent government reports.) In this chapter we are concerned with fuel for transportation, so we focus on direct spending on fuel for private cars and the share of fuel in public transport costs, rather than on kerosene and other fuels for lighting and heating.[3] The data for this chapter are extracted from the Ghana Living Standards Survey (GLSS). The most recent round of the survey is round five (GLSS5). However, limited data are currently available from this survey (for example, spending on food items was unavailable at the time of writing). We are therefore using data from round four (GLSS4). The GLSS4 was conducted between April 1998 and March 1999. For GLSS4, information for a particular household was collected on patterns of household consumption and expenditure on food and non-food items over a period of 35 days. The survey year was divided into 10 cycles of 36 days each, interviews were conducted on the first 35 days of each cycle, and the team used one day (i.e., the 36th day) to travel to the next set of enumeration areas (EAs). The data set includes annualized information on patterns of household consumption and expenditure on food and non-food items. The total sample is 5,998 households.

Our focus is data on the total expenditure on transport, which includes purchases of gasoline for private vehicles and expenditure on public transport. In addition to expenditure on transport, other items included in the computation of total expenditure in the GLSS survey are expenditures on food, education, water, electricity and garbage disposal, remittances, miscellaneous expenditure, and rental payments. Items are classified as food expenditure, frequent non-food expenditure, and less frequent non-food expenditure.

A simple test of whether a given tax is progressive or regressive is available by comparing whether the budget shares for the consumption of that particular good are higher or lower among high- and low-income earners (see Chapter 6). In Figure 11-3, we calculate the direct and indirect share of fuel for transportation in expenditure for each income decile in Ghana. Direct expenditure is the amount spent on fuel for private vehicles. Indirect expenditure is the fuel cost of public transportation. Fuel accounts for approximately 26% of public transport costs in Ghana, and so the indirect fuel cost is calculated as 0.26 times the total cost of public transport (Coady et al. 2006;

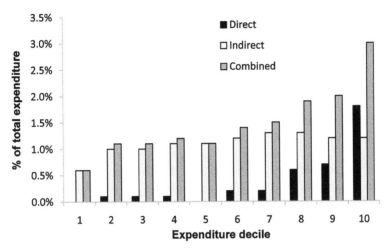

FIGURE 11-3 Direct and Indirect Expenditure on Fuel for Transportation by Expenditure Deciles.

Source: GLSS4.

personal communication with the NPA).[4] Therefore the combined budget share for fuel expenditure for transportation is given by:

$$\frac{\text{Expenditure on fuel for private vehicles} + 0.26 \times \text{Public transport expenditure}}{\text{Expenditures on all goods and services}}$$

We also calculate the Suits coefficient, which is similar to the Gini coefficient (Suits 1977). Upwards sloping graphs and positive Suits coefficients indicate progressivity; conversely, downwards sloping graphs and negative Suits coefficients indicate regressivity. From our data, we calculate the Suits coefficient to be 0.605 when we only take into account the direct impact of fuel taxes and 0.217 when we also include the indirect impact through public transport expenditure. That is, the direct effect of fuel taxes is progressive, and the progressivity remains even when we take into account indirect effects, though these are not as strong. We found the Suits coefficient to be –0.244 for expenditure on food, indicating that the indirect subsidy for food does benefit the poor in Ghana. This supports a Van Heerden et al. (2006) finding that subsidizing food and taxing gasoline could bridge the income gap between the poor and the rich.

In Ghana, as in many low-income countries, private car ownership is small, but growing. In 1991–1992 (GLSS3), 1.48% of households owned a car; in 1998–1999 (GLSS4) 3.77% did so; car ownership had risen to 7.97% by 2005–2006 (GLSS5). When we look at expenditure on fuel by expenditure deciles for GLSS4 (Figure 11-3) we see that direct expenditure on transportation fuel is negligible for low-expenditure deciles and it is only in the highest expenditure deciles that a greater share of household expenditure is on direct transportation fuel and not indirect. Not surprisingly, direct taxation on petroleum products for private vehicles is highly progressive.

Consequences of the 2009 Fuel Price Increases

The immediate consequence of the government's removal of the fuel subsidy in June 2009 was for fuel prices to jump approximately 30% at the pump. These fuel price increases had been anticipated for several weeks before the actual increase, with the Ghanaian government seemingly uncertain as to whether and when they would actually put up the prices. Regular gasoline increased from GHc 0.857 (US$ 0.60) to GHc 1.11 (US$ 0.78) per liter; diesel, kerosene, and liquefied petroleum gas (LPG) underwent similar increases. Although kerosene prices are lower than gasoline and diesel, kerosene was not spared the price increases.

Road transport fare increases followed quickly. Because fuel accounts for approximately 26% of public transport costs in Ghana, as fuel prices increased by 30% we might expect public transport prices to increase by about 8%. Yet what we saw were large increases, with public transport prices in Accra going up on average 17%. In Ghana the transport unions are very strong and have been involved in negotiations over how much public transportation prices should increase as a result of the reduction in subsidy. The initial negotiation was for a 10% rise in public transport costs. This was to allow the providers of public transport to absorb both the immediate increase in fuel prices and earlier increases in other costs, such as for spare parts, that have gone up in the past but for which the transport unions have not had a ready opportunity to increase the prices. Because of the strong resistance of the public to increases in public transport fares, increases in other costs are not typically immediately passed on to the consumer. The fuel price increase gave the unions the opportunity to pass on the cumulative increases in other transport costs. A 10% price increase would increase the 20-pesewa fare (about 15 cents, a typical fare for short journeys in the city) to 22 pesewas.[5] However, following the argument that there would be a shortage of change, specifically one-pesewa coins, the unions argued that the fare should increase to 25 pesewas, an increase of 25%. Similarly, other fares typically increased more than 8%. The increase in fuel prices facilitated the increases in public transport prices over and above the 8% justified by the higher fuel costs.

Conclusion

Although economists have highlighted the typically regressive nature of fuel subsidies in low-income countries, particularly those for petroleum and diesel that are used by individuals in their private vehicles, transportation fuel subsidies remain a common feature in many countries. In Ghana and other African countries, fuel subsidies have proven easy to introduce—often implicitly rather than explicitly, such as when international oil prices increase and nominal domestic petroleum prices remain unchanged, or when exchange rates change. But they have proven particularly difficult to remove. In Ghana, in part this is because increases in fuel subsidies have tended to be incremental, a consequence of the gradual increase in crude oil prices, whereas the periodic removal of subsidies has tended to result in large discrete jumps in the price of fuel at the pump. Individuals therefore do not necessarily realize that the fuel subsidy is increasing. However, when the subsidy is removed the impact is visible, immediate, and most likely felt most strongly by the poor, for whom fuel, transport, and food, are a large part of their expenditure. Although in the past Ghana did attempt to offset large increases in fuel prices with programs targeted to help the poor, particularly during the 2005 price increases, in 2009 the overwhelming perception of many Ghanaians was that the poor were worse off.

Ultimately in Ghana, pressure from the IMF, and the new government's inherited fuel debt of US$ 350 million deficit—a combination of subsidies for the power utilities and subsidies for petrol consumption—meant that the new administration had little choice but to raise fuel prices. Despite these increases, Ghana's debts due to fuel subsidies in the past remain high. For example, the Tema Oil Refinery, which supplies approximately 70% of the country's petroleum (Coady et al. 2006), had debts in mid-2009 of approximately GHc 114.6 million.[6] Consequently, in September 2009 the country was not able to import crude oil for refining because of these previously incurred debts, and so had to import refined petroleum products.

Fuel taxes are also environmental taxes, having the effect of reducing the demand for fuel and therefore reducing emissions (see, e.g., Stiglitz 2006; Sterner 2007; Hsu et al. 2008). Fuel subsidies are therefore likely to have the opposite effect—increasing the demand for fuel and therefore increasing emissions. Where fuel taxation is progressive, there is a double benefit from raising fuel taxes: a reduction in income inequality and a reduction in climate-harming emissions (Baranzini et al. 2000; Sterner 2007). The additional tax revenue generated from fuel taxes can be targeted and redistributed to the poor to further reduce income inequality. As noted by Baranzini et al. (2000), the lump-sum redistribution of fiscal revenues to the population could benefit the poor since the lowest incomes will receive a higher amount, relative to their income, compared to the highest income households.

Although arguably countries in Africa should not be asked to take on the burden of reducing their emissions, and although fuel usage tends to be relatively inelastic in the short term (Sterner 2007), anecdotal evidence suggests that the increases in fuel prices will encourage more efficient use of petroleum products. For example, Ghana's *Daily Graphic* reported on June 8, 2009 that the government announced fuel-saving measures following the price increase, aiming to reduce expenditures on fuel for ministers and officials (and thereby fuel consumption) by 30%. In part this will be achieved through reducing the extent to which official cars and fuel are being used for unofficial purposes.

Fuel subsidies have other consequences. The problem of petrol smuggling across Ghana's borders has long been recognized as a consequence of subsidized, and therefore lower, prices in Ghana relative to its neighbors, such that Ghana is in effect subsidizing fuel consumption in the neighboring countries. For example, at the end of 2008, after prices were reduced almost 40%, petrol was being sold at the pump for the equivalent of GHc 8.36 in Côte D'Ivoire, GHc 8.44 in Burkina Faso, GHc 6.32 in Togo, but just GHc 3.62 in Ghana (*The Chronicle* 2009).

Whether or not the fuel subsidy is progressive or regressive, large increases in fuel prices in the short run directly and negatively affect the poor through increases in public transportation costs and increases in food prices, often disproportionate to the fuel price increase. Allowing incremental increases in fuel prices rather than implementing large jumps of 30% and more, as have followed the last two changes in government in Ghana, would reduce the negative shock to poorer households and allow them to adjust gradually to the changes.[7] Replacing general fuel subsidies with specific subsidies, such as for kerosene, can retarget the subsidies towards items most used by the poor. However, even such approaches can have unintended consequences. For example, differential tax rates on fuel prices in Tanzania have resulted in adulteration of diesel with cheaper kerosene.

Notes

1   The "implicit subsidy" means that the country sells gasoline to consumers at a price below its economic value (i.e., the cost of producing it), and the subsidy increases as the cost of production increases, but the domestic price remains constant—or, as the domestic currency depreciates against major foreign currencies while the cost of production remains constant.
2   Policy prescriptions by the World Bank and IMF support have called for Ghana to remove its petroleum subsidies (Amegashie 2006).
3   For a detailed analysis of the impact of fuel subsidies in Ghana that includes kerosene, please see Coady and Newhouse (2006).
4   26% is comparable to other country estimates, including 21% for Costa Rica, 21% for Mali, and 19% for South Africa (see Chapter 5; Chapter 13; Mabugu et al. 2009). Mutua et al. (see Chapter 12) use an estimate of 30% for Kenya, based on observations that in Europe the figure is 17% for urban buses and 28% for long-haul, and that salaries typically are much lower in Africa.
5   100 pesewas = one Ghana cedi.
6   OMCs provide the remaining approximately 30% of Ghana's petroleum.
7   Exceptions could be made for oil price shocks in the world market.

CHAPTER 12

Distributional Effects of Transport Fuel Taxes in Kenya: Case of Nairobi

JOHN MUTUA, MARTIN BÖRJESSON, AND THOMAS STERNER

E nergy is a key to economic growth and development as well as improvement of quality of life, and world energy demand has increased tremendously. The major source of energy is currently fossil fuels, which account for about 81% of the world's total primary energy supply, but hydro power, nuclear, and biomass are also significant sources. Oil use accounts for 43% of total final consumption (IEA 2006). Most of the oil is used in the transport sector. To limit the environmental damage of the transport sector and generate income to state budgets, fuel taxation is widely used; however, the appropriate level of fuel taxation is a debatable issue, and taxes vary dramatically across countries. Great Britain's gasoline tax of 50 pence/liter in 2000 (about US$ 2.80/U.S. gallon) is the highest among industrial countries, while United States tax level of about 40 cents/gallon is among the lowest (Parry and Small 2005).

In Kenya, petroleum accounts for 22% of the total primary energy supply, 67% of which is consumed in the transport sector. The rest is consumed mainly in industry and power generation (Aligula 2006).[1] In recent years, 2003–2007, Kenya has recorded a substantially increased demand for transport fuels (both diesel and gasoline). Total consumption of petroleum products, about 3.61 million metric tonnes in 2009, is projected to more than triple in line with Vision 2030, Kenya's blueprint for economic and investment policy. Gasoline and diesel demand were 0.462 million metric tonnes and 1.42 million metric tonnes (year 2009), respectively (KNBS 2010). The balance of about 1.73 million metric tonnes is from jet fuel, illuminating kerosene, fuel oil, and heavy diesel oil (see Table 12-A1 in Appendix 12-1). The gasoline tax is 40, 43, and 28 US cents/liter for motor gasoline super, motor gasoline regular, and automotive gas oil (diesel), respectively (KRA 2007). Motor gasoline super, motor gasoline regular, and gas oil in Kenya are mainly used as transport fuels. Their tax rates therefore take into account transportation as their main use. Other products such as fuel oil (used for power generation), illuminating kerosene (mainly used for cooking and lighting), and heavy diesel oil have lower tax regimes.

The steady increase of motor vehicles in the rapidly growing Kenyan cities makes transport emerge as a key contributor to regional air pollution. By the end of 2009, over 1.15 million vehicles were already licensed to ply Kenyan roads, with the number increasing by 40,000 per year on average, leading to increasing emissions of both local and global air pollutants. Ambient lead concentrations in urban centers in Kenya in 2004 were recorded in the range of 0.4–1.3 µg/m³. Kenya is one of the last countries to finally phase out lead additives in gasoline. Just a few years ago the content in gasoline was 0.4g/liter, and only 4% of the gasoline consumed in the country in April 2004 was unleaded. The proportion increased to 30% in 2005. This implies that about 245 million liters of leaded gasoline were consumed in 2004, releasing tonnes of lead, 75% of which ended up in air, soils, and plants, while the rest remains in engine block and lubrication oil (UNEP 2006). In adherence to the Dakar Declaration of 2002, the government of Kenya phased out unleaded gasoline by end of 2005 (UNEP 2010).

Regulatory functions in the Kenyan petroleum sector have previously been shared among various players, including the Ministry of Energy, Ministry of Finance, and Provincial Administrations and Local Authorities. The Petroleum Institute of East Africa (PIEA), a voluntary membership institution patronized by major oil companies, plays a key role in capacity building and awareness creation. For many years Kenya did not have a clear policy governing all the energy sectors. The challenges posed by this policy vacuum led to a new policy: Sessional Paper on Energy No. 4 of 2004 and the Energy Act 2006, through which the Energy Regulatory Commission (ERC) now regulates all energy sectors (petroleum, electricity, renewable energy and biomass). The policy aims to achieve sustainable energy supply.

Various instruments have been applied in the case of Kenya to achieve certain fiscal policy objectives as well as environmental sustainability objectives. Kenya uses fiscal instruments to pursue environmental objectives, albeit in a limited way. Subsidies are more prevalent; fees and royalties are too low to induce much behavioral change, and taxes and levies are rare (Ikiara et al. 2007). Some of the instruments previously used include: an increase in the excise duty on petroleum products for gasoline and for gas oil; a levy of 1 Kenyan shilling (Kshs) per liter of kerosene to raise revenue for use in water harvesting and other projects; a surcharge on imported secondhand vehicles; and an increase in the excise duty on motor vehicles to decongest roads. Others include reduction of import duties, excise duties, and value added taxes (VAT) on diesel and residual fuels, liquefied petroleum gas (LPG), kerosene, and imported timber and other wood products to reduce deforestation. The most important taxes related to transport are the excise duty and road maintenance levy, which finance activities of the Kenya Roads Board. These taxes are comparable to those in other developing countries and have not seen any changes in real terms in the last three years. Studies have shown that there is seemingly inelastic short-run demand for gasoline coupled with pressure to address urban traffic congestion (Mutua et al. 2009).

The current trend of emissions from gasoline and diesel fuels is a threat to the well-being of citizens as well as contributing to global warming and climate change. The fact that the public transport system in Kenya has performed poorly over the years has further deteriorated the pollution levels from the transport sector since a majority of high- and middle-income households prefer private transportation. This has led to increase in ownership of private cars and hence high vehicle densities. With the high levels of fuel consumption and high emissions levels, there is a great need to discourage consumption. Since Kenya is also a signatory to many climate change and environment-related programs, such as the Kyoto Protocol, revising the fuel tax regime is an issue to consider. Environmentally related fiscal policies also have the potential to raise more revenue for the state and possibly lower the domestic debt since it cushions the government from going into deficits, while at the same time reducing environmental damage.

A common argument against transport fuel taxes is that because fuel demand is inelastic the environmental benefit of fuel taxes is small. Sterner (2007), however, concludes that the long-run price elasticity of gasoline is high, although in the short run it may be quite inelastic. This has implications for policymakers, who often depend on observable short-run progress.

Even in Kenya, distributional effects of fuel taxation are considered important, and it is commonly argued that gasoline taxes are regressive. The debate in Kenya today with regard to fuel taxation seems to assume that the current taxes such as the fuel levy and excise duty are too high (Mutua et al. 2009). Ikiara et al. (2007) also argue that there has been increased pressure from stakeholders and consumer organizations to reduce some of the fuel taxes; however, this is not supported by evidence-based research. The debate was intensified by the oil price volatility which led to high prices of gasoline— up to Kshs. 110 (US$ 1.37) in November 2008, following the increase of crude petroleum prices in the international market to US$ 147 per barrel. Only the road maintenance levy for road construction and rehabilitation increased. After that spike, the price of gasoline dropped to about Kshs. 75 (US$ 0.94) by March 2009. The government has moved in to reduce taxes on oil used in power generation, but transport fuel taxes were not touched. Motorists generally argue that the tax burden is too high and also that once it has been collected it is not used well to finance road development and rehabilitation works. They also argue that fuel taxes hurt the poor. Evidence-based research is necessary to evaluate these claims and beliefs. The present study investigates the distributional effects of Kenya's transport fuel taxes to come up with policy recommendations for developing countries currently facing unprecedented vehicular growth.

Methodology and Data

As in previous chapters, to establish whether the fuel taxes are regressive or progressive distributional effects are estimated by calculating the budget share of transport expenditures in total household expenditures for each category of population classified by household income deciles. Suits coefficients are then calculated to obtain a single, comparable measure of the distributional burden.

The use of total expenditures rather than annual income is a way of approximating the lifetime income (see Chapters 1 and 2). In short, this approach is argued to provide a better measure, since an individual often belongs to one income group temporarily but another when considering the entire lifetime (e.g., students or pensioners). For comparison—and to show how substantial the difference between the two approaches can be—Suits coefficients based on the temporary income approach are also reported. Furthermore, besides owners of motorized vehicles who consume fuel directly, we take into account those who pay indirectly for fuel by using public transport. In this chapter we calculate both the burden from direct and indirect consumption.

In our analysis, we use data from a 2004–2005 urban household travel survey done by the Kenya Institute for Public Policy Research and Analysis (KIPPRA). A representative sample of about 2,200 households in Nairobi were interviewed, including low, middle and upper income households.

Empirical Analysis

Table 12-1 summarizes the statistics used in the analysis. Notable is the average monthly income of Kshs. 15,700 while the average household expenditure is Kshs. 18,200, illustrating that households spend more than they earn. The

TABLE 12-1 Summary Statistics.

Variable	Observations	Mean	Std	Minimum	Maximum
Average income (Kshs)	2,105	15,662	15,351	1,500	50,000
Distance to CBD (km)	2,071	5.84	4.46	0	20.16
Household expenditure on transport (Kshs)	2,054	3,603	4,474	0	50,000
Household size	2,105	3.54	1.96	1	14
Expenditure on leisure (Kshs)	2,105	1,859	4,597	0	70,833
Household monthly expenditure (Kshs)	2,105	18,200	21,261	700	287,850
Age (years)	2,104	35.8	9.79	6	55

Source: Authors' computation from 2004–2005 urban public transport survey (UPTS) data, KIPPRA.

extra income for an individual could be from transfers and gifts from relatives as well as initiatives in cooperatives and rotating savings and credit associations. At the aggregate level, however, this discrepancy must surely be a sign that income is underreported. The mean monthly expenditure on transport is Kshs. 3,600 and on leisure Kshs. 1,860. The mean distance from the central business district (CBD) is about six kilometers, while the maximum is 20 km. The mean household size is four persons, with the average age of the household head about 36 years. Average car ownership is 0.14 per household.

Expenditure Shares

To analyze the distributional effect of fuel taxes, data from the household survey is used to calculate the budget shares spent on transportation by income deciles (corrected for household size). Figure 12-1 below shows the distribution of household income per month and number of households in the survey. The horizontal axis shows the number of households against the average household income per month in the vertical axis. Note that about 50% of the households earn Kshs. 10,000 per month or less, while fewer than 15% of the households have incomes of Kshs. 50,000 per month or higher.

Figure 12-2 presents the resulting public transport expenditures as shares of total expenditures for households for each income decile. As can be seen, the three lowest income household deciles spend about 10% or less of their total household expenditures on public transport. For middle income households the share is higher, near 14%. For the income deciles with the highest household incomes, the public transport expenditure share of total expenditures

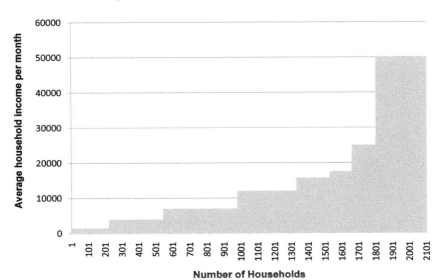

FIGURE 12-1 Household Income Distribution.

Source: UPTS data, KIPPRA, 2004–2005.

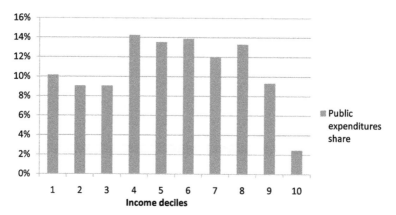

FIGURE 12-2 Share of Total Household Expenditures Spent on Public Transport per Household Income Decile (1 = Lowest Income, 10 = Highest Income).

Source: UPTS data, KIPPRA, 2004–2005.

drops considerably, and the highest income decile spend only about 2.5% of the total household expenditures on public transport.

The distribution of expenditures on private transport is completely different from that on public transport. Figure 12-3 gives the private transport expenditures as shares of total expenditures for households for each income decile. For most households, the share of total expenditures spent on private transport is very low; the six lowest income deciles all have private transport expenditures lower than 2% of total expenditures. The two income deciles with the highest incomes have significantly higher shares: the second highest income decile's share is 6.93% and the highest income decile's share is 12.66%.

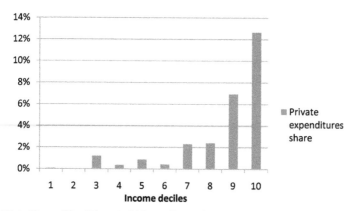

FIGURE 12-3 Share of Total Household Expenditures Spent on Private Transport per Household Income Decile (1 = Lowest Income, 10 = Highest Income).

Source: UPTS data, KIPPRA, 2004–2005

Public transport is by, e.g., bus or trains, while private transport is mainly by cars and motorcycles. Note that only a portion of the transport expenditures can be considered to be fuel costs. For European conditions, it has been estimated that about 17% of the operating costs for urban buses stem from fuel costs, and the corresponding share for line-haul buses has been estimated to 28%. The largest shares of the operating costs are the driver costs (salary etc.), which are 46% and 31% for urban and line-haul buses respectively (Jobson 2007). Since salaries are generally lower in developing countries like Kenya than in European countries, it could be argued that the corresponding fuel share of the total public transport costs should be somewhat higher. Indeed, a study done in Kenya (Aligula et al. 2005) shows that about 30% of public transport expenditures and 80% of private transport expenditures originate from fuel-related costs. Considering that no driver salaries or administrative costs, etc., are included in the expenditures for private transport, the fuel share of the private transport expenditures is, as expected, considerably higher than for the public transport expenditures.

In Figure 12-4 it is assumed that 30% of the public transport expenditures and 80% of the private transport expenditures stem from fuel costs. Figure 12-4 shows the fuel expenditures from both public and private transport as shares of total expenditures for each income decile. From the figure, it is clearly the case that transport fuel taxes in Kenya are not regressive but progressive.

Distributional Effects of Transport Fuel Taxes

In our analysis for the Suits index, we have accumulated the income and transport expenditures for every income decile. In the analysis of tax burden

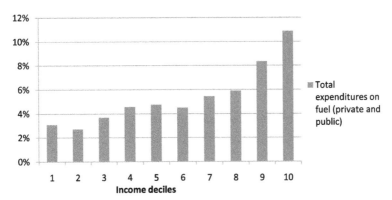

FIGURE 12-4 Share of Total Household Expenditures Spent on Transport Fuel per Household Income Decile (1 = Lowest Income, 10 = Highest Income).

Assumes that fuel costs make up 30% of public transport expenditures but 80% of private transport expenditures (Aligula et al. 2005).
Source: UPTS data, KIPPRA, 2004–2005.

for both private transport fuel, we find that transport fuel taxes in Kenya are progressive, not regressive. Figure 12-5 illustrates this finding with concentration curves for public transport and private transport Taxes by summarizing the accumulated percentage of tax burden from private expenditures on transport and a combined burden for both private and public fuel expenditures. In both cases, the burden is larger in the high income households. A tax increase in the transport sector will thus mainly affect the high income earners because of its progressivity.

The obtained Suits coefficients are 0.483 and 0.464 for the tax burden calculated based on lifetime income approach and ordinary income approach respectively (see Table 12-2). This result concerns only the tax induced from private consumption of fuel. When also including public transport expenditures, the coefficients become less progressive—0.225 and 0.171, respectively—but these

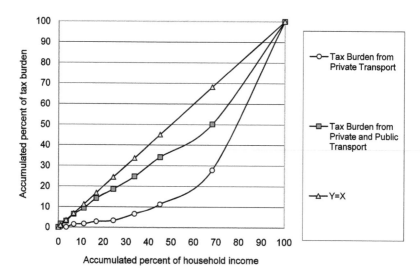

FIGURE 12-5 Concentration Curves for Public Transport and Private Transport Taxes.

Source: UPTS data, KIPPRA, 2004–2005.

TABLE 12-2 Progressivity of Kenya's Private and Public Transport Fuel Taxes.

Tax Progressivity of Private Transport Taxes	
Type of income defining tax burden	Suit Index(S)
Lifetime income approach	0.483
Temporary income approach	0.464
Tax Progressivity for Combined Private and Public Transport Fuel Taxes	
Lifetime income approach	0.225
Temporary income approach	0.171

Computation done from the KIPPRA UPTS, 2004–2005.

are still high values. From the foregoing, we can conclude that fuel taxes in Kenya are progressive. The degree of progressivity varies, but the progressivity becomes weaker if one also includes the indirect component.

Conclusion and Policy Recommendations

Changing transportation patterns in developing countries are very important for carbon emissions and for the climate. Rising income will typically create a tendency for expansion of private transport. The public transport system is in most cases seen as an inferior service. Therefore, these countries need to come up with policies that promote public transport and mass transit and discourage excessive private transportation. The best instrument for this would be a tax on fuel. In addition, many developing countries face huge domestic debt and budget deficits; the potential to raise more revenue through a fuel tax provides an attractive way to bridge the financing gap.

This chapter has estimated distributional effects of fuel taxes by expenditure and income deciles, calculating tax burdens and Suits indices for private transport fuel taxes and combined distributional effects of both private and public transport fuels. The analysis of distributional effects of gasoline taxes has shown that they are strongly progressive. The lowest income household deciles spend somewhat less than 10% of their total household expenditures on public transport. In the middle income households the share is higher, nearly 14% for several of the middle income household deciles. For the income deciles with the highest household incomes, the public transport expenditure share of total expenditures drops considerably, and for the highest income decile only about 2.5% of the total household expenditures constitutes public transport expenditures. Low and middle income deciles spend very little of their total expenditure on private transport: 70% of the lower income households all have a budget share for private transport of less than 2%. Meanwhile, the highest income decile has a share of private transport expenditures around 12%. When the impacts for public and private transport are combined, the study finds that the total shares of transport fuel expenditures to total household expenditures are lower for low-income than high-income households.

From the foregoing analysis and conclusion, the following policy recommendations can be drawn from the study.

First, the public transport system needs to be improved and mass transit use encouraged, so private ownership of vehicles and gasoline consumption can be reduced. This could be through improvement in the railway system and public bus/metro system. Second, taxes on high gasoline consumption vehicles not used for public transport should be revised, reducing per capita consumption of gasoline and hence achieve emissions abatement. The taxes generated can furthermore be used to finance the necessary investments in public transport and infrastructure.

Appendix 12-1

TABLE 12-A1 Fuel Consumption, Energy Usage, and Vehicle Ownership, 2005–2009.

Petroleum products consumption (000s metric tonnes)	*2005*	*2006*	*2007*	*2008*	*2009*
Automotive gas oil (gas oil)	892.4	1,081.9	1,116.5	1,141.1	1,416
Motor spirit premium (motor gasoline super)	334	404	367	381	462
Illuminating kerosene	307	281	265	245	333
JET/turbo Fuel	561	595	641	562	571
Liquefied petroleum gas (LPG)	49	96	77	84	75
Fuel oil	547	878	615	690	729
Heavy diesel oil	26	41	40	30	24
Total consumption of liquid fuels, (000s metric tonnes)	**2715.9**	**3375.8**	**3121.8**	**3133.2**	**3609.4**
Energy consumption in tonnes of oil equivalent					
Total consumption of liquid fuels, (000s, metric tonnes)	2,716	3,038	3,122	3,133	3,611
Total energy consumption, (000s, metric tonnes)	3,155	3,509	3,627	3,614	4,002
Energy per capita consumption (kilograms of oil equivalent)	90	97	98	98	102
Number of vehicles	749,680	787,036	872,360	994,191	1,156,004

Source: Kenya National Bureau of Statistics (KNBS), 2010.

Note

1   Aligula (2006) estimates on shares of energy consumption in Kenya were based on the Kenya Integrated and Household Budget Survey 2005–2006 and the National Energy Consumption time-series data from the economic surveys by the Kenya National Bureau of Statistics. These are reliable data sets in estimating energy consumption shares.

## CHAPTER 13

Assessing the Impact of Oil Price Changes on Income Distribution in Mali: An Input-Output Approach

KANGNI KPODAR

For many developing countries, the increase in oil prices over the past few years has made structural reform of the domestic petroleum pricing system a critical component of their macroeconomic policies. From a relatively low price of US$ 26 a barrel in January 2001, oil prices—which have shown an upward trend, albeit volatile—hover currently around US$ 100 a barrel. Although in some countries, this large oil price increase may have been partly offset by exchange rate movements (notably, the weakening of the U.S. dollar against the euro), it has also had major socioeconomic impacts. Many governments have been reluctant to pass on to consumers a rise in international oil prices because of the potential for social resistance to a policy that could hurt the poor. However, by not passing on higher prices, lowering fuel taxes, a revenue loss would result, jeopardizing social spending.

In Mali, oil subsidies are implicit and result from low, government-controlled petroleum prices. Given the budgetary cost of such a policy, nearly 2% of GDP in 2004,[1] it is important to assess the level and distribution of the effects of oil price hikes on household welfare to determine if the subsidies are serving their intended purpose.

Mali is large, poor, and landlocked—and highly dependent on imported oil products. In 2003, income per capita was about US$ 260. The main sources of energy are wood, charcoal, and petroleum oil products, with the latter accounting for 61.5% of the country's total energy use. Its total supply of petroleum products was 544,000 tons in 2001, of which diesel represented 51%; gasoline, 17%; fuel oil, 18%; and kerosene, 8%. Thus, Mali is one of the low-income countries most vulnerable to oil price shocks. Its vulnerability, measured by the ratio of net oil imports to GDP, reached 5.4% in 2001, compared with an average of 3.34% for all oil importers and 3.97% for landlocked oil importers. Mali's vulnerability mainly results from the fact that the country is not an oil producer. That vulnerability is also a product of geography: its nearest ports are nearly 1,000 km from the capital, and about one third of its electricity is generated from oil.

To take into account both the direct and indirect effects of oil price increases on household welfare, we combine household survey data with a sectoral input-output data set. Households in Mali use kerosene for lighting and cooking and gasoline for transport; thus, higher oil prices lead directly to a decrease in their real incomes. Higher oil prices also affect households indirectly, through a rise in the prices of goods and services of other sectors that use oil products as intermediate goods. Depending on how much of their budget households spend on oil products and the input-output linkages between the petroleum sector and other sectors, the effect of oil price increases on household income may be substantial, although it varies by income group.

The chapter is organized as follows. The following section reviews the major factors behind oil price increases and highlights the implications of low, government-controlled prices. The methodology and data used to estimate the impact of oil price hikes on household expenditures are presented next. We then report the result of the policy simulation and conclude with policy recommendations.

How Domestic Oil Prices Are Linked to International Oil Prices Changes

The authorities control domestic oil product prices by setting a monthly ceiling price. The ceiling price is calculated on the basis of four main factors: crude oil prices, the exchange rate, taxation (an excise tax and a value-added tax), and domestic margins. Below, we discuss the consequences of each factor in Mali's context.

The first fact is the price of crude oil. Since 2001 buoyant emerging market demand and limited supply have caused oil prices to rise significantly (Figure 13-1). Mali imports only refined petroleum products, not crude oil. Therefore, supplier prices include crude oil price, refinery costs, and in some cases an import tax on petroleum products collected by transit countries.

The next factor is Mali's exchange rate. Because world crude oil prices are denominated in dollars and the CFA (Communauté Financière Africaine) franc is pegged to the euro, the recent appreciation of the euro against the dollar has reduced Mali's oil bill. The CFA franc appreciated against the dollar by 65% between 2001 and 2007, leading to a lower rate of increase in oil prices denominated in CFA francs compared with prices in dollars (Figure 13-1).

Distribution and retail margins also affect prices. In 2005, retail margins represented only 5.2% to 8% of pump prices, depending on the oil product. Transport costs accounted for a larger share (9% to 13.2%) of oil prices, probably because of long and costly routes of importation (Table 13-1).

Finally, there is taxation. A value added tax (VAT) of 18% is applied to all oil products. However, distribution of petroleum products is exempt from the VAT,

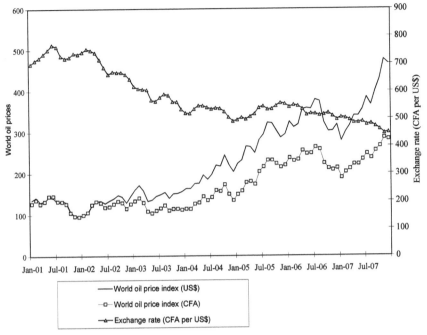

FIGURE 13-1 Changes in the International Oil Price, 2001–2007.

Source: International Financial Statistics (2008).

TABLE 13-1 Mali: Structure of Fuel Pump Prices, June 2005 (in Percent of Pump Prices).

	Kerosene	*Gasoline*	*Diesel*
Supplier price	60.1	34.3	49.3
Tax	21.6	48.6	31.5
of which: TIPP	3.9	30.9	12.1
VAT	13.3	13.3	13.1
Transport costs	13.2	9.1	11.7
Margins	5.2	8.0	7.5
Total	100.0	100.0	100.0

Notes: Data for imports from Dakar.
Source: Malian Ministry of Finance.

so oil products pay VAT at the border on a tax base that includes the cost, insurance, and freight (CIF) value of imports, customs duties, and a specific excise.

The *taxe interieure sur les produits pétroliers* (TIPP) is the excise tax levied on oil products in Mali. The TIPP represents an important revenue source for the government, with gross receipts amounting to CFAF 30 billion in 2004 (over 1% of GDP). It is adjusted monthly by decree of the Minister of Economy and Finance, and the rate differs by product and route of importation.

Given that kerosene is consumed primarily by low-income groups, it is taxed at lower rates than other oil products, with the TIPP accounting for 3.9% of kerosene pump price compared with 30.9% for gasoline (Table 13-1). The TIPP rate is also lower for diesel than for gasoline because the former is often used for transport and electricity production.

In addition, imports from distant ports such as Cotonou and Lomé[2] receive favorable tax treatment, owing to the government's desire to diversify supply routes. However, given the cost of the policy of differentiating the TIPP, the government decided to substantially reduce the differentiation among routes from June 2005.

The VAT and TIPP have different implications on the government budget through petroleum revenues, which represent over 20% of the total tax revenue. In 2004, the VAT accounted for 45% of petroleum revenues in 2004, the TIPP 40%. An increase in international oil prices would lead to a rise in the VAT revenues (unless the government changes its tax policy). In contrast, an increase in international price has no effect on the TIPP revenues since the TIPP is an excise tax. However, the TIPP revenues may decline if a substantial increase in domestic oil prices leads to a reduction in oil consumption. When the government changes the tax policy, for example by lowering the TIPP rates to soften domestic oil price increases, it loses potential tax revenues not only because of lower TIPP rates but also because of lower VAT revenues since the VAT rate is applied on the price with the TIPP included. Alternatively, if the TIPP rates are increased following a drop in international oil prices so that pump prices remain unchanged or decrease more slowly, the result would be an increase in petroleum revenues.

Currently, the authorities use the following formula to set pump prices:

$$\text{Pump price} = \left[\text{CIF Import Price} * \text{DT} + \text{TIPP}\right] * \text{VAT}$$
$$+ \text{Transport cost} + \text{Domestic margin}$$

where the CIF import price includes the supplier price and transport cost to the border. The supplier price depends on crude oil price, exchange rate between the euro and the dollar, and refinery cost. DT represents the duty tax.

In Mali, the government has primarily used the TIPP rates to hit targeted pump prices given the change in international oil prices. As a result, an increase in international oil prices may cause the country to lose a significant amount of fiscal revenue depending on the extent to which the TIPP rates are lowered to offset partially the impact of international oil price hikes on consumer prices. Figure 13-2 shows that in Mali, there was less than full pass-through from January 2002 until the beginning of 2003, which may explain why the drop in world prices during 2003 was not passed through to the pump. From early 2005 onward, domestic oil prices have been rising less rapidly than international prices. As the gap increases, the cost of keeping pump prices low has placed a significant burden on public finance and may not be sustainable if the world oil price continues its upward trend.

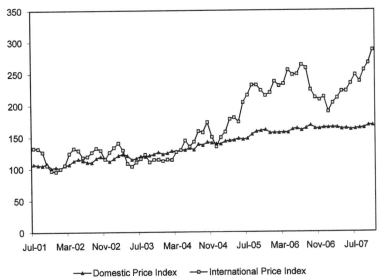

FIGURE 13-2 Mali: Pass-Through of World Oil Prices, 2001–2007. (Index, Jan 2002 = 100).

Source: IMF (2007).

An important implication of the pass-through mechanism is its effects on resource distribution (equity). Households will be affected differently by oil price increases and thus by a policy aimed at controlling oil prices. Is that policy favorable to the poor? Or may the impact of that policy on the poor be improved? To address this issue, it is necessary to assess the effect of oil price increases on households, which is the purpose of the next section.

The Consequences of Rising Oil Prices for Households

Rising oil prices have both a direct and an indirect effect on household real incomes. Oil price increases directly affect household real incomes through the higher prices for the oil products they consume, but they also indirectly raise the prices of goods produced by other sectors through their input-output linkages with the petroleum sector. The magnitude of the direct effect will be higher for those households who allocate a high proportion of their budget to oil products. Similarly, the indirect effect will be higher for households who devote a significant share of income to goods whose costs are highly influenced by oil prices.

Methodology: the Input-Output Approach

First we look at direct expenditure effects. The budget shares give a first-order indication of the magnitude of income effects resulting from price changes.

For a given product, its budget share corresponds to the price elasticity of real income or total spending assuming the volume of demand constant.

$$b_i = \frac{\partial \log Y}{\partial \log P_i}$$

where Y is the level of income or expenditure, b_i is the share of spending on good i in total expenditure and P_i is the price of good i.

The budget shares of oil products determine the direct effect on consumers of the increase in oil prices.

$$\partial \log Y_{dir} = \sum_{t=1}^{m} b_t * \partial \log P_t$$

where $\partial \log Y_{dir}$ is the direct expenditure effect (expressed in percentage), b_t is the budget shares of the oil product t, P_t is the change in the price of oil product t, and m is the number of oil products consumed by households.

To evaluate the indirect effect, we need to calculate the price rises resulting from the oil price increases of all other final goods purchased by households. The share of household expenditures on each of these final goods multiplied by their respective price changes gives a first-order estimate of the increased cost for purchasing the identical basket of goods before and after the oil price rises. The formula of the indirect effect is then similar to that of the direct effect:

$$\partial \log Y_{ind} = \sum_{i=1}^{n-m} b_i * \partial \log P_i$$

where $\partial \log Y_{ind}$ is the indirect expenditure effect (expressed in percentage), b_i is the budget shares of final goods other than oil products, P_i is the change in the price of good i, n and m are respectively the number of final goods and the number of oil products.

To compute the change in commodity prices following a change in oil prices, we follow the input-output approach adopted by Coady and Newhouse (2005).[3] We then consider two categories of sectors: the *noncontrolled sectors* and the *controlled sectors*. In noncontrolled sectors, higher input costs are pushed fully through to output prices (mostly goods not traded), or lead to lower factor payments or lower profits (e.g., traded goods for which output prices are determined by world prices).[4] On the other hand, in controlled sectors, output prices are controlled by the government. In these sectors, output prices remain unchanged, so that higher input costs would reduce factor payments, profits, or government tax revenue.

For noncontrolled sectors, the relationship between user and producer prices is given by:

$$P_{nc}^u = P_{nc}^p + t_{nc} \tag{13.1}$$

where P_{nc}^u is the price paid by consumers, P_{nc}^p is the price received by producers,[5] and t_{nc} is the tax imposed by the government.[6] Changes in the user prices are then given by:

$$\Delta P_{nc}^u = \Delta P_{nc}^p + \Delta t_{nc} \qquad (13.2)$$

For controlled sectors, producer prices are determined by the government. The change in consumer prices is equal to the change in producer prices plus the change in tax.

$$\Delta P_c^u = \Delta P_c^p + \Delta t_c \qquad (13.3)$$

where ΔP_c^u is the change in consumer prices, ΔP_c^p is the change in producer prices, and Δt_c is the change in tax on controlled-sector commodities. As we are not looking for the effect of a change in tax policy, we assume that tax remains constant: $\Delta t_c = \Delta t_{nc} = 0$.

The technology of production is captured by an input-output coefficient matrix A, with a_{ij} denoting the cost of input i in producing one unit of output j. Also, a_{ij} represents the change in the production cost of a unit of j resulting from a unit change in the price of input i. This implies a Leontief production function, where the firm's demand for inputs is relatively insensitive to the changes in input prices.[7]

Using the input-output coefficient matrix and assuming that factor prices are constant, the change in producer prices is derived as:

$$
\begin{array}{cccc}
\Delta P_{nc}^p & = & A_3' \cdot \Delta P_c^p & + & A_4' \cdot \Delta P_{nc}^p \\
(n-p,1) & & (n-p,p) \cdot (p,1) & & (n-p,n-p) \cdot (n-p,1)
\end{array}
\qquad (13.4)
$$

where A_3' is the matrix of the input requirements from the p controlled sectors for the production of one unit of output in each $n-p$ noncontrolled sectors, n being the total number of sectors. A_4' is the square matrix of the input-output coefficients of the $n-p$ noncontrolled sector.

By rearranging (13.4), we obtain the following equation that gives the effect of a change in controlled prices (for instance, a change in oil prices) on the prices of noncontrolled sectors:

$$\Delta P_{nc}^p = (I - A_4')^{-1} \cdot A_3' \cdot \Delta P_c^p \qquad (13.5)$$

Data and Descriptive Statistics

We use data from the 2000–2001 household survey of 4,966 households, 63% of which are in rural areas. A two-stage random sampling procedure was used to draw the sample of households. First, the country was divided into 12,000 districts,[8] from which 750 were randomly selected. Second, 10 households were randomly selected within each district, resulting in a sample of 7,500 households. However, only 4,966 households were completely surveyed.

By regressing oil spending shares on the level of expenditure per capita in logarithm and a dummy variable for various pieces of equipment, the results suggest that household income levels and equipment ownership help explain oil consumption (Table 13-2). Expenditure per capita is negatively correlated with kerosene budget shares and positively correlated with gasoline and diesel budget shares. Holding constant the level of expenditure per capita, agricultural households that own a tractor or a cultivator tend to consume more oil products. Similarly, households that possess a car or a moped tend to have higher petroleum consumption. Data also reveal that a household's access to electricity is negatively associated with kerosene consumption.

The 1998 Input-Output Table

Because the 2000 input-output table is not available, we use the input-output table for 1998 and assume that the structure of the economy did not change dramatically between 1998 and 2000. This assumption seems reasonable because, to our knowledge, the Malian economy did not experience any major technological shocks during that period.

The petroleum sector is strongly linked to other sectors in the economy and has the highest sum of domestic input coefficients. The most oil-intensive sector consists of services intended for the agricultural sector,[9] with a per unit requirement of 0.29, followed by the electricity and transport sectors, which have, respectively, 0.27 and 0.19 per unit requirements. A sector's dependence

TABLE 13-2 Equipment Ownership and Oil Consumption.

	Total	Kerosene	Gasoline	Diesel
Expenditure per capita	+	−	+	+
Equipment				
Tractor	+	ns	+	+
Cultivator	+	ns	+	+
Mill	+	ns	+	+
Generator	+	−	+	ns
Car	+	−	+	+
Moped	+	−	+	+
Pirogue (a canoe-like boat)	ns	+	−	ns
Access to electricity	+	−	+	+
Power-driven pump	+	−	−	+

Notes: Ordinary least squares. Sign of coefficients are reported. "ns" means nonsignificant.
Source: Mali 2000–2001 household survey.

on oil products determines the increase in the cost of production following a rise in oil prices.

Although firms consume a major share of oil products, household fuel consumption is also significant. Roughly 20% of oil products are consumed by households, 76% are used by firms as intermediate goods, and 4% are exported. Among the sectors, transport and metallurgy consume the largest shares of oil products (Figure 13-3). Surprisingly, the public sector seems to consume more oil products than the electricity sector. The mining sector and the export-oriented agricultural sector consume relatively small shares of oil products; as a result, these two sectors benefit less than other sectors from low domestic oil prices.

Results

In this section, we simulate a 34% rise in oil prices corresponding to the change in international oil prices expressed in domestic currency between 2001 and 2005. We assume that international price changes are passed through to domestic prices. First, we present the distribution of the direct expenditure effects.[10] We next examine the indirect distributional effects caused by the increase in other commodity prices. Finally, we highlight the distribution of the total effect and assess the progressivity of petroleum taxation in Mali.

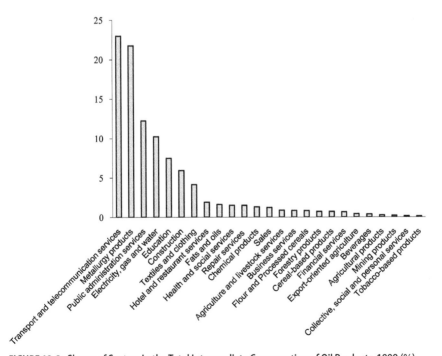

FIGURE 13-3 Shares of Sectors in the Total Intermediate Consumption of Oil Products, 1998 (%).

Direct Effect

Household fuel expenditure is predominantly for kerosene and gasoline, which account for 95% of total spending on fuel. On average, 1.45% of household expenditures is allocated for kerosene and 1% for gasoline (Table 13-3). Diesel and other sources of energy represent a relatively small share of household consumption.

The consumption of fuel products differs significantly across households according to their expenditure levels (Table 13-3). The poorest households have the highest average budget share for kerosene (2%). As expenditure per capita increases, budget shares for kerosene decrease. Other fuel products and sources of energy are disproportionately consumed by wealthy households. However, the bottom quintile seems to consume more diesel than the second and third quintiles, perhaps because diesel is often used in agricultural activities, in which poor households are mainly concentrated.

Given the pattern of budget shares, the rising kerosene price has a slightly regressive effect on income, whereas an increase in the gasoline price is almost progressive. As we mentioned previously, budget shares give an indication of the price elasticity of households' real income. The bottom quintile will experience the biggest decrease in income following a rise in the kerosene price and the lowest decrease in income when the gasoline price goes up. The distributional effects of the rise in diesel prices are almost negligible for the second and third quintiles.

The direct expenditure effect of the rise in oil prices is modest, and its distribution follows a nonlinear pattern. Indeed, the lowest-quintile households are more affected than households in all other quintiles except those in the top one (Table 13-4). A 34% rise in the prices of all oil products leads to a 0.9% decrease in real income for the bottom quintile, while the income of households in the top quintile drops by 1%. Intermediate quintiles experience a smaller decrease in income than the top and bottom quintiles. Although

TABLE 13-3 Household Budget Shares of Energy Spending by Product (% of total Spending).

Quintile	Kerosene	Gasoline	Diesel	Charcoal	Electricity
Top	0.88	1.98	0.10	0.91	1.47
Fourth	1.30	1.07	0.06	0.54	0.45
Third	1.47	0.67	0.01	0.29	0.16
Second	1.54	0.68	0.01	0.10	0.07
Bottom	2.04	0.61	0.04	0.06	0.01
All	1.45	1.00	0.04	0.38	0.43

Notes: Shares are calculated using data from the 2000–2001 household survey. Quintiles are based on the national distribution of consumption per adult equivalent.
Source: Mali, 2000–2001 household survey.

TABLE 13-4 Direct Expenditure Effects of Oil Price Increases.

Quintile	Expenditure Effect (percent of spending)	Expenditure Per Capita	Nominal Expenditure Effect	Subsidy Share (%), All Oil			
		(Thousands of CFA francs)	(Thousands of CFA francs)	Products	Kerosene	Gasoline	Diesel
Top	1.0	316	3.6	43.5	19.2	59.0	71.1
Fourth	0.8	152	1.3	19.9	22.4	18.3	17.0
Third	0.7	100	0.8	14.1	23.0	8.3	2.5
Second	0.8	70	0.5	11.5	17.6	7.6	3.3
Bottom	0.9	42	0.4	11.2	17.8	6.7	6.1
All	0.9	136	1.3	100.0	100.0	100.0	100.0

Notes: Shares are calculated using data from the 2000–2001 household survey. Expenditure effects are obtained by multiplying the sum of oil product spending shares (kerosene, gasoline, and diesel) by the oil price increase (0.34). Nominal expenditure effect is equal to the expenditure per capita multiplied by the expenditure effect. Quintiles are based on the national distribution of consumption per adult equivalent. Average expenditure per capita is based on annual per adult equivalent consumption and is in thousands of CFA francs.

Source: Mali, 2000–2001 household survey.

poor people lose less in nominal terms than other income groups (Table 13-4, column 3), the drop in their income relative to the level of expenditure is high.

Regardless of the oil product considered, the evidence is clear that rich households benefit more from price subsidies than poor households (Table 13-4). Assuming that all oil prices are subsidized at the same rate, the bottom three quintiles receive less than their population shares while the top quintile receives more than double its population share. When we assume that only the price of kerosene is subsidized, the share of subsidies accruing to the three bottom quintiles increases, but the bottom and the second quintiles still receive less than their population shares. Obviously, gasoline and diesel subsidies disproportionately benefit the richest households. These results suggest that the removal of oil subsidies is progressive. They are consistent with the results of other studies (e.g., Coady and Newhouse 2005) that find that oil subsidies are not the most efficient means of protecting the poor against higher oil prices.

Indirect Effect

Although poor households devote a relatively small share of their budgets to transport and electricity, they are likely to be affected indirectly if food prices are highly sensitive to oil prices. Table 13-5 classifies goods consumed by households into nine broad categories. To identify which commodities are more important for poor households than for wealthy households, we divide

TABLE 13-5 Share of Household Expenditure on Different Goods and Services by Household Income Quintile (%).

Household Expenditures	Household Income Quintile					All	Relative Measure (Bottom/Top)	Rank
	Top	4th	3rd	2nd	Bottom			
Food	67.5	75.4	80.2	82.1	84.0	77.8	1.24	1
Education	0.5	0.4	0.3	0.4	0.5	0.4	0.91	2
Oil products	3.0	2.4	2.2	2.2	2.7	2.5	0.90	3
Clothing	5.7	5.3	5.3	5.5	5.0	5.4	0.88	4
Health	2.4	2.4	1.9	1.8	1.6	2.0	0.66	5
Furniture and household accessories	1.4	1.4	1.2	0.9	0.8	1.1	0.57	6
Other goods and services	6.5	4.9	4.2	3.5	2.8	4.4	0.42	7
Transport and communication	4.9	2.8	1.6	2.2	1.7	2.6	0.35	8
Housing, water and energy	8.0	5.1	3.2	1.4	1.0	3.7	0.12	9

Notes: Shares are calculated using data from the 2000–2001 household survey. Relative measure is derived as the budget share of the bottom quintile divided by the budget share of the top quintile. Quintiles are based on the national distribution of consumption per adult equivalent.
Source: Mali, 2000–2001 household survey.

the budget share of the bottom quintile by the budget share of the top quintile. Although this approach does not take into account intermediate quintiles, it may provide useful information about commodities for which price increases are more harmful to the poor. For instance, the poor spend higher budget shares on food products than rich households; therefore, poor households are likely to be more negatively affected by food price increases. Oil price increases will hurt the poor slightly less than the rich, while housing, water, and energy price increases will primarily affect rich households.

Sectors experience significant price changes depending on their input-output linkages with the petroleum sector (Figure 13-4). Following a 34% rise in oil prices, services to the agricultural sector experience the largest price increase (11%), followed by metallurgy products (7.44%) and transport and telecommunication (7.22%). These increases were estimated using the input-output matrix under the assumption that cost increases caused by higher oil prices are fully passed through to output prices, except to the prices of electricity and public services, which are assumed to be controlled by the government.

The bulk of the indirect expenditure effect comes through increases in food expenditures, although food price hikes are relatively small (Table 13-6). Expenditure effects for foods are the largest, mainly because of their high budget shares. Likewise, the expenditure effects for transport and telecommunication are also large, albeit because of high price increases rather than high budget shares.

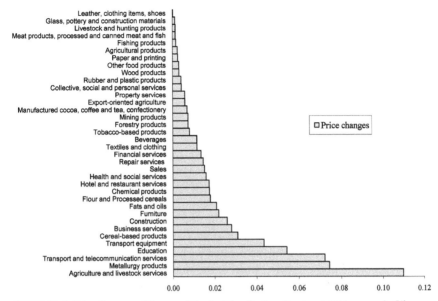

FIGURE 13-4 Price Changes of Commodities in Other Sectors Due to a 34% Increase in Oil Prices (%).

Source: Author's calculations based on the 1998 input-output matrix.

TABLE 13-6 Indirect Price and Expenditure Effects by Sector (%).

Sector	Price Change	Budget Share	Expenditure Effect
Flour and processed cereals	1.80	13.40	0.24
Fatty substances	2.09	3.80	0.08
Other food products	0.24	31.10	0.08
Transport and telecommunication services	7.22	0.90	0.07
Textile and clothing	1.17	4.50	0.05
Chemical products	1.75	2.20	0.04
Cereal-based products	3.10	1.10	0.04

Notes: Shares are calculated using data from the 2000–2001 household survey. The expenditure effect is obtained by multiplying the price change by the budget share. Only sectors with an average expenditure effect greater than 0.03% are presented.

Source: Mali, 2000–2001 household survey.

Indirect expenditure effects of oil price increases are modest and slightly regressive, with the poor experiencing the biggest expenditure rises (Table 13-7). This is the case because of the relatively high food budget shares of poor households and the assumption that the electricity price is fixed for wealthy

TABLE 13-7 Indirect Expenditure Effects by Quintile (%).

Quintile	All	Subsidy Share
Top	0.86	39.8
Fourth	0.78	19.4
Third	0.76	15.4
Second	0.83	13.7
Bottom	0.88	11.8
All	0.82	100.0

Notes: Quintiles are based on the national distribution of consumption per adult equivalent. Numbers are derived by aggregating the expenditure effects in Table 13-6 for each household.

households. As for the direct effect, the intermediate quintiles experience smaller indirect expenditure effects than the bottom and the top quintile.

As regards the oil subsidies associated with the indirect effect, rich households receive the highest share. Indeed, the share of oil subsidies accruing to the top quintile is three times larger than that accruing to the bottom quintile.[11]

Total Effect

Total expenditure effects yield a conclusion similar to the direct effects. Oil price increases have a limited impact on household expenditures, with rich households benefiting disproportionately from oil subsidies because they consume a larger share of oil products (Table 13-8). In addition, the highest incidences of rising oil prices are borne by the bottom and the top quintiles. Finally, the relative share of the direct effect is roughly equal to that of the indirect effect. Although intermediate oil consumption (76% of total) is nearly four times more than final household oil consumption (20% of total), indirect and direct effects are similar because a large proportion of intermediate oil consumption is in the electricity sector, where prices are assumed fixed, in the mining sector, where production is exported, or is consumed by the public administration.

TABLE 13-8 Total Direct and Indirect Expenditure Effects by Quintile (%).

Quintile	All	Subsidy Share
Top	1.86	41.7
Fourth	1.61	19.6
Third	1.50	14.7
Second	1.59	12.5
Bottom	1.79	11.5
All	1.67	100.0

Note: Quintiles are based on the national distribution of consumption per adult equivalent.

Wealthy households benefit not only from oil subsidies but also from controlled electricity prices (Table 13-9). When we remove the assumption of a fixed electricity price, a 34% rise in oil prices leads to a 10.72% increase in the electricity price. The distributional effects show that the removal of electricity subsidies is slightly progressive. However, the urban poor appear to be affected negatively by a higher electricity price and therefore need to be protected.

We also assess the progressivity of petroleum taxation using the Suits index, which with a value of 0.5 suggests that petroleum taxation in Mali is progressive as it places a greater burden on higher rather than lower income groups. As a result, a drop in the petroleum tax to ease the pass-through of international oil price hikes will reduce the tax burden for higher income groups more than that of lower income groups. However, this is not true for all quintiles: the Suits index measures the average progressivity of a tax across the entire income range but may hide some heterogeneity. The percentage of tax burden borne by the lowest quintile (12.4%) is larger than their share of total expenditure (11.4%), suggesting that the petroleum tax is actually regressive for the bottom quintile, but progressive for the second through the fourth quintile (Table 13-10).

TABLE 13-9 Indirect Expenditure Effects by Quintile, Rural and Urban: Distribution Effects of Controlled Electricity Price (%).

Quintile	Noncontrolled Electricity Price (1)	Controlled Electricity Price (2)	Difference (1)–(2)
Top	1.24	0.86	0.38
Bottom	1.04	0.88	0.17
All	1.06	0.82	0.24

TABLE 13-10 Distribution of Petroleum Tax Burden (in Thousands of CFA Francs per Year, and Average per Household, Unless Otherwise Noted).

Quintile	Total Expenditure on All Goods and Services		Petroleum Tax Burden			
	Level	Cumulative Share (%)	Direct[1]	Indirect[2]	Total	Cumulative share (%)
Top	2760.2	100.0	30.3	8.5	38.7	100.0
Fourth	1732.8	65.5	12.0	12.2	24.2	61.6
Third	1445.9	43.9	7.8	6.2	14.0	37.6
Second	1155.9	25.8	6.3	5.2	11.4	23.7
Bottom	910.2	11.4	6.2	6.3	12.5	12.4

Notes: [1] Weighted average petroleum tax rate multiplied by fuel expenditure. The weighted average tax rate for a quintile is the tax rate for each oil product (in Table 13-1) multiplied by the share of each oil product in fuel consumption. [2] Weighted average petroleum tax rate multiplied by indirect fuel expenditure (which is equivalent to the fuel cost of producing other goods and services).

This is consistent with the results of the input-output approach. Simulations suggest that reducing taxes on kerosene and diesel improves the average progressivity of the petroleum tax in contrast to lower gasoline tax.

Conclusion and Policy Implications

Oil price increases have a clearly adverse yet modest impact on household expenditures in Mali. This result suggests that a 34% rise in oil prices leads to a 1.67% average increase in household expenditures. The indirect effect caused by the rise in the prices of other goods and services is calibrated through input-output linkages with the petroleum sector; it represents roughly 50% of the total effect.

Although the lowest and the highest expenditure quintiles are most adversely affected by oil price rises, the benefits of blanket subsidies accrue largely to the non-poor. Mali therefore stands to reap major gains by trying to target subsidies to achieve its poverty reduction objectives. The impact of fuel prices on household budgets had a U-shaped relationship with expenditure per capita. The households in the bottom quintile experienced a 1.79% rise in expenditures, which is smaller than the impact on households in the top quintile but larger than the impact on households belonging to the intermediate quintiles. However, given that higher income groups consume the largest share of oil products, they benefit the most from implicit subsidies on oil consumption. Likewise, control of the electricity price to smooth the impact of oil price increases benefits rich households rather than poor households.

However, these results should be interpreted as representing the maximum short-run impact of the oil price increase. In the medium and long run, households and firms will adjust their demand for oil products, leading to smaller expenditure effects. The adjustment is made through the price elasticity of demand for oil products. That is, to the extent that consumers and producers reduce their demand for oil, either by switching to a different fuel or by switching to other goods or products, the long-run expenditure effect will be smaller than the short-run impact. For developing countries, the long-run elasticity is estimated at about 0.25 (Gately and Huntington 2001), which implies that about 25% of the impact calculated above might be offset by adjustment in domestic demand. In practice, these second-order effects are complex, and estimating their magnitude requires a computable general equilibrium model with a set of supply and demand elasticities.

The policy implications of our results are clear: Neither oil subsidies nor electricity subsidies are effective in protecting the poor. Nevertheless, given that rising oil prices harm the poor, particularly in the short run, it is necessary to try to mitigate the impact through well-targeted social safety nets using some of the resources generated through a subsidy reform. First, a kerosene subsidy or the use of coupons could help to soften the impact of oil price shocks on the poor. These subsidies should be made available on a temporary basis and in a transparent manner to discourage the direct diversion of subsidized kerosene to other purposes. Second, the government could generate additional revenue through subsidy reform to finance social spending, such as education and health programs, which are more beneficial to the poor. Third, because the elimination of electricity subsidies hurts rich households but also the urban poor, a subsidized rate could be granted up to modest levels of consumption. This would prevent richer households from taking advantage of those subsidies and would lessen poor households' incentive to switch to kerosene. Also, the withdrawal of electricity subsidies should be accompanied by a rural electrification program to expand the access of rural households to electricity; this may reduce their consumption of kerosene. Finally, the removal of subsidies can be politically difficult. One way to soften the impact is to clearly explain the rationale to the public and to introduce higher prices gradually according to a clear timetable. In the medium run, it would be desirable to allow market mechanisms to determine domestic oil prices and to promote the efficient use of fuel.

Notes

1   Of the increase, 0.9% is due to the cost of tax differentiation by route of importation, 0.71% represents the tax exemption received by the mining sector, and 0.23% constitutes the loss of petroleum revenue in 2004, assuming that oil tax rates remained constant between 2003 and 2004.
2   Mali imports oil products from the ports of Dakar, Abidjan, Lomé, and Cotonou.
3   See Kpodar (2006) for a detailed presentation of the model.
4   Foreign goods are assumed to be perfectly competitive with domestically produced traded goods.
5   Producer prices are a function of intermediate goods costs and factor prices.
6   Note that t is a trade tax when we consider traded goods but a domestic tax when we consider nontraded goods. A tax on domestic production of traded goods does not affect user prices but reduces producer prices.
7   However, because this assumption ignores substitution effects, the adverse income loss resulting from price increases might be overstated.
8   In rural areas, a district is defined as a geographic area where 800 to 1,000 persons are living, whereas in urban areas the number of persons in each district ranges from 1,000 to 1,500.
9   They include coffee bean hulling, protection and treatment of crops, and tool and agricultural machine rental.
10 For convenience, we assume that a rise in expenditure resulting from higher product prices is equivalent to a decrease in real income.
11 In contrast to subsidy shares linked with the direct effect, those related to the indirect effect depend on the subsidy rate. From March 2003 to March 2005, international oil prices (in CFA francs) rose by 37.42%, while domestic oil prices increased by 16.64%. First, we compute the change in the prices of other goods

following a 37.42% rise in oil prices. Second, we do the same thing for a 16.64% rise in oil prices. Finally, we calculate the difference between household expenditures under the two scenarios, which is equivalent to the amount of subsidies received by each household. Since petroleum products are aggregated in the input-output table, we are unable to compute the subsidy shares by quintile for each oil product. However, we could assume that subsidies associated with indirect expenditure effects are negligible when only kerosene is considered.

## CHAPTER 14

An Analysis of the Efficacy of Fuel Taxation for Pollution Control in South Africa

MARGARET CHITIGA, RAMOS MABUGU, AND
EMMANUEL ZIRAMBA

The study of fuel consumption is important because of its implication for pollution. The rapid rate of urbanization, the accompanying rapid increase in human population and of vehicles, and the subsequent expansion of economic activities in major towns and cities in Africa have led to increased demand for fossil fuels such as gasoline, resulting in increased emissions of carbon pollutants. This increased fuel consumption has direct bearing on climate change both at national and international levels. The present study investigates whether fuel expenditures are regressive or progressive in South Africa. This would then suggest whether a fuel tax would be an effective and desirable instrument for pollution control. The study then investigates alternative price reforms for different types of fuel to ascertain their distributional effects on different households. Such a comparison can aid policymakers looking at which types of fuels can be taxed more without inducing much pain on poor households. The study uses expenditure incidence computations as well as the Suits index to assess the progressivity (or regressivity) of fuel taxes.

Liquid fuels and gas were the most important source of energy for final consumption in South Africa for year 2004. They accounted for 30% of the total final consumption of energy as compared to coal (which accounted for 28%), electricity (which accounted for 26%), and combustibles, renewable, and waste (which accounted for 15%).

A large proportion of liquid fuel and gas, 84%, is consumed in the form of motor gasoline, diesel, and gas, while the remaining is consumed as kerosene, liquefied petroleum gas (LPG), and residual fuel oil. In terms of sectoral consumption of both liquid/petroleum and gas fuels in 2004, the transport sector was by far the major user of petroleum products (81%), while the industry sector and the residential sector each used 5%. The agriculture sector used 6% and the commercial and public services sector used the remaining 3%.

Emission Profile, Energy, and Environmental policy

South Africa's economy is very energy- and carbon-intensive, and coal is the second most important export commodity. Its greenhouse gases (GHGs) are well above the average in per capita terms, ranking it among the top ten countries contributing to global warming (Davidson et al. 2002). By the mid-1990s South Africa was already the 18th largest source of GHGs. South Africa contributes 1.6% to global GHG emissions and 42% to the total GHG emissions emitted in Africa (Davidson et al. 2002). At the same time the country is also GHG intensive in that it produces relatively low GDP values for a tonne of CO_2. The energy sector in South Africa contributes the most GHG emissions—about 89% of total emissions, with carbon dioxide being the most significant contributor (DEAT 2000). Carbon dioxide accounts for more than 80% of the three GHGs in the national inventory.

Use of coal by South Africa's industrial and power sectors is the primary source of the country's air pollution. Several industrial centers—South Durban, the Vaal Triangle, and Milnerton, in Cape Town—are considered air pollution "hotspots." More than 90% of South Africa's electricity is generated from the combustion of coal, which contains approximately 1.2% sulfur and up to 45% ash. Despite harmful environmental effects, coal-fired power stations are not required to use coal scrubbers to remove sulfur, as use of cleaner coal technology would significantly raise consumers' electricity costs. In addition to power generation, coal combustion in stoves and coal-heated boilers in hospitals and factories contributes to low-level coal-related atmospheric pollution.[1]

In addition to coal combustion, motor vehicle emissions also contribute to air pollution in urban centers. Emissions from vehicles (airplanes, ships, trains, and road vehicles) contribute 44% of the total national nitric oxide emissions and 45% of the total national volatile organic compound emissions (VOC). VOCs combine with nitric oxide and carbon monoxide, in the presence of sunlight, to form photochemical smog, which contains gases that are toxic to plants and animals, a problem in urban areas. Agricultural activities contribute the most to methane emissions (48% of the national total) and nitrous oxide emissions (78% of the national total) (State of the Environment: South Africa 2000). The effects of pollution caused by older vehicles and the lack of emissions control technology has been compounded by the historical absence of vehicle emissions legislation.[2]

Since 1970, South Africa has consistently consumed the most energy and emitted the most carbon dioxide per dollar of GDP among major countries in Africa. In 2002, South African energy intensity measured 11,359 Btu per 1995 US$, higher than most other important African energy-producing countries such as Algeria, Egypt, and Morocco. South Africa's energy intensity surpasses that of several other rapidly industrializing countries, such as India and China, as well as the United States. In 2002, South Africa's carbon dioxide intensity was approximately 0.8 metric tons per thousand US$ 1995—larger than all other African countries and the United States. In 2002, of the 4.5 quadrillion Btu (quads) of primary energy consumed in South Africa, 74.0% was coal, 20.9% was oil, 2.6% was nuclear, and 2.0% was natural gas. While South Africa accounted for 35.6% of the total primary energy consumed in all of Africa in 2002, the country was responsible for only 1.1% of total primary world energy consumption. However, South Africa emitted 306.3 million metric tons of carbon dioxide from coal consumption, amounting to 90.6% of Africa's energy-related carbon emissions and 3.4% of world energy-related carbon dioxide emissions. Reliance on coal-based energy sources explains South Africa's relatively larger carbon dioxide emissions in comparison with many other industrializing countries.

South Africa's four oil refineries are another major contributor to energy-related air pollution. The refineries, located in the northern suburbs of Cape Town and in the southeastern coastal city of Durban, emit high levels of sulfur dioxide and several other harmful pollutants.

Over the past 25 years, primary energy consumption in South Africa's residential and commercial sector has risen only gradually. The slow increase of primary energy consumption in the residential sector is in part due to South Africa's reliance on fuelwood, which is not accounted for in primary/commercial energy consumption estimates. Fuelwood is still the largest source of household energy in remote rural areas, estimated to meet the daily energy needs of more than one third of South Africa's population. Use of fuelwood is believed to contribute to deforestation in South Africa. Furthermore, it is usually burned in enclosed spaces without adequate ventilation, leading to increased risk of respiratory health problems and other ailments.

According to UNICEF, in 2000 respiratory infections from air pollution were the fourth-largest cause of death in children under five in South Africa (more than 6,000 deaths per year). The effects of air pollution on children are made worse by poverty, including lack of access to potable water, sanitary facilities, and health care. Residents of low-income communities have been forced to relocate due to pollution from refineries and their waste sites. Currently, there is significant public pressure on Dallas-based Caltex, operator of the Milnerton refinery, to upgrade operations and to reduce emissions at the site. In addition, Shell and BP, operators of the 165,000 barrels per day Sapref refinery, are evaluating possible upgrades to their facility and to a 53-mile underground pipeline system, which contribute to air and ground water pollution respectively.

It is likely that South Africa will have to face up to negotiations on international environmental issues in the near future. The National Climate Change Response Strategy for South Africa (2004) is the most recent policy statement on the management of GHG emission reduction in South Africa. Other documents[3] also discuss GHG emissions. These documents show the importance of this problem to South Africa.

Methods Used for Analysis

Generally the literature presents different approaches used to assess the economic incidence of taxes on energy products. We can cluster these techniques into three main categories: the macroeconomic method, assessing how energy taxes affect macroeconomic variables such as the GDP, exchange rate, inflation, employment, and so on; the microeconomic approach, using a micro database like a household survey and sometimes an input-output table to estimate the direct and indirect effects of energy price increases on households' well-being; and third, the linked micro and macroeconomic method, which combines both micro and macro level analysis in a single framework to fully account for the direct and indirect effects of a shock on the economy, as well as their distributional impacts on sectors and households. In this study, we use a household survey to analyze household expenditures on energy products and how increased fuel prices influence them. To evaluate the price reforms, supply and use tables are used to gauge the change in consumption resulting from a change in price. This allows estimation of the direct and indirect effects of the price changes on households using the following formula:

$$\Delta C_i = \left(\sum_p c_{p,i}^1 - \sum_p c_{p,i}^0 \right) \Big/ \sum_p c_{p,i}^0$$

where $c_{p,i}^0$ and $c_{p,i}^1$ are the user cost of product p by consumer i before and after the shock. For oil and oil products (op) $c_{op,i}^1 = 2 \cdot c_{op,i}^0$, and for non oil products (np) $c_{np,i}^1 = c_{np,i}^0$, with $\{op, np\} \in p$. The value of $c_{p,i}^0$ is given by supply and use tables (Fofana et al. 2009).[4] Thereafter, welfare is then measured using the equivalent variation measure. The budget share for each expenditure decile can be calculated as follows:

$$E_{shd} = (FE/TE) \times 100 \qquad (14.1)$$

Using national data from South African Consumer Expenditure Surveys we will test the commonly held view that fuel taxes are regressive and hence impose burdens mainly on low-income households. As stated earlier, the regressivity hypothesis suggests that a fuel tax would not be a politically attractive instrument for pollution control. To summarize we will use the Suits index (1977).

Evaluation of Price Reforms Methods

Higher prices of fuel products increase the cost of living through direct and indirect effects. The direct effects of fuel price rises are on petroleum products (notably gasoline, LPG, and kerosene) purchased directly by households. The much more complex indirect effects of fuel price increases pass through the input cost of products consumed by households. For example, an increase in diesel cost increases transport cost, thereby increasing food costs. The magnitude to which such effects affect the various groups of households depends on the relative share of fuel- and fuel-input-intensive products in household budgets.

The calculation of the full price impacts should integrate the chain of indirect effects along with the direct effects, as well as the share of each product in the total household budget. This requires both an input-output table and a household expenditure survey (see Chapters 8 and 9).

Results of the Expenditure-incidence Analysis—Suits and Empirical Methods

The basic data source for this analysis is the year 2000 Income and Expenditure Survey (IES). The survey has a representative sample of 26,264 households drawn from all provinces of South Africa and detailed household level data on consumption patterns, with some data on household income and taxes.

Household disposable income is defined as regular income plus other income, measured on an annual basis. Total expenditures are the sum of expenditures on most household activities such as housing, food, beverages (alcoholic and nonalcoholic), clothing and footwear, health, household fuel for private transport and domestic uses, and recreation and entertainment.

We use the expenditure measure to assess the distributional impact of fuel taxation. We assign households to deciles by total expenditure. Each decile has about 2,626 households. Fuel expenditure shares or fuel expenditure-to-total-expenditure ratios within each decile are then calculated to illustrate the distribution of fuel expenditure patterns. We also calculate similar ratios for transport-related expenditures.

Measuring the incidence of a tax requires a number of simplifying assumptions. First, we need to determine the unit of analysis. This involves deciding on whether to allocate tax liabilities to individuals or to households. In this study the unit of observation is the household (because the only data available is at that level). Second, we must make assumptions about the direction of the shifting of the tax under analysis. Following earlier studies, we assume that the supply of petroleum products is perfectly elastic since South Africa is a price taker in the world market. This assumption implies that the imposition of a tax on fuel does not affect the producer price of the good, and thus the entire tax burden falls on consumers (Musgrave et al. 1974). Third, we need to determine how to rank people or households by some measure of "well-being." Annual household spending has been used to sort households in this study.

Due to the highlighted problems with reported annual income, this study uses an expenditure-based measure of lifetime income, which is annual household expenditure. All households were classified to expenditure deciles.

Table 14-1 shows the ratio of fuel expenditure (for own transport services) in total household expenditure. The budget share of fuel generally increases

TABLE 14-1 Fuel expenditure/Total Expenditure, by Expenditure Decile, 2000.

Expenditure Deciles	Fuel expenditure/Total expenditure (%)
1	0.03
2	0.03
3	0.05
4	0.11
5	0.27
6	0.50
7	0.74
8	1.30
9	2.74
10	3.39

Source: Authors' tabulations using 2000 IES.

as income increases. The lowest expenditure decile devotes 0.03% of total expenditures to fuel. The highest decile devotes 3.39% of total expenditures to fuel. The expenditure-based calculations suggest that the distribution of fuel expenditure is generally progressive, with higher income households devoting the highest budget shares to fuel.

Households also make use of fuel indirectly in other transport-related activities. This is done through the use of public and hired transport. Such transport takes various forms, including buses, trains, rented vehicles, taxis, and furniture removal for the transportation of goods. The total indirect fuel expenditure was computed by adding household expenditures on bus travel, train travel, rented vehicles, and furniture removal and transportation of goods. These expenditures are reported in the survey. Figure 14-1 shows the proportion of expenditures devoted to transport-related activities in total household spending. The first decile devotes 2.27% of total expenditures on transport. The share of such expenditures in total expenditure increases with income until the seventh decile.

Figure 14-1 plots the share of transport-related expenditures in total household expenditure. This figure shows interesting results. The curve is hump-shaped, indicating that the middle-income households spend a larger share than the rich or poor on fuel. This result indicates that transport-related services are a necessity for middle income households. A similar observation was made by Santos and Catchesides (2005).

Table 14-2 shows the proportion of expenditures devoted to fuel and transport-related activities, which generally increases with income. The lowest decile devotes 2.58% of total expenditures to fuel, while the highest devotes 3.94% of expenditures to fuel and transport-related activities. The budget shares show progressivity of fuel expenditures, with the ninth decile devoting 5.24% to fuel and transport-related expenditures.

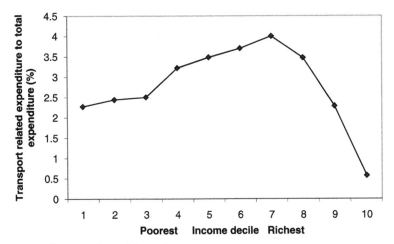

FIGURE 14-1 Transport-Related Expenditures.

Source: 2000 IES.

TABLE 14-2 Fuel and Transport-related Expenditures/Total Expenditure, by Expenditure Decile, 2000.

Expenditure Deciles	Transport Expenditures/Total Expenditure (%)
1	2.58
2	2.77
3	2.73
4	3.36
5	3.89
6	4.22
7	4.65
8	4.93
9	5.24
10	3.94

Source: Authors' tabulations using 2000 IES.

The expenditure-based measure of fuel tax incidence shows that fuel taxes are progressive: even though the percentage share for the tenth decile is lower than for deciles 6 to 9, it is still higher than those for the poorest five deciles. This point confirms Poterba's (1991) result that using an expenditure-based measure (as a proxy for lifetime income) will result in less regressivity. When all forms of fuel use are taken into account fuel expenditures are progressive.

Figure 14-2 plots the fuel tax liability (concentration) curves for South Africa when we consider direct as well as indirect fuel use. Both curves are convex from below, indicating progressivity. The concentration curve for

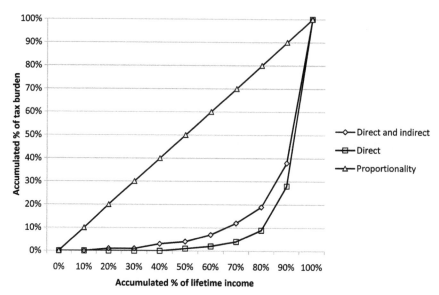

FIGURE 14-2　Concentration Curves for South Africa.

Source: Authors' calculations based on the 2000 IES.

direct fuel use is more convex than when both uses are considered together, meaning progressivity is reduced when we take indirect fuel use into account. This is confirmed by the Suits coefficients reported in Table 14-3. Both coefficients are positive, but the index value decreases when we combine indirect fuel use with direct fuel use—reducing progressivity. This is explained by the fact that indirect fuel use separately appears to be progressive until the seventh decile when it becomes regressive.

We use the Suits index (Suits 1977) to estimate the progressivity of South African fuel taxes for the year 2000. Table 14-3 shows that direct taxes on fuel are progressive (with a Suits index of 0.284). When indirect effects are considered, the Suits Index of 0.164 indicates that fuel taxation is still progressive. These results are shown in Figure 14-2.

Results Based on Evaluation of Price Reforms Methods

The discussion above focused on the pattern of expenditures for fuel among deciles. Next, we simulate price changes to different types of fuel to assess the

TABLE 14-3　Suits Index Estimates of Fuel Tax, 2000.

Tax type	Suits Index
Direct tax	0.284
Direct and indirect tax	0.164

impact on households. Specifically, we increase transport fuel price, transport services price, and lastly paraffin prices. For these simulations we collapse the deciles into quintiles for ease of analysis, still using the 2000 IES. We simulate three price changes. Simulation 1 is a doubling of the price of fuel used for transport. In simulation 2 we increase the cost of transportation by 15%. finally, in simulation 3 we double the price of paraffin. We analyze the impact of these changes on household expenditures and on welfare as measured by the equivalent variation.

Household Quintile Expenditure Shares of Various Fuels

Table 14-4 shows the expenditure shares of energy and major petroleum products in household expenditure for quintile expenditure groups. It should be noted that adding up the groups and including expenditure on paraffin means that the first quintile has higher expenditure than quintiles 2, 3 and 4. However, the last quintile still has the highest expenditure. Fuel expenditure is still regressive, as in the case with deciles. This means that the poorest groups rely more on paraffin as a source of fuel, while the richest groups rely on transport fuel. This same pattern is observed as we disaggregate households into rural and urban, although, notably, rural households rely more on paraffin than urban households (see Table 14-5).

The analysis of transport service expenditure by quintile groups reveals the same pattern as for the deciles. The-middle income groups spend the most on this service. For the whole of South Africa, average expenditure on transport services is 4.5% (Table 14-6). This expenditure is higher in urban than rural areas.

The next section analyzes the impacts of the three simulations—transport fuel prices, transport costs, and paraffin prices—as an indication of how a fuel tax would affect South African households.

TABLE 14-4 Shares of Energy and Petroleum Products Expenditure in Total Household Expenditure by Quintile in South Africa.

	All energy sources	Petroleum Products				
		All	Paraffin	Gas	Other Domestic Fuel	Transport fuel
Quintile 1	8.2	3.9	3.8	0.1	0	0
Quintile 2	7.9	3.1	2.9	0.2	0	0.1
Quintile 3	7.5	2.5	1.9	0.2	0	0.5
Quintile 4	7.7	2.6	0.8	0.2	0	1.5
Quintile 5	8.8	4.8	0	0.1	0.1	4.5
All	8.5	4.2	0.5	0.1	0.1	3.5

Compiled using the 2000 IES.
Source: Fofana et al. (2009).

TABLE 14-5 Shares of Petroleum Products Expenditure in Total Household Expenditure for Quintile Groups in Urban and Rural Areas.

	All Energy Sources	Petroleum Products				
		All	*Paraffin*	*Gas*	*Other Domestic Fuel*	*Transport Fuel*
Urban	8.4	4.2	0.3	0.1	0	3.7
Quintile 1	8.8	4.6	4.3	0.2	0	0
Quintile 2	8	2.9	2.6	0.1	0	0.1
Quintile 3	7.4	2.3	1.7	0.1	0	0.5
Quintile 4	7.6	2.3	0.7	0.1	0	1.5
Quintile 5	8.6	4.6	0	0.1	0	4.5
Rural	9.1	4.4	1.5	0.3	0.3	2.3
Quintile 1	7.8	3.5	3.4	0.1	0	0
Quintile 2	7.8	3.4	3.1	0.2	0	0.1
Quintile 3	7.6	2.9	2.1	0.3	0.1	0.5
Quintile 4	7.8	3.7	1.3	0.6	0.1	1.7
Quintile 5	11.1	6	0.3	0.3	0.6	4.8
All	8.5	4.2	0.5	0.1	0.1	3.5

Compiled using the 2000 IES.
Source: Fofana et al. (2009).

TABLE 14-6 Shares of Transport Expenditure in Total Household Expenditure.

	All	Areas	
		Urban	Rural
Quintile 1	4.2	4.9	3.4
Quintile 2	4.7	5.9	3.2
Quintile 3	5.8	6.6	4
Quintile 4	5.4	5.5	4.6
Quintile 5	4.3	4.4	2.7
All	4.5	4.6	3.5

Source: Fofana et al. (2009).

Increasing Transport Fuel Prices

If we double the price of fuel in a bid to encourage efficient use of this energy source, the quintiles bearing the highest burden will be the richest households (Table 14-7). The richest households tend to own their own vehicles and thus spend more on transport fuel than the poorer households. In terms of welfare, the households that see the largest fall in welfare are the richest, with the very poorest quintile least negatively affected. Such a tax is thus progressive.

TABLE 14–7 Estimated Percentage Changes in Household Expenditure and Welfare for Fuel Price Increase.

| | *Household Expenditure* | | | *Welfare (Equivalent Variation)* | | |
| | *All* | *Areas* | | *All* | *Areas* | |
		Urban	*Rural*		*Urban*	*Rural*
Quintile 1	0	0	0	0	0	0
Quintile 2	0.1	0.1	0.1	−0.1	−0.1	−0.1
Quintile 3	0.5	0.5	0.5	−0.3	−0.3	−0.3
Quintile 4	1.6	1.5	1.8	−1	−1	−1.2
Quintile 5	4.6	4.6	5.4	−3.1	−3	−3.5
All	3.6	3.8	2.6	−2.4	−2.5	−1.7

Source: Fofana et al. (2009).

Increasing Transport Costs

Looking at direct purchases of transport fuel may exclude those households that do not have their own vehicles. Yet in South Africa, during the apartheid period, most poor workers were located at the outskirts of cities and towns; till this day, many commute to their working places using public or minibus taxi transport. We simulate a 15% increase in transport costs to see the effects of expenditure and welfare. The results are reported in Table 14-8. In this case the middle-income households bear the brunt of such an increase—in both rural and urban areas. The fall in welfare is highest for quintile 3. Thus such a policy would require more careful design to prevent income distribution from becoming even more skewed than it already is across South Africa.

TABLE 14–8 Estimated Percentage Changes in Household Expenditure and Welfare for Transport Cost Increase.

| | *Household Expenditure* | | | *Welfare (Equivalent Variation)* | | |
| | *All* | *Areas* | | *All* | *Areas* | |
		Urban	*Rural*		*Urban*	*Rural*
Quintile 1	0.56	0.66	0.48	−0.52	−0.61	−0.45
Quintile 2	0.64	0.8	0.48	−0.59	−0.74	−0.45
Quintile 3	0.8	0.91	0.59	−0.74	−0.84	−0.54
Quintile 4	0.78	0.8	0.7	−0.72	−0.74	−0.65
Quintile 5	0.58	0.6	0.36	−0.53	−0.55	−0.33
All	0.62	0.65	0.5	−0.58	−0.6	−0.46

Source: Fofana et al. (2009).

TABLE 14-9 Estimated Percentage Changes in Household Expenditure and Welfare for Paraffin Price Increase.

	Household Expenditure			Welfare (Equivalent Variation)		
		Areas			Areas	
	All	*Urban*	*Rural*	*All*	*Urban*	*Rural*
Quintile 1	3.8	4.3	3.4	−2.5	−2.9	−2.3
Quintile 2	2.9	2.6	3.1	−1.9	−1.8	−2.1
Quintile 3	1.9	1.7	2.1	−1.3	−1.2	−1.4
Quintile 4	0.8	0.7	1.3	−0.5	−0.4	−0.9
Quintile 5	0	0	0.3	0	0	−0.2
All	0.5	0.3	1.5	−0.3	−0.2	−1

Source: Fofana et al. (2009).

Increasing Paraffin Prices

Here, to see clearly the impacts, we double the price of paraffin. Doubling the price of paraffin leads to a regressive pattern of expenditure. This is in line with the shares as shown above, where the poorest households have the highest paraffin expenditure. The changes in household expenditures required are quite big and particularly so for the poor and the rural households (see Table 14-9). Table 14-9 shows the welfare loss using the equivalent variation measure. Poor households suffer the most, with the poorest in both urban and rural bearing the highest burden from such an increase. However, the urban poorest quintile is the worst affected of all the quintiles.

Conclusions

Our analysis of household survey data demonstrates that a fuel tax is not as regressive as might be perceived. Using household income and expenditure data from South African IES of 2000, we have tested the commonly held view that gasoline taxes are regressive and hence impose burdens only on low-income households. Poterba (1991) argued that households who would be most heavily burdened by the fuel tax are those who spend more than 10% of their budget on fuel. In our case, none of the income deciles spends up to 10% on fuel. The analysis of indirect fuel use shows that middle-income groups spend more on fuel than other income groups. Such an analysis shows some progressivity of fuel tax as the budget share of indirect fuel increases until the seventh decile. Our results suggest that a fuel tax would not impose an excess burden on the poorest households, as has been argued in the literature. When all forms of fuel use are taken into account, fuel expenditures are in effect progressive. This suggests that fuel tax would be an effective and desirable instrument for pollution control.

Evaluation of alternative price reforms shows that increasing the price of fuel in a bid to reduce emissions is progressive for fuel tax. However, when one takes transport costs into account, it is not as progressive particularly between the middle-income and the richest households. With paraffin expenditure, the pattern turns out to be regressive. These results are in line with the individual item expenditure shares of the different quintiles.

Acknowledgments

Emmanuel Ziramba acknowledges a research grant received from the Centre for Environmental Economics and Policy in Africa (CEEPA), financed by the Swedish International Development Cooperation Agency (SIDA), and the International Development Research Centre (IDRC), for parts of this work. Ramos Mabugu acknowledges funding from the Financial and Fiscal Commission for parts of this work, and substantial contributions to methodology used by Ismael Fofana.

Notes

1   www.eia.doe.gov
2   In addition to air pollution, energy-related marine pollution is a growing environmental problem, threatening rare mollusk, fish, seabird and mammal populations, as well as pristine beaches. In July 2004, a collaborative $11 million project aimed at reducing land-based pollution of the western Indian Ocean was launched by several African nations, including South Africa, Kenya, Mozambique, and several island nations. The three-year project, funded by the Global Environment Facility (GEF) of the World Bank and the Norwegian government, aims to curb sewage, chemicals, and other toxins that pollute regional rivers and coastal waters. South Africa is especially vulnerable to oil spills due to the high volume of oil transported around the country's coasts by ships en route from the Middle East to Europe and the Americas.
3   Such as the National Environmental Management Act, Air Quality Act (2004), the 2003 White Paper on the Promotion of Renewable Energy and Clean Energy Development, the 2004 White Paper on the Renewable Energy Policy, the Integrated Waste Management Strategy (2000), the 1998 Energy White Paper, and a draft Policy Paper recently launched by the National Treasury on a framework for considering market-based instruments to support environmental fiscal reform in South Africa.
4   See Fofana et al. (2009) for details on the methodology used.

CHAPTER 15

Fuel Taxation and Income Distribution In Tanzania

Adolf F. Mkenda, John K. Mduma, and
Wilhelm M. Ngasamiaku

In her 2007–2008 Budget speech, the Minister of Finance of the United Republic of Tanzania, Hon. Zakia Meghji, proposed an increase in excise taxes on petroleum products as a measure to increase government revenue. The suggestion entailed a 33% increase in the fuel levy and a 7% adjustment of excise duty on fuel to account for inflation. This proposal met with unprecedented outrage from members of parliament, the mass media, and large sections of the public. Their objections were largely based on two arguments: that such a tax increase was inflationary and regressive. In response to the uproar, the Minister of Finance withdrew the tax increase on kerosene, viewed as more regressive than the tax increase on petrol and diesel. In this rare turnaround, the Minister further reduced annual motor vehicle fees for motorcycles and other vehicles with engine capacity below 1,500 cc. while increasing such fees on luxury cars. As this episode shows, higher taxation of fossil fuel, an increasingly common policy option for African governments, is very unpopular. One of the reasons for objecting to this tax is its perceived regressive effect on distribution.

This chapter reports the assessment of the first order distributional implications of taxation on fuels in Tanzania, which means that we only focus on the immediate tax incidence of fuel taxation on the consumers of fuel and the likely immediate impact on the cost of transport. We assume that consumers of fuel bear the full brunt of tax incidence. This assessment does not take into account the full multiplier impact of fuel taxation on the entire economy, which must include the extent to which higher prices of fuel would affect the cost of production and thus possible increases in inflationary pressure. To do this, a general equilibrium approach would be necessary, an approach we cannot employ for lack of detailed data and time. And yet because parliamentary arguments against higher fuel taxation in Tanzania included the objection to the perceived first order distributional impact of such taxes, it is useful to shed light on the extent to which such objections are supported by empirical evidence. The aim of this chapter is to establish whether fuel taxation in Tanzania has a regressive or progressive distributional impact.

Fuel taxation in Tanzania is used primarily to raise government revenues. No doubt this is an important policy objective particularly because the tax to GDP ratio in Tanzania is lower than for other comparable countries, and because Tanzania still depends heavily on foreign aid for fiscal solvency. The presumption here is that the price of fuel is sufficiently inelastic to ensure imposition of tax-yielding substantial revenue.

Fuel taxation is also important for environmental reasons in that it leads to less consumption of these fuels and thus reduces pollution. Transport-related pollution is one of the six major environmental problems that the National Environmental Policy (Tanzania 1997, p. 21) has singled out for urgent attention. The Policy states that the transport sector shall focus on the following environmental objectives:

1 Improvement in mass transport systems to reduce fuel consumption, traffic congestion, and pollution;
2 Control and minimization of transport emission gases, noise, dust, and particulates; and
3 Disaster/spill prevention and response plans and standards shall be formulated for transportation of hazardous/dangerous materials.

Table 15-1 gives some measures of pollution in Dar es Salaam, the largest urban center in Tanzania. The emission levels of sulfur dioxide, suspended particulate matter (SPM), and lead concentration exceed World Health Organization (WHO) maximum recommended levels. The high level of air pollution in Dar es Salaam is mainly due to the increased number of vehicles, a situation made worse by the fact that most cars are imported from secondhand markets and continue to use leaded fuel. This problem has drawn the attention of policymakers; a punitive tax of 95% of the cost, insurance, and freight (CIF) has been imposed on imported used cars over 10 years old, and the tax rate is graduated in accordance with engine size. The government of Tanzania has now passed the Environmental Management Act to promote the use of

TABLE 15-1 Air Pollution in Dar es Salaam.

Pollutant	*Level in Dar es Salaam in µ/m³ per Hour (Average)*	*WHO Recommended Maximum in µ/m³ per Hour*
Sulfur Dioxide	127–1385	350
Nitrogen Dioxide	18–53	200
Suspended Particulate Matter (SPM)	98–1161	230
Lead Concentration	0.60–25.6	1.5

Source: Msafiri (2005).

economic instruments such as taxation for the purpose of protecting the environment. The extent to which fuel taxation reduces consumption and thus cuts down pollution depends to a large extent on the price elasticity of fuel. Indeed, one of the objections of higher fuel taxation has been that these fuels have no substitutes, so taxing them would only prove inflationary rather than reducing fuel consumption. This claim deserves further empirical research.

The rest of this chapter is organized as follows. The next section reviews approaches for assessing tax progressivity. Two commonly used indices, the Kakwani index and the Suits index of progressivity are explained and their relationship explored. A stochastic dominance test for assessing progressivity of tax is also explained. The subsequent section briefly explains the empirical strategy and assumptions used for assessing tax incidence from household budget survey data. We then present empirical results and relate the results to some other findings. The final section concludes the chapter.

Measurement of Tax Incidence and Progressivity

In this chapter we use both Suits and Kakwani indices (see Suits 1977 and Kakwani 1977a). As discussed in the introduction, these can give contradictory results. Another approach, the dominance approach for assessing tax progressivity, makes use of the stochastic dominance approach to determine whether one tax structure has a concentration curve that is above the other tax structure. The intuition for this is straightforward. If tax concentration curve A is above concentration curve B, then for any cumulative share of the population ordered by income the cumulative share of the tax in A is higher than in B. This means that the poorest segment of the population in A pays a bigger share of tax than the poorest segment of the population in B, making taxes under B more progressive than the taxes under A. Specifically, Yitzhaki and Slemrod (1991) have proved that if the concentration curve of one commodity is above the concentration curve of another commodity, then a small tax decrease in the first accompanied by a tax increase in the second (with revenue remaining unchanged) increases social welfare, where social welfare is based on minimum and uncontroversial assumptions. Exploiting the social welfare function makes it possible to infer welfare superiority of a more progressive tax, assuming that issues of efficiency remain unchanged. In practice, a tax structure with a lower concentration curve is more progressive and thus preferred to a tax structure with a higher concentration curve.

More insight on the progressivity of a tax can be gained by using both concentration curves and Lorenz curves. When a concentration curve for tax is below the Lorenz curve of income it means that households whose income is below any given threshold pay the percentage share of tax that is lower than their percentage share of income. This can be interpreted as a sign of progressivity. On the other hand, when the concentration curve of a tax is above the

Lorenz curve of income, households with income below any given threshold pay the cumulative percentage share of tax that is higher than their cumulative percentage share of income, which can be viewed as a sign of regressivity. In fact, Sahn and Younger (2003) preferred combining concentration and Lorenz curves to assess the progressivity of indirect taxes on the grounds concentration curves alone produce graphs too close to each other to make much sense. Besides, the use of both the concentration and Lorenz curves affords the opportunity to infer the income elasticity of the commodity based on the position of the concentration curve relative to the Lorenz curve and the 45-degree diagonal line (see Kakwani 1977b). Income elasticity tells us whether a commodity is a necessity, a luxury, or inferior. This information is useful for determining on which good to impose a tax.

Assessment of tax progressivity presumes correct measurement of tax incidence, that is, the estimation of the actual distribution of the economic burden of the tax. It is challenging to calculate tax incidence for a commodity tax because of tax avoidance and tax evasion and also because the actual burden of an indirect tax may be shared across producers, traders and final consumers. Besides, imposition of a tax normally induces a behavioral reaction that is difficult to capture empirically. In a developing country such as Tanzania, the pervasiveness of the informal sector increases incidence of tax evasion, making the measurement of indirect tax incidence even more difficult. Following Sahn and Younger (2003) we adopt the following approach. First we specify household expenditures after tax as follows:

$$e_{i,j} = p_j \left(1+t\right) x_{i,j} \tag{15.1}$$

where

$e_{i,j}$ = stands for the after tax expenditure by household i on commodity j.
p_j = price of commodity j.
t_j = *ad valorem* tax on commodity
$x_{i,j}$ = amount of commodity j consumed by household i.

If we let $T_{i,j}$ denote the total amount of tax paid by household i due to consuming commodity j, then equation (15.1) implies that

$$T_{i,j} = t_j p_j x_{i,j} = \frac{t_j}{1+t_j} e_{i,j} \tag{15.2}$$

We then assume that tax incidence is proportional to expenditure, which means that if there is only one tax rate we can use the concentration of the observed expenditure on commodity j in place of the amount of tax paid in assessing the progressivity (see Sahn and Younger 2003). In this way, we do not need to calculate the actual tax paid for each commodity; it is sufficient to use the expenditure by household on the commodity of interest.

The post-tax household expenditure by commodity is available in the household budget survey data, the last of which was collected in Tanzania in 2007. From this data the following categories of fuel and fuel-related products are chosen to assess the progressivity of tax: petrol and diesel, kerosene, electricity, charcoal, firewood, and transport. Progressivity of taxes on these commodities is assessed by calculating the Suits and Kakwani indices and their concentrations, and the Lorenz curves of the households' expenditure on these commodities.

Results and Implications

Table 15-2 reports the estimated Suits and Kakwani indices for the following commodities: charcoal, firewood, kerosene, electricity, petrol and diesel, and household expenditure on transport. Transport is included in the analysis

TABLE 15-2 Suits and Kakwani Indices for Selected Categories of Fuels in Tanzania.

Type of Fuel	Suits Index	Inference on Progressivity from the Suits Index	Kakwani Index	Inference on Progressivity from the Kakwani Index
Charcoal (national)	0.13*	**Progressive**	0.057*	**Progressive**
Firewood (national)	−0.34*	Regressive	−0.35*	Regressive
Kerosene (national)	−0.20*	Regressive	−0.22*	Regressive
Electricity (national)	0.52*	**Progressive**	0.34*	**Progressive**
Petrol/diesel (national)	–	–	0.52*	**Progressive**
Transport (national)	–	–	−0.0098*	Regressive
Charcoal (urban)	−0.09*	Regressive	−0.13*	Regressive
Firewood (urban)	−0.39*	Regressive	−0.50*	Regressive
Kerosene (urban)	−0.26*	Regressive	−0.25*	Regressive
Electricity (urban)	0.33*	**Progressive**	0.16*	**Progressive**
Petrol/Diesel (urban)	–	–	0.50*	**Progressive**
Transport (urban)	–	–	−0.064*	Regressive
Charcoal (rural)	0.07*	**Progressive**	0.021	Not Significant
Firewood (rural)	−0.23*	Regressive	−0.28*	Regressive
Kerosene (rural)	−0.15*	Regressive	−0.21*	Regressive
Electricity (rural)	0.71*	**Progressive**	0.53*	**Progressive**
Petrol/Diesel (rural)	–	–	0.46*	**Progressive**
Transport (rural)	–	–	−0.080*	Regressive

* indicates that the estimated value is statistically different from zero. Test of significance for Suits index was carried out by the authors using the bootstrapping method. Test of significance for the Kakwani index is based on the DASP program in STATA in which the Kakwani is estimated as the difference between the Gini coefficient and the concentration index (see Lambert 2001).

because when taxes are imposed on petrol and diesel, the cost of transport is likely to go up. At the national level, both the Suits and Kakwani indices show that taxes on firewood and kerosene are regressive. The Kakwani index further shows that a tax on transport cost is also regressive, although the value of the index (which is statistically significant) is very small. Both the Kakwani and Suits indices show that taxes on charcoal and electricity are progressive, and the Kakwani index further shows that taxes on petrol and diesel are also progressive. Overall, these results at the national level are as one would expect. Households that use electricity are few and relatively well-off.

In urban areas, taxes on the following are shown to be regressive: charcoal, firewood, kerosene and transport. Significantly, a tax on charcoal—progressive at the national level—is found to be regressive in urban areas. This makes sense because in rural areas charcoal is used by relatively well-off households, whereas in urban areas well-off households use electricity instead. Taxes on electricity and on petrol and diesel are progressive in urban areas.

A tax on charcoal in rural areas is found to be progressive using the Suits index, but the estimate of the Kakwani index is not significantly different from zero. Taxes on firewood and kerosene are found to be regressive in rural areas, and the Kakwani index indicates that tax on transport is also regressive. Taxes on electricity, petrol, and diesel are all found to be progressive, signifying that these commodities are mostly consumed by the relatively well-off households.

Figure 15-1 depicts the concentration curves for firewood, charcoal, kerosene, electricity, transport, and petrol and diesel for the whole country, together with the Lorenz curve. As discussed above, a tax on a commodity whose concentration curve is above the other is more regressive than a tax on a commodity whose concentration curve is below. Figure 15-1 therefore shows that a tax on firewood is more regressive, followed in descending order of progressivity by kerosene, transport, charcoal, electricity, and petrol. In other words, a tax on petrol is more progressive, followed by electricity, charcoal, transport, kerosene, and firewood. This ranking is consistent with the ranking obtained using the progressivity indices in Table 15-2.

Figure 15-1 can also be used to assess the income elasticity of these fuel products. Kakwani (1977b) has proved that when the concentration curve is above the 45-degree diagonal line, the income elasticity of the commodity is less than zero, signifying that it is an inferior good. When the concentration curve is between the 45-degree diagonal line and the Lorenz curve, income elasticity is between zero and one, which means that the commodity is a necessity.

When the concentration curve is below the Lorenz curve, income elasticity is above one, meaning that the commodity in question is a luxury. Using this insight we can see from Figure 15-1 that firewood is a slightly inferior commodity: its concentration curve is marginally above the 45-degree diagonal line. The concentration index[1] for firewood is –0.015 and is barely significant. The concentration curve of kerosene is between the 45-degree line and the Lorenz curve, showing that kerosene is a necessity. The concentration curve for

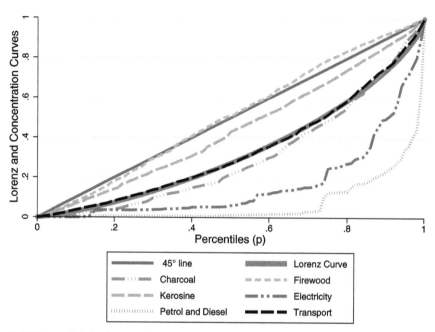

FIGURE 15-1 Relative Progressivity Using Concentration and Lorenz Curves, All Areas.

transport tracks the Lorenz curve very closely, showing that its income elasticity is close to one; thus it is on the borderline between a necessity and a luxury. Charcoal is marginally a luxury good, with its concentration curve just slightly below the Lorenz curve, but electricity and petrol/diesel are luxury goods.

Figure 15-2 depicts the concentration and Lorenz curves for urban areas. A tax on firewood is shown to be rather regressive, followed in descending order of progressivity by taxes on kerosene, charcoal, transport, electricity, and finally petrol and diesel, the latter being the most progressive. Firewood is clearly an inferior good in urban areas, where its concentration curve is clearly above the 45-degree diagonal line; its concentration index is found to be –0.16 and is statistically significant. This finding is highly plausible, as only very poor households use firewood in urban areas. Kerosene, charcoal, and transport are necessary goods in urban areas, while electricity, petrol and diesel are luxury goods.

Figure 15-3 reports the concentration curves for rural areas, which suggests that a tax on firewood is the most regressive of the fuel taxes, followed in descending order of regressivity by taxes on kerosene, transport, charcoal, petrol/diesel, and electricity. On this score there is much similarity with urban areas. However, the income elasticity of these fuels shows slight variation from the urban pattern. Figure 15-3 shows that firewood is a necessity in rural areas: the concentration curve is slightly below the 45-degree diagonal line, and its concentration index is found to be 0.05 and significantly different from zero.

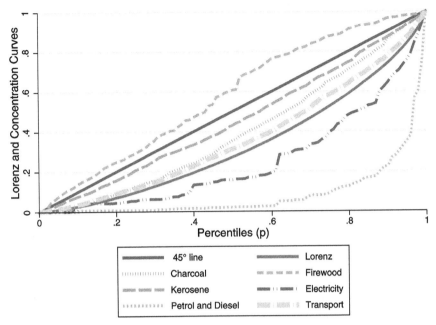

FIGURE 15-2 Relative Progressivity Using Concentration and Lorenz Curves, Urban Areas.

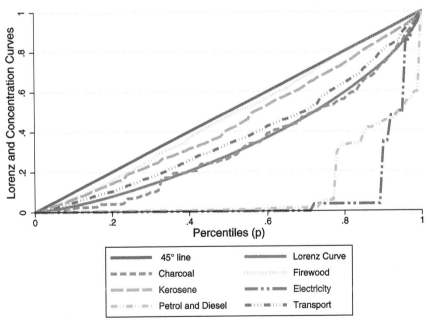

FIGURE 15-3 Relative Progressivity Using Concentration and Lorenz Curves, Rural Areas.

Kerosene and transport are also found to be necessary goods in rural areas. Charcoal is on the borderline between a necessary and a luxury good, given that its concentration curve closely tracks the Lorenz curve. This is in fact consistent with the finding in Table 15-2, where the Kakwani index for charcoal, which is given by twice the area between the concentration curve and the Lorenz curve, is found to be statistically insignificant.[2] This finding makes sense in that charcoal in rural areas is not as commonly used as firewood, and it is slightly more popular among relatively well-off households. Electricity and petrol/diesel are also found to be luxury goods in rural areas.

The results reported here are roughly what one would expect. To some extent they are in keeping with findings by other studies on Tanzania using different data sets and findings originating from other African countries. For example, Sahn and Younger (2009) have used data from two unofficial surveys—the Human Resources Development and ERB/Cornell surveys—to assess progressivity of certain taxes using the dominance of concentration curves and the extended Gini coefficient. They found that a tax on kerosene is neutral while taxes on petrol and diesel are progressive. The surveys they used did not contain information on firewood, charcoal, and electricity. Surprisingly, however, Sahn and Younger (2009) found that a tax on transport services is either neutral or progressive, a finding which is at odds with our finding, but consistent with their findings from Ivory Coast, Ghana, Guinea, Madagascar, South Africa, and Uganda in the same study.

Other studies on other African countries have found results that are broadly similar to ours. Chen et al. (2001) found that a petroleum tax in Uganda is progressive except for a tax on kerosene which is regressive. Younger (1993) also found that a tax on kerosene in Ghana is regressive but taxes on other petroleum products are progressive or proportional. Similarly, Younger et al. (1999) found that in Madagascar, a tax on petrol is highly progressive but a tax on kerosene is regressive.

Our findings suggest that taxes on petrol, diesel, and electricity are progressive if we only focus on the first order direct impact on the consumer. However, to the extent that tax incidence on petrol and diesel can be shifted to the consumer of transport services, fuel taxation would be regressive. Moreover, the analysis in this chapter does not take into account the behavioral changes that taxes on fuel may induce to consumers, and thus possible changes in consumption due to taxation are not taken into account. In any case, in spite of the fact that taxing electricity is progressive, such a tax is not necessarily desirable if it would induce a switch from consumption of electricity to that of charcoal or firewood, which might inflict bigger local environmental destruction. Furthermore, a full appraisal of the desirability of a tax must take into account both efficiency considerations and distributional concerns. This chapter focuses on the distributional issues only. Finally, since fuel is used as an intermediate input in many production processes, welfare impact of fuel taxation would not only have to take into account the efficiency and distributional considerations but also the possible overall inflationary impact to the economy. Because of these limitations, the results of this chapter cannot be taken as conclusive in terms of the distributional impact (because inflation also tends to redistribute) or general desirability of fuel taxation. It only sheds useful light on the first order impact of fuel taxation on distribution.

Conclusion

This chapter has used a fairly robust approach to assess first order distributional effects of fuel taxation in Tanzania. Taxes on firewood, kerosene, transport, and charcoal have a regressive impact on the distribution, while taxes on electricity, petrol, and diesel are progressive. Using the concentration and Lorenz curves, the chapter determines that firewood is largely an inferior good, particularly in urban areas, while transport and kerosene are necessary goods. Charcoal is a borderline luxury good in rural areas but a necessity in urban areas. Electricity, petrol, and diesel are luxury goods. There is basis, therefore, for asserting that taxing petrol, diesel, and electricity would not have a regressive impact on distribution. The repercussions of taxation may multiply, however, with some regressive tendencies. For instance, this chapter shows that taxing transport is regressive. This means that if a tax on fossil fuel is shifted onto higher transport costs, the multiplier effect of such a tax would be regressive with respect to transport. To capture the entire distributional impact of fuel taxation it is important to use a general equilibrium approach. The results of this chapter are therefore only a first order approximation.

APPENDIX 15-1

In estimating and bootstrapping the confidence interval of the Suits index, 58 observations were deleted because they had zero amounts in total consumption. Thus, the results presented here are based on a sample of 10,406 observations. The study bootstrapped 1,000 samples to estimate the confidence interval. Graphical inspection (density plotting) reveals that bootstrapped suit indices are fairly normally distributed (Figures 15-A1 to 15-A5).

TABLE 15-A1 Bootstrapped Confidence Intervals of Suits Index.

Suits index (Bootstrapped Mean and 95% Confidence Interval)

	Lower	Mean	Upper
Charcoal (national level)	0.10	0.13	0.17
Firewood (national level)	−0.36	−0.34	−0.32
Kerosene (national level)	−0.22	−0.20	−0.17
Electricity (national level)	0.45	0.52	0.59
All fuels (national level)	−0.16	−0.14	−0.12
Charcoal (urban areas)	−0.13	−0.09	−0.06
Firewood (urban areas)	−0.46	−0.39	−0.33
Kerosene (urban areas)	−0.29	−0.26	−0.22
Klectricity (urban areas)	0.25	0.33	0.40
All fuels (urban areas)	−0.16	−0.13	−0.11
Charcoal (rural areas)	−0.03	0.07	0.16
Firewood (rural areas)	−0.26	−0.23	−0.21
Kerosene (rural areas)	−0.18	−0.15	−0.12
Electricity (rural areas)	0.56	0.71	0.84
All fuels (rural areas)	−0.20	−0.18	−0.16

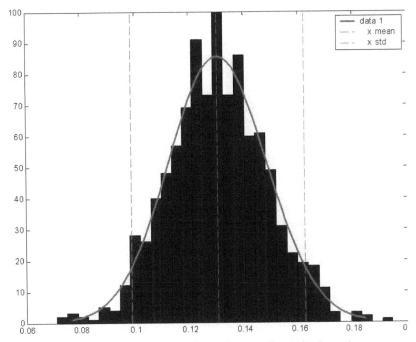

FIGURE 15-A1 Bootstrapped Suits Index for Expenditure on Charcoal in Tanzania.

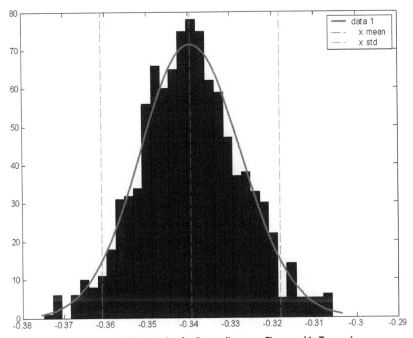

FIGURE 15-A2 Bootstrapped Suits Index for Expenditure on Firewood in Tanzania.

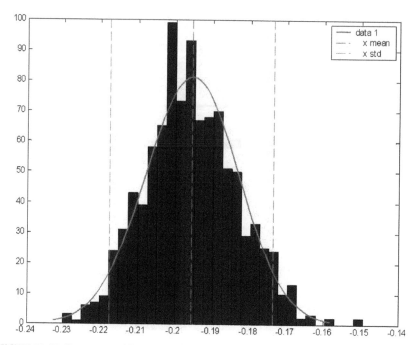

FIGURE 15-A3   Bootstrapped Suits Index for Expenditure on Kerosene in Tanzania.

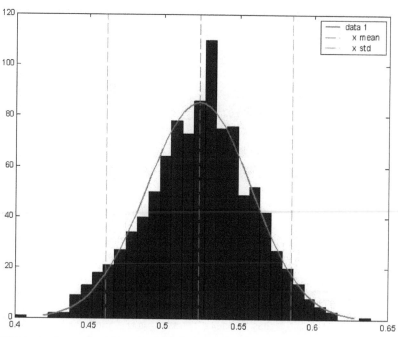

FIGURE 15-A4   Bootstrapped Suits Index for Expenditure on Electricity in Tanzania.

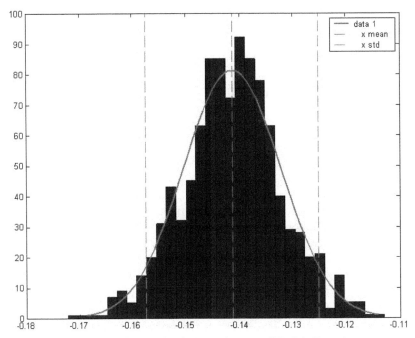

FIGURE 15-A5 Bootstrapped Suits Index for Expenditure on all Fuels in Tanzania.

Notes

1 The concentration index is given by twice the area between the concentration curve and the 45-degree diagonal line (see Kakwani 1977b). It ranges from 1 to −1. If the concentration curve overlaps the 45-degree line, the concentration index becomes zero.

2 It should be remembered that the Kakwani index is given by twice the area between the Lorenz curve and the concentration curve. When the Lorenz curve overlaps the concentration curve the Kakwani index would have a value of zero.

## CHAPTER 16

Distributional Effects in Europe

Thomas Sterner and Emanuel Carlsson

Europe is one of the most affluent of continents and of course the original home of the automobile society, even if the U.S. has taken over the leading position a long time ago. If global emissions are to be cut to meet even moderately ambitious climate goals, this will mean that global consumption of fossil fuels must decrease quite drastically. This must apply all the more so to Europe which, after the U.S., has one of the highest rates of carbon emissions per capita in the world, with a particularly large share of these emissions coming from the transport sector. It seems quite inevitable that we will need considerably higher prices for all fossil fuels to drive both behavioral and technical changes necessary to bring about such a transformation.

Europe has a self-proclaimed role as a leading agent when it comes to climate change. When the U.S. pulled out of the Kyoto Treaty, the agreement was saved by joint efforts from the EU and Japan. The European Union Emissions Trading System (ETS) has attracted much attention as the first mechanism that has really created a carbon price for industrial sectors. However, it is worth pointing out that the high taxes on fossil transport fuels such as gasoline and diesel in the EU are considerably higher than the EU ETS prices and corresponding fuel taxes in the U.S. Sterner (2007) analyzed the importance of these taxes in lowering emissions of greenhouse gases and showed that not only are they more important than the ETS, they have even had a significant effect on global concentrations of carbon dioxide. Clearly, these fuel taxes originally were not generally designed for environmental or climate change mitigation purposes, but their effect on the environment is still noteworthy.[1] The experience of fuel taxes in countries within Europe or Japan is in fact a full-scale demonstration of just how powerful such economic instruments can be.

Raising fuel taxes, however, meets stiff opposition. The past decade saw protests in a number of European countries. For instance, truckers and fishermen calling for tax cuts and lower fuel prices were perhaps the most prominent example (see, for example, *Guardian* 2008). An often heard argument is that fuel taxes are strongly regressive. Regressivity was found in some early studies in the U.S. during the 1980s and 1990s (Poterba 1991). Although it might seem natural to assume that most modern economies would see those same effects, in fact this empirical matter must be studied in each country and time period separately.

Only few earlier studies have focused on Europe. For instance, Santos and Catchesides (2005) concluded that the British fuel tax is regressive as long as only car-owning households are included but more or less neutral if one also includes those without a car (which would seem to be the more relevant analysis). Asensio et al. (2003) showed that for Spain, the largest burden from fuel taxation falls on middle-income earners. Bureau (2011) finds that fairly simple schemes of revenue recycling can make an otherwise regressive French fossil fuel tax progressive. In addition, he shows that reductions in traffic congestion that follow an increased tax would benefit the poor more than the rich, thus removing the regressive effect even further.

Our study fills a gap by comparing seven European countries. Five of them are chosen due to their size and importance in the political and environmental context—France, Germany, Italy, Spain, and the United Kingdom. To broaden the analysis, we also include Sweden and Serbia—two small countries with very different characteristics. The following section discusses fuel taxation in Europe, the subsequent one presents the data, followed by our distributional results for a number of European countries before we conclude.

Fuel Taxation in Europe

Fuel consumption is determined by demand functions that depend primarily on income and prices—in particular, fuel prices. As discussed in Chapter 1, average long-run elasticities are likely to be in the range of –0.7 to –0.8 for price, and unity or slightly above for income. The average tax on gasoline for Europe is very high compared to the U.S. and many other non-European countries. This is interesting from the viewpoint of climate policy, since it is clearly related to lower levels of fuel use and carbon emissions (see Sterner 2007).

There is, however, also likely to be some causation that goes in the opposite direction. One might wonder why European countries are more prone to raise fuel taxes than the U.S. It might be tempting to conclude that these countries are more environmentally oriented—but that would be overly simplistic. It is more likely that there is considerable path dependency. In countries where the population is used to high levels of (fuel) taxation, people and institutions adapt: they not only buy smaller vehicles, but town planning adapts by providing more services close to where people live. Likewise people select to live closer to work, bus companies are found to be profitable, and authorities run metros, railways, and other public transport. In such a context, officials face less resistance to new fuel tax increases and so forth. Other mechanisms of adaptation such as the profitability and desirability of research into smaller and more fuel efficient engines and so forth might be encouraged (see Hammar et al. 2004). As an anecdotal illustration, take Italy, which has always had the highest fuel taxation in Europe. This might be a reflection of a determined and aggressive environmental fiscal policy, or it could be that Italian goverments found other taxes harder to collect, while fuel taxation is fairly popular in a country where Fiat wants to avoid competition from German and other more fuel-thirsty cars. Whatever the original reason, the effect is of course that fuel use is low in Italy.

In the UK it was during the rule of Margaret Thatcher that Britain introduced the "Fuel Price Escalator," which transformed the UK from a country with fairly cheap fuel to one of the countries with the highest tax levels in Europe. Thatcher appears to have been one of the first European politicians who understood the gravity of climate change.

Data

We focus in this chapter on seven European countries whose tax levels are shown in Table 16-1 below. Columns 2 through 5 give tax rates and prices in US$ cts/liter, while columns 6 and 7 show the tax as a percentage share of the fuel price. As we can see, gasoline taxes are now well harmonized, with the slight exception of Spain. As for diesel, the disparity is bigger, with the UK having almost twice the tax of Spain. The share for Serbia is considerably lower.

The data we use to estimate the incidence of fuel taxes come from seven European household budget surveys (Table 16-2). From these we use information on households' disposable income, total expenditures, and specific expenditures on gasoline, diesel, public transportation, and taxis. This information is available by income deciles for all countries except Italy, where it was only possible to acquire the information by expenditure deciles, and Spain, where

TABLE 16-1 Prices and Excise Taxes on Fossil Fuels (in US$ cts/liter and as %) in Seven European Countries, 2008.

	Gasoline Tax	Diesel Tax	Gasoline Price	Diesel Price	Gasoline Tax as %	Diesel Tax %
France	121	93	198	186	61.1	50.0
Germany	128	100	205	195	62.4	51.3
Italy	116	94	202	196	57.4	48.0
Serbia					34.9	19.9
Spain	90	67	162	165	49.4	40.6
Sweden	118	104	191	202	61.8	51.5
United Kingdom	123	124	196	215	61.7	57.7

Note: Only national excise taxes included, excluding value added tax (VAT).

Sources: IEA Energy Prices and Statistics, for Serbia: UNECE (2007a) and GTZ (2007).

TABLE 16-2 Sources of Data.

Country	Name of Survey	Provider	Year
France	Enquête Budget de Famille	Institut national de la statistique et des études économiques	2006
Germany	Wirtschaftsrechnungen	Statistisches Bundesamt Deutschland	2006
Italy	I consumi delle famiglie	Istituto nazionale di statistica	2006
Serbia	Living Standard Measurement Study	Republički Zavod Za Statistiku, Republike Srbije	2007
Spain	Encuesta de Presupuestos Familiares	Instituto Nacional de Estadística	2006
Sweden	Hushållens utgifter	Statistiska Centralbyrån	2004–2006
United Kingdom	Family Spending	Office for National Statistics	2006

only income quintiles were available (the point estimates were not significant for deciles). The data are furthermore adjusted according to an equivalence scale, that is, adjusted for each household's size and composition (number of children and adults, etc.).

In addition to increasing fuel prices, fuel taxes also change prices of other goods and services. In order to come up with a *total tax* burden, one would have to include this *indirect* effect in the analysis, in addition to the *direct* effect of households purchasing fuel at the pump. A more ambitious attempt to do so would be to use an input-output model together with detailed information of fuel use for every industry in the economy. Such an analysis is done for India Chapter 8 by Datta, who finds auto fuel taxes to be progressive in India. However, doing such an analysis for all seven countries in this chapter would require input-output tables for each country. We choose a strongly simplified approach and only include indirect fuel use through public transportation and taxis, which are affected when fuel taxes are increased (see Metcalf 1999). To do so, we needed the share of fuel tax costs in expenditures on public transport. Such data are hardly available at the national level; for illustrative purposes we have therefore chosen to estimate them.[2]

The shares differ due to various nation-specific factors, e.g. the rate of fuel tax, the type and price of the fuel used, and other costs in public transport and taxis. Moreover, in some countries (like the UK and France), fuel subsidies are given for public transportation and taxi companies. Table 16-3 summarizes the estimated shares used in our calculation.

TABLE 16-3 Share of Taxation on Fuel in Expenditures on Bus and Taxi for Seven European Countries.

Expenditure variable	Type of fuel	France	Germany	Italy	Serbia*	Spain*	Sweden	United Kingdom
Bus	Gasoline	8.3%	8.8%	6.3%	7.0%	6.3%	9.2%	12.4%
	Diesel	5.6%	4.0%	4.6%	4.0%	4.6%	9.5%	5.9%
Taxi	Gasoline	**	4.6%	5.0%		5.0%	2.1%	3.9%
	Diesel		3.3%	3.7%		3.7%	1.9%	4.1%

Note: We did not find data for these countries and therefore made some assumptions. For *Serbia*, the share of fuel in public transport expenditures is assumed to be 20%. For *Spain* we used the same shares as for Italy, since that country had the closest income level. In addition to the shares above, we have also used information on the share of each fuel type for bus and taxi for all seven countries (from national databases on the vehicle stock). Using these weights, we calculated the indirect tax burden for each decile and country. Figures for Serbia include both bus and taxi. **In France, taxis are exempt from fuel tax for up to 5,000 L per year.

Source: Authors' calculations based on information from various national transport statistics and sources.

The Distributional Effect of Fuel Taxes in Europe

Starting with the reported income approach[3] (see Figure 16-1) an upward sloping line would suggest progressivity and a negatively sloping line would suggest regressivity. The results are quite mixed, but the first general impression is that effects are small, and in many countries the tax appears reasonably close to proportionality. This applies in particular to deciles 3 through 8. The richest two deciles have in several countries slightly lower shares but this does not apply to Serbia. It seems there is some overall regressivity for Sweden and United Kingdom, while Serbia shows progressivity. In Germany the middle income earners seem to bear the largest burden.

Visible in Figure 16-1 is also a difference in the level of burden *across* countries. Households in France seem to bear a relatively low burden while the fuel tax in United Kingdom, Germany and Spain constitutes a larger share of each decile's average disposable income. Partly this reflects the level of taxation which is highest in the UK. Taking the national averages across all deciles, France has an average burden of 0.78% and the United Kingdom 2.20%.

We turn next, in Figure 16-2, to exactly the same data—but now as shares of total expenditure rather than income. This allows us to bring in Italy (not included in Figure 16-1 due to missing income data) where, as in the UK, the total fuel taxes are high and hence incidence on average also. Otherwise figures are fairly similar, but the overall impression is one of greater progressivity in these figures. This is also confirmed when we look at the overall Suits (1977) measures in Table 16-4 (as well as Figures 16-A1 and 16-A2 of the concentration curves, which can be found in Appendix 16-1). The Suits coefficients

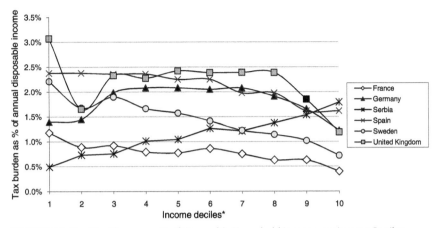

FIGURE 16-1 Fuel Tax Shares as a % of Disposable Household Income, per Income Decile.[a]

Source: Authors' calculations on data from seven national statistical authorities (see Table 16-2).
[a] Includes both direct fuel purchase and indirect use through public transport.

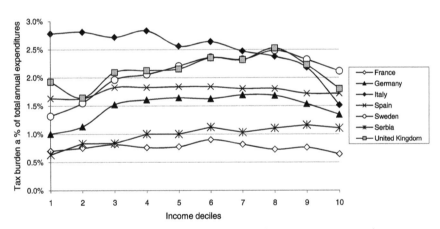

FIGURE 16-2 Fuel Taxes as a % of Total Household Expenditures,[a] per Income Deciles.

Source: Authors' calculations on data from seven national statistical authorities (see Table 16-2).
[a] Includes both direct fuel purchase and indirect use through public transport.

are all negative—that is, they illustrate regressivity when we use a traditional income measure. (The exception is Serbia—a lower income country with a smaller share of the population having access to cars (see Table 16-4), where it is perhaps unsurprising that fuels have more of a "luxury" character and hence fuel taxes there are more progressive.) However, when expenditures are used as a proxy for lifetime income, we find that the picture is much more mixed. In both Germany and Sweden fuel taxes are actually now progressive instead

TABLE 16-4 Suits Coefficients for Seven European Countries.

Type of Income	Suits Coefficient for Each Respective Country						
	France	Germany	Italy	Serbia	Spain	Sweden	UK
Total, temporary	−0.157	−0.067		0.172	−0.086	−0.178	−0.125
Direct, temporary	−0.155	−0.066		0.187	−0.086	−0.171	−0.123
Total, lifetime	−0.024	0.008	−0.110	0.051	−0.002	0.064	−0.004
Direct, lifetime	−0.021	0.009	−0.110	0.066	−0.002	0.072	−0.003
GDP/capita	30,150	31,571	29,406	5,713	27,542	31,264	31,585
Gini coefficient	27	27	32	30	31	24	32
Gas consumption	224	354	292	132	213	545	396
Population Density	111	232	195	112	86	20	248
Cars per capita[y]	495	546	195	161	455	455	439
Urban %[z]	77	75	68	52	77	84	90

Sources: Progressivity coefficients: Authors' calculations on data from seven national statistical authorities (see Table 16-2); GDP/capita: (PPP-adjusted, current international US$) IMF (2009); Gini coefficients: For all but Serbia, 2006 from Eurostat (2009), and for Serbia, 2003 from CIA (2009); Motor gasoline consumption: (liters/capita, 2005) WRI (2009a). Population density (people per square kilometer, 2005) WRI (2009a); [y] Passenger cars per 1,000 People: 2003 for all countries but Serbia, WRI (2009b) (1999), and for Serbia, UNECE (2007b); [z] Urban population as % of total population: WRI (2009c).

of regressive. Still, for practically all the countries except Italy, values are very small, and the tax is best characterized as proportional on average.

Table 16-4 allows us to study the difference between direct and total effects, which are however in all cases very small. The addition of indirect effects thus does not turn out to make a big difference in these countries (as it does in some very low income countries). Table 16-4 also provides some additional comparative data on the countries, income, inequality as measured by Gini, population density, vehicle density, and degree of urbanization. Yet these variables show limited variation, with a few exceptions such as income (which is much lower in Serbia). Sweden is an outlier when it comes to low population density but still has a high degree of urbanization and high gasoline demand per capita. These factors might contribute to making a fuel tax somewhat more regressive—although the regressivity in Sweden still only applies to the case when yearly income and not lifetime income (expenditures) are used. Sweden is also the country where the overall Gini coefficient is the lowest—reminding us that general economic policies are much more important for equality than the results of a specific tax such as on fuel.

Conclusion and Discussion

Fuel taxation is an effective and potentially important instrument in dealing with climate change. Those who disfavor fuel taxes often claim they are strongly regressive. Earlier studies have shown that this depends on the country studied and on the details of the methods used (whether lifetime or temporary income is used, or substitution or other adaptations are allowed for in the analysis, and so on). There is a tendency to progressivity in low-income countries but regressivity in high-income countries. We use fairly simple methods to study a larger selection of European countries. We find that when using traditional income as a measure, there is indeed some regressivity in most of the European country studies. However, it is so small that the tax can for practical purposes be considered broadly proportional or neutral. In the lowest income country, Serbia, the tax is instead progressive. The inclusion of indirect consumption through public transport has little or no effect on results. The use of lifetime income (as proxied by total expenditures) does however more or less reverse results, and the tax is now found to be neutral or proportional (or weakly progressive in some countries and weakly regressive in others). As shown by West and Williams (2004), inclusion of adaptive mechanisms such as elasticities—and in particular individual elasticities for each decile and the use of theoretically preferable measures such as equivalent variation instead of ordinary consumer surplus measures—tend to weaken regressivity or make results even more progressive.

The reader should note that there are also distributive effects of changes in pollution, transitional effects on various markets, changing property prices, and so on (see Fullerton et al. 2008). As an example, Bento et al. (see Chapter 3) study effects through the used car market, but that line of inquiry is outside the scope of this particular study. Most important is the possibility of using proceeds of a gasoline tax in some way. Either the budget will be strengthened or public expenditures increased; also, other taxes could be lowered. Depending on which combination of policies is chosen, small distributional effects could quite easily be corrected if the policymaker so desires. Ahola et al. (2009) analyze the distributional consequences of using increased fuel taxes to lower the VAT on food in Sweden using the same data as in this chapter. Employing the West and Williams (2004) index—which can be seen as a variation on the Suits index to evaluate tax reforms—they show that a balanced tax reform with increased fuel taxes, combined with a lowering of VAT on food (for the corresponding amount of fiscal revenue), would indeed be an overall progressive tax reform. The European experience has demonstrated two things: first, on the whole, fuel taxes are effective at reducing fossil fuel emissions in the long run; second, the distributional consequences have been, in the larger scheme, fairly negligible. Fuel tax increases do of course cause protests, but on the other hand this applies to all taxes. Seen from a U.S.

perspective, the more interesting and striking fact is that such high tax levels do not engender more protests.

Acknowledgments

Valuable comments from Francisco Alpizar, Ashokankur Datta, and an anonymous referee are gratefully acknowledged.

Appendix 16-1

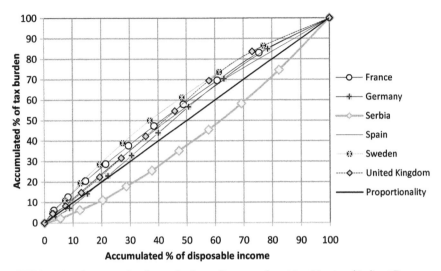

FIGURE 16-A1 Concentration Curves for Seven European Countries: Direct and Indirect Tax Burden Plotted against Annual Disposable Income.

Source: Authors' calculations on data from seven national statistical authorities (see Table 16-2).

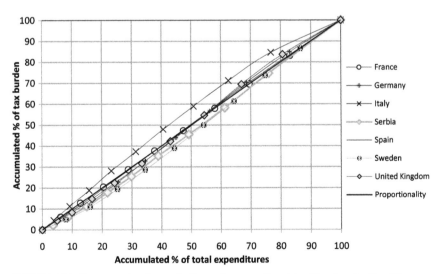

FIGURE 16-A2 Concentration Curves for Seven European Countries: Direct and Indirect Tax Burden Plotted against Total Expenditures.

Source: Authors' calculations on data from seven national statistical authorities (see Table 16-2).

Notes

1   The stated motives for gasoline taxes vary considerably. In some countries they are just a convenient tax base. In others they contribute to road building and maintenance. These vary geographically; Parry and Small (2005) question their level for such purposes. Historically, climate externalities have played a small role (if any) in motivating gasoline taxes—but the taxes play a big role in reducing emissions of GHGs.

2   This estimation was possible for France, Germany, Italy, Sweden, and UK. In a first step, we calculated the fuel tax per vehicle km by multiplying the fuel tax per liter with an assumed level of average fuel consumption per km (0.4 L/km for buses and 0.1 L/km for taxis). The share of fuel tax in public transport expenditures was then given by dividing the fuel tax per km by the revenue per km for either buses or taxis. For buses, information on total revenues and vehicle km were directly available from national statistical agencies, while the revenues per km for taxis were calculated from a small sample of companies in each country.

3   Burden is defined as the share of fuel (direct and indirect) in disposable income.

## CHAPTER 17

Who Pays Taxes on Fuels and Public Transport Services in the Czech Republic? *Ex Post* and *Ex Ante* Measurement

MILAN ŠČASNÝ

A fter the so-called Velvet Revolution in 1989, the Czech economy started to be transformed from a centrally planned system into a market economy. This movement has brought many changes in the economy, but also in behavioral patterns of consumers. Indeed, passenger transport is not an exception. Its performance has changed with respect to volume and structure.

First, we experienced a boom in the number of passenger cars. Their numbers increased by 70% during 1993–2008, from 274 up to 467 cars per 1,000 people (Table 17-1). Despite a growing public transport infrastructure over the last 20 years, the number of passenger cars per kilometer of road network increased by 73%, from about 50 to 88 cars. Transport performance also increased in terms of volume; while the performance of passenger transport had been less than 70 billion pkm (passenger km), by 2003 it was more than 90 billion pkm, and in 2008 it reached the 100 billion level. Except for air transport (which increased by almost 500%), individual car transport grew most (by 48% during 1993–2008), keeping its share of total passenger transport performance within the range of 71% to 78%. In contrast, the share of public road transport continuously decreased from about 13% in 1993 to around 9% during the last five years. The most dramatic change, however, appeared in railways. Their share of passenger transport performance shrank from 12% at the beginning of the 1990s to the 7% level where it has stayed since 2002. Transport performance in volume—as measured by pkm per 1000 inhabitants—remained constant only for public road transportation covering buses and urban public transportation systems. On average, each person travelled about 900 km per year by them, with some lower performance at the end of the 1990s. In contrast, the use of personal cars increased from less than 4,800 pkm in 1993 to almost 7,000 pkm in 2008.[1]

TABLE 17-1 Performance in Passenger Transport in the Czech Republic.

		1993	1995	2000	2005	2008
Transport performance in passenger transport						
Total (in billion pkm*)		68.9	73.2	86.4	93.7	99.3
Car transport	%	71.1	74.4	74.0	73.3	72.9
Public road	%	13.2	10.5	10.8	9.2	9.4
Railway	%	12.4	10.9	8.5	7.1	6.9
Air	%	3.3	4.1	6.8	10.4	10.8
Transport performance in passenger km per inhabitant						
Car transport	pkm	4,743	5,275	6,228	6,707	6,914
Public road	pkm	880	742	911	841	893
Railway	pkm	828	776	711	651	650
Air	pkm	218	293	570	951	1,027
Passenger cars and light-duty vehicles up to 3.5 tons						
per 1,000 inhabitants		274	295	335	387	467
per km of road network		50.6	54.8	62.1	71.3	87.7

Source: CDV (2009). * pkm = passenger kilometer.

All of these trends have led to an increasing environmental burden whether measured by pollution, use of fossil fuels and space, noise annoyance, or congestion (CDV 2009). Despite the negative external effects of transportation (Verhoef 1994) and growing demand for investment in transportation infrastructure, regulation of passenger transport by market-based instruments has not become stricter. Indeed, the excise tax that presents about 40% of the final price of fuel in the Czech Republic has not been permanently nominally adjusted for inflation—even when the inflation rate reached levels of 8% to 10% per year. Because the rate of excise tax on fuels was increased slightly and only a few times over the last 15 years, its real rate was continuously decreasing. Except for when world oil prices escalated in 2000 and 2001, the final price of fuels has remained more or less constant during the entire period of 1995–2008, that is, at 80% of its 1993 level.

New impetus in more than just transport pricing was not taken by the implementation of the 2003/96/EC Directive that set minimum tax rates for petrol and diesel at levels even smaller than actual rates at the time of Directive implementation in the Czech Republic. Nor did several attempts of the Czech government to introduce an environmental tax reform, all of which failed to be introduced into the Czech tax system (see Brůha and Ščasný (2006a) and Ščasný and Máca (2010) for a review).

However, higher taxation on fuels has still remained on the political agenda in the Czech Republic. Environmentalists primarily use green rhetoric to support higher fuel taxation, while others call for any additional revenues for the State Fund on Transportation Infrastructure to be used to support new investment. Meanwhile, some economists consider higher taxation on fuels a good source of public finance stabilization. Others argue quite strongly against such an increase, especially due to the fear that a fuel tax would hurt poorer households.

The aim of this chapter is to address these conjectures and fears by examining the distributional impacts of transport taxation for the Czech Republic. We first measure tax progressivity of several taxes *ex post* by both the Suits (1977) index and the newly developed Jinonice index covering the last 16 years, a period of economic and social transformation. Then we use a micro-simulation model and the results from a properly estimated household demand system to analyze the distributional impacts of fuel taxation and of taxation on public transport services. We conclude that although higher fuel taxation as analyzed in this chapter increases household expenditures and reduces welfare by 0.7% and 1.5% of total expenditures on average, the effect on income distribution and tax progressivity is very small. Both of the impacts on household expenditures and welfare are distributed evenly across income deciles, although these impacts vary more if we examine tax incidence in the household segments defined according to family status and size of residence. Since we simulate the impacts for each household included in our rich database separately, the distributional effects can be evaluated in every detail and for various household segments. We believe that such results can provide useful information to decisionmakers for targeting possible mitigating measures or redesigning policy.

The rest of the chapter is organized as follows: The next section describes our data. We then summarize methodology and methods, starting with a description of our micro-simulation model, continuing with household demand estimations, and ending with inequality and progressivity measurements. A measure of tax progressivity *ex post* follows, and then the subsequent section measures the distributional impacts *ex ante* for several policy scenarios on fuel and transport service taxation. The final section offers conclusions.

Data

In our analysis we use a comprehensive micro database from the Household Budget Survey (HBS) collected regularly by the Czech Statistical Office. The HBS includes information about household annual expenses on several hundred consumption items, income from various sources, possession of durable goods, home characteristics, and other socioeconomic data about household members. However, with the exception of certain food and clothing items, the survey never asked about consumption of nondurable goods expressed in physical units, such as liters of fuels used or kilometers driven. Households included in the survey are selected using the nonprobability quota sampling technique. Annual samples have on average 2,700 to 3,000 observations. The variable we call "PKOEF" reflects how each household is represented in the entire Czech population, allowing us to compute aggregate statistics and make country-wide predictions. Our dataset as compiled from HBS covers the period 1993–2008 and includes 44,432 observations. We use the 2005 sample of 2,877 observations for *ex ante* simulations using our micro-simulation model.

One should be cautious when interpreting the results for 2007–2008 reported below. The Czech Statistical Office changed its sampling strategy in 2006, mainly to include households of the unemployed as well as households living in municipalities with fewer than 20,000 inhabitants—groups that were underrepresented in the earlier samples. Proper interpretation should therefore consider the periods 1993–2006 and 2007–2008 separately.

Interestingly enough, transportation expenditures of Czech households were quite stable over the entire period of economic transformation during 1993–2008. During the last 15 years, the households surveyed spent annually on average CZK 11,400 to 13,400 on motor fuels, CZK 2,100 to 2,500 on urban public transport, CZK 1,700 to 2,300 or CZK 800 to 1,400 on buses or rail, respectively.[2] Thanks to the continually increasing wealth of Czech households, the share of transportation expenditures decreased; for instance, expenditures on fuels decreased from 5.5% of total household expenditures in 1993 to 4.8% in 2008. The share of expenses on other transportation modes remained more or less constant at the level of 1%, 0.9%, and 0.5% for city public transport, bus, and rail, respectively (see Table 17-2). The share of those who did not consume rail at all remained stable over the period, however, the share of households with some expenses decreased from 65% to 60% for city public transport and from 75% to 62% in the case of buses. Simultaneously, the share of those with some expenditure on fuels increased from about 62% to 74% during the period of 1993–2008.

However, on average, the consumption of motor fuels rose from 1993 to 2003, when it was almost 60% more than in the year 1993. The reason for increasing consumption with simultaneously constant real expenses over that period can be found in a significantly larger inflation rate (9% to 10% during 1993–1998) compared to the growth rate in real fuel price.

As real expenditures on fuels were constant over the period 1993–2008, the only factor that led to an increase in total fuel consumption was an increasing penetration of personal vehicles within the households. Indeed, while only slightly more than 50% of the households owned a personal vehicle at the beginning of the 1990s, since 2001 that figure has risen to around 60% (see Table 17-1).

Ownership of a personal vehicle and consumer expenditures on transportation might, however, have different tendencies within different household groups. We therefore look at household segments defined by net total income per household member (income deciles) and then by family status and size

TABLE 17-2 Household Expenditures on Transportation during 1993–2008.

Year	N	Expenses per Family Member (1,000s CZK)	Had a Vehicle	Fuel Consumption (Liters)	Had Expenses On … (In %)				Expenditures as % of Total (If Any)			
					Fuels	Urban Transport	Bus	Rail	Fuels	Urban Transport	Bus	Rail
1993	2,922	183	53	217	63	67	75	49	5.55	1.14	1.04	0.42
1994	2,409	187	51	233	62	64	75	52	5.39	1.07	0.98	0.41
1995	2,387	202	52	262	64	64	74	53	5.02	1.01	0.93	0.44
1996	2,456	215	53	277	66	63	73	53	4.82	1.03	1.05	0.47
1997	2,441	226	56	293	67	65	69	53	4.88	0.99	1.03	0.46
1998	2,377	219	56	303	67	64	68	53	4.67	1.10	0.93	0.54
1999	2,457	227	58	311	68	66	66	53	4.69	1.04	0.96	0.54
2000	2,994	223	59	286	69	65	66	52	5.25	1.06	0.97	0.55
2001	3,045	224	61	313	70	64	65	51	5.13	1.00	1.03	0.56
2002	3,038	225	61	328	71	63	65	48	4.71	0.97	0.98	0.60
2003	2,760	235	62	344	71	64	63	47	4.72	1.02	0.78	0.45
2004	2,883	237	63	337	72	64	64	46	4.89	1.02	0.94	0.50
2005	2,877	232	62	333	71	64	61	45	5.30	1.07	0.94	0.54
2006	2,752	243	62	339	73	61	65	45	5.12	0.94	0.90	0.51
2007	2,963	255	61	343	74	60	63	46	4.82	0.94	0.89	0.53
2008	2,934	245	61	352	74	59	62	46	4.83	0.94	0.88	0.55

of residence. The reason of this choice is based on our expectation that having children might involve specific transportation patterns, and living in a large municipality can determine infrastructure availability. We therefore expect wider heterogeneity in consumer patterns of the household segments according to the following definitions. Based on the factor analysis provided by Brůha and Ščasný (2005), we define overall 13 household segments. Specifically, we identify farming families, families with retired person(s), and families with one (EA1) or two (EA2) economic-active persons with or without children (+).[3] For each of these groups, we further define whether the household is living in a municipality with fewer than or more than 20,000 inhabitants (labelled "small" or "large"). For the households of pensioners, we delineate three sizes of households. For instance, the label "EA2+_large" indicates a household with two economically active persons with at least one child living in a town or a city with more than 20,000 inhabitants.

The budget shares on fuels were in the range of 2.5% to 4.5% of total expenditures, and their shares increased across the higher deciles over the entire period analyzed (Figure 17-1), indicating that expenses were larger for richer households not only nominally but also relative to a household's wealth. We also observe the same increasing trend, though weaker, for the budget shares on public road transport (urban transport and buses). Increasing budget shares across deciles only hold, however, if we consider all households, i.e., even those that did not spend anything on the items listed. If we only consider those on the market with non-zero expenditures, the budget shares are more or less constant across income deciles (see Table 17-3).

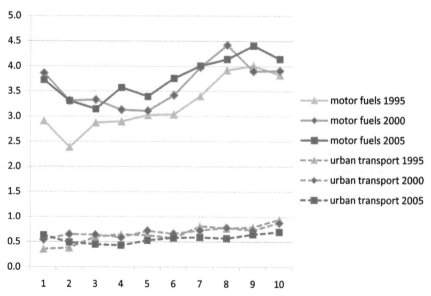

FIGURE 17-1 Budget Shares on Fuels and Urban Transport by Deciles, 1995, 2000, 2005.

TABLE 17-3 Household Expenditures on Transportation for Income Deciles and Household Groups, 2005.

Decile	Had Vehicle (%)	Had Expenses on... (% of households)				Expenses as % of Total (for Expenses > 0)			
		Fuels	Urban Transport	Bus	Rail	Fuels	Urban Transport	Bus	Rail
1	61	69	55	67	44	5.4	1.0	1.8	0.6
2	51	58	59	62	47	5.7	1.3	1.1	0.6
3	51	59	58	57	36	5.3	1.0	0.7	0.6
4	58	68	60	60	41	5.3	1.0	0.8	0.5
5	56	70	58	65	47	4.9	1.0	0.6	0.6
6	62	71	65	64	50	5.3	1.0	0.9	0.5
7	66	71	66	60	48	5.6	1.1	1.2	0.4
8	71	79	67	57	45	5.2	1.2	0.8	0.6
9	72	82	71	61	48	5.4	1.0	0.8	0.5
10	72	82	76	58	49	5.1	1.1	0.7	0.5
Household groups									
Farmer_small	82	90	32	75	34	5.7	0.3	1.4	0.6
Farmer_large	82	88	43	70	44	6.2	0.6	1.8	0.6
Retired_small	48	65	27	73	32	5.4	0.4	0.5	0.4
Retired_medium	44	58	37	67	36	5.2	0.2	0.5	0.4
Retired_large	35	44	64	47	45	4.3	1.0	0.5	0.4
EA1_small	43	60	34	77	28	6.7	0.7	2.5	0.7
EA1+_small	82	86	47	70	45	6.6	0.7	1.4	0.5
EA2_small	84	93	46	72	38	6.7	0.7	1.5	0.5
EA2+_small	95	97	56	83	49	6.7	0.6	1.5	0.7
EA1_large	35	49	67	54	43	5.3	1.7	1.2	0.9
EA1+_large	65	72	77	58	49	5.0	1.1	0.8	0.5
EA2_large	83	89	74	61	44	5.4	1.0	0.8	0.4
EA2+_large	84	89	77	63	54	4.9	1.2	0.9	0.5

Table 17-3 reports averages of several transport-related variables for several household groups for 2005. We found, for instance, that the wealthier the household, the more likely the household owns a personal vehicle, consumes more fuel, and has expenses on city public transport and rail. This, however, does not hold true for expenses on bus transport, which did not vary across the deciles. If we consider the households that had non-zero expenses on fuel or one certain transportation service only, the share of these expenditures on total expenditures is similar across all income deciles and over the entire period.

As expected, the share of those with expenditures on urban public transport is greater in households living in larger municipalities, while households living in smaller towns and villages spent more on buses. The use of rail is least frequent, especially in the single families and families of pensioners living in smaller municipalities. The average values of personal vehicle ownership, the share of those who had some expenditures on fuels or public transport, and the share of these expenditures on total expenditures of a given household segment are reported in detail in the lower part of Table 17-3. Whether price responsiveness or tax incidence differs across these household segments will be further examined in the next section.

Methodology

We assess distributional effects of transport price changes by using DASMOD, a static micro-simulation model of the Czech Republic (see Brůha and Ščasný 2006b; Ščasný and Brůha 2007). The way this model has been constructed allows us to predict price changes for several types of energy—electricity, solid fuels, natural gas, and centrally supplied heat—as well as for motor fuels and three types of public transportation, namely urban public transport, intercity buses, and rail.

The DASMOD micro-simulation model predicts changes in expenditures and consumption of these non-durable consumption items when new quantity demanded q is calculated as follows:

$$q_k^1 = q_k^0 (1 + \Delta P_k)^{(1 + \eta_{kk})} \times \prod_{c \neq k} (1 + \Delta P_c)^{\eta_{kc}} \times (1 + \Delta Y)^{\eta_k^y} \qquad (17.1)$$

where the subscript denotes the commodity, the superscript denotes policy (with 0 for the initial state), P is a vector of prices, Y is total income, and η_{kc} is the elasticity of demand for good k with respect to the price of good c.[4]

One specific feature of the DASMOD model is the possibility of simultaneously predicting changes in energy or fuel taxation on one hand, and changes in taxation on the factor of labor on the other. The model's flexibility with respect to policy scenarios is ensured by the possibility of simulating the impacts of various revenue recycling options, including a revenue-neutral environmental tax reform. Specifically, the model allows simulating the revenue recycling via personal income tax rate cuts, changes in tax deductibles or provision of tax credit, or changes in obligatory social security contributions paid by employer, employee, or both. Moreover, a provision for social transfers—income tested or not—to mitigate adverse social effects of a policy on certain households can also be part of a policy scenario for making a particular tax incidence prediction. To ensure consistency in the tax incidence evaluation, the model first predicts the labor taxes for set parameters of labor taxation (that can actually differ from those recorded in the database), and then predicts the labor taxes for considered changes in the parameters of labor taxation as proposed in the evaluated policy scenario.

Using predictions about changes in paid energy, fuel, and labor taxes—including proposed social benefits—the model predicts effects on public finance, specifically the effect on public revenues and the dead-weight loss.

This model was built from micro-level data from the HBS. A recent version uses 2005 data, but basically it can be run on a dataset for any year of the HBS. The predictions for all endogenous variables are performed at the lowest possible level, i.e., we predict changes for each household included in the model database separately. The predictions for each individual household are then clustered according to predefined household groups and segments such as deciles, and aggregated for the entire Czech population by using the PKOEF variable.

The DASMOD model offers several options for optimization. It can compute the implicit rates or other parameters of labor taxation, or the level of social benefits to reach revenue-neutrality, or keep the household budget, or keep overall welfare at the macro level unchanged (see Ščasný 2006).

Another specific feature of this model allows us to compute the effects of policy both on household expenditures and the cost-of-living index (CLI). In fact, both of these indicators provide useful but different information for decisionmakers. Changes in household expenditure patterns may inform a policymaker about expected fiscal impacts and environmental effects. Because the change in energy expenditures may determine investment in energy saving, the predictions of household energy expenses can serve as a useful indicator for the possible targeting of social mitigation measures or for considering a support measure that enhances energy saving installations within households.

In comparison, the CLI-based measures provide information about welfare loss or benefit induced by intended policy. Overall changes in welfare inform a policymaker about economic (Pareto) efficiency and desirability of a proposed policy. In the most recent version of DASMOD, the welfare changes are approximated by the geometric mean between the Paasche and Laspeyres CLIs.[5] Effects on the environment are evaluated through the impact on consumption of environmentally-related goods such as use of electricity or fuels. To provide a more comprehensive picture of the involved benefits, the external costs attributable to the changes in physical consumption of environmentally harmful goods such as energy or fuels are quantified by using unit damage factors as quantified for the Czech emission sources by the ExternE method. Benefits from the change in environmental damage are then added to the welfare impacts calculated for the price changes.

Responsiveness and Household Demand Estimation

One can hardly expect that under new conditions from a prospective policy, a household will consume as much as it consumed without this policy. As a matter of fact, proper evaluation of distributional effects of policy should not ignore the behavioral responses of individuals on price and wealth changes connected to a policy. To consider the behavioral responses of the consumers, key parameters of household demand have to be plugged into any simulation or prediction model.

Empirical studies that provide such estimates may utilize various econometric models, data, and approaches which can result in a wide range of elasticity estimates.[6] The estimates of elasticities may also vary for the natural reason that people are just different. As a consequence, the response to price and income changes might differ for different people (see Ščasný (2006) for a discussion). Therefore, in order to consider heterogeneity in consumer behavior and across different household segments, the DASMOD model can be run with several sets of elasticities estimated separately for various household groups.

Specifically for Czech households, the elasticities of demand for motor fuels were estimated for the first time by a time-series technique and using macro data over the period 1993–2003 for income deciles by Ščasný and Brůha (2003). Their subsequent empirical research (Brůha and Ščasný 2005) provides the elasticity estimates for families of pensioners, families of farmers, and seven household groups of economically active persons (EA) then further differentiated by income level. They specifically estimate a log-linear specification of demand given as

$$d_{it}^n = \sum_j \varepsilon_{ij}^n p_{jt} + \eta_i^n y_i^n + \sum_k \beta_{ki}^n x_{kt}^n \qquad (17.2)$$

with the coefficients for price of j type of goods (p), for total income (y) (that is replaced in the 2004 work by total expenditures to solve intertemporal problems of household), and for durable goods and other relevant exogenous variables (x). They specifically estimate a system for n goods (electricity, gas, coal, and motor fuels) by ordinary least squares (OLS) for i household groups and by using the means for each year during t years (1993–2003). Because there are only few yearly observations, the standard tests on residuals have a weak power, and the used data may be contaminated. Ščasný and Brůha (2003) therefore run the Least Absolute Deviation (LAD) estimator for each regression to check whether the results would be drastically different. In most cases, the OLS and LAD estimators give very similar results, and in all other cases of striking differences, the OLS point estimations lie in the 95% confidence interval of the LAD estimation. To address the issue of the correct functional specification of the energy demand functions the Box-Cox regression is run for each regression (for each good and each household group). The likelihood ratio statistics do not reject the null that $\lambda = 0$ for any regression on a 5% confidence level supporting the choice of using the log-linear specification.

Table 17-4 reports the elasticity estimation results for motor fuel demand. The average of price elasticity is –0.504 and the null about equality of the price elasticity estimates for each decile, and the average value is not rejected on a 5% level. This is not the case for the estimates of income elasticity, however, which vary more across the deciles with their average of about +0.75. Income elasticities do not vary considerably across groups defined by family and wealth status. Motor fuel demand is not price sensitive and is highly income sensitive for the two lowest income deciles, while the demand for fuels does not respond to income changes in the two highest deciles at all. Households of pensioners have the strongest response to price changes, the richest households the weakest. The cross-price elasticities of motor fuel demand with respect to prices of other transportation services were not significant.

Both of the empirical studies just mentioned, however, suffer from using macro data (see Halvorsen 2006b for criticism), are unable to provide consistent aggregates, and do not examine possible effects of transportation

TABLE 17-4 Elasticity Estimation from Macro Data by Using Time-Series Technique.

Groups Defined by Income Deciles			*Groups Defined by Family and Wealth Status*		
Decile	*Income elasticity*	*Price elasticity*		*Income elasticity*	*Price elasticity*
1st	3.025	0.475‡	Pensioners	0.64	–0.82
2nd	1.540	–0.370‡	Farmers	0.63	–0.55
3rd	0.554	–0.514	EA <1.6	0.6	–0.52
4th	0.631	–0.638	EA 1.6–1.9	0.62	–0.61
5th	1.746	–0.339	EA 1.9–2.2	0.65	–0.72
6th	1.410	–0.424	EA 2.2–2.5	0.62	–0.57
7th	0.690	–0.69	EA 2.5–3.0	0.67	–0.77
8th	0.553	–0.608	EA 3.0–4.0	0.61	–0.53
9th	0.128‡	–0.53	EA >4.0	0.53	–0.14(‡)
10th	0.446‡	–0.418			
Weighted average	0.75	–0.5	Weighted average	0.63 for EA or 0.62 for the rest	–0.65 for EA or –0.67 for the rest

Note: Significance level at 5% and smaller; ‡ = estimate is not significant even at more than 10%. The number behind EA indicates the net income bracket defined as a multiple of the "minimum living standard" set annually for the household and each family member by the Czech government; also used for provision of income-tested social benefits. Averages weighted by relative consumption of motor fuels by each respective household segment.

substitutes. All of these problems are addressed in the research of Brůha and Ščasný (2005, 2006b), who estimate the coherent demand system, namely the Almost Ideal Demand System (AIDS) (Deaton and Muellbauer 1980). Four main improvements have been made to previous empirical work. First, they estimate an AIDS system on micro data by using about 14,700 observations from HBS 2000–2004. Second, the AIDS consists of the budget shares for motor fuel, but also for urban public means of transport, for buses, and for rail. Third, the AIDS is estimated separately for 13 distinct household groups defined by a combination of social status, presence of children in family, and size of residence (see the previous section on data for a detailed definition). Fourth, zero expenditure is treated with the two-stage Heckman-style correction (1979) to avoid biased estimates. Moreover, to avoid a problem with biased estimates due to too many zeros in expenditures, the AIDS did not include the expenditures on urban public means of transport for all four groups of households living in small towns and villages.

The AIDS model gives the share equations in an n-good system ($i = 1,...n$) for given year t and household h as (Deaton and Muellbauer 1980):

$$w_{it}^h = \alpha_i + \sum_{j=1}^n \gamma_{ij} \ln p_{jt} + \beta_i \ln\left(\frac{x_t^h}{P_t^h}\right) \qquad (17.3)$$

while Brůha and Ščasný (2005a, 2006b) estimate the following specification of the AIDS:

$$w_i = \alpha_{i0} + \sum_h \alpha_{ih} x_h + \varepsilon_i + \sum_j \gamma_{ij} \log(p_j) + \left[\beta_{i0} + \sum_h \beta_{ih} x_h\right]\log\left(\frac{y}{P}\right) \qquad (17.4)$$

where w_i is the expenditure shares on the ith commodity, p_j are prices, y is the total expenditures, x_h are household characteristics—which enter both the intercept and expenditures—and ε is the unobservable random effect. The AIDS obeys a set of parameter restrictions due to additivity, homogeneity and symmetry:

$$\sum_i \alpha_i = 1, \text{ where } \alpha_i \equiv \alpha_{i0} + \sum_h \alpha_{ih} x_h + \varepsilon_i$$

$$\sum_i \beta_i = 0, \text{ where } \beta_i \equiv \beta_{i0} + \sum_h \beta_{ih} x_h, \sum_i \gamma_{ij} = 0$$

$$\sum_i \gamma_{ij} = 0$$

$$\gamma_{ij} = \gamma_{ji}$$

where relevant, the intercept α also contains the inverse Mills ratios from the Probit estimation, since this may mitigate the estimation bias from the zero expenditure problem (Heien and Wessels 1990). Although, as shown by

Shonkwiler and Yen (1999), this procedure can be biased in the case of a large number of censored observations, Brůha and Ščasný follow the Heien and Wessels approach and estimate the AIDS for all households including those with zero expenditures to avoid losing too many observations.

There are two possible approaches to defining the Stone index *P*. The first is to approximate the index by an empirical index which does not depend on the parameters (as in Chapter 4), or to use a nonlinear estimation technique. Brůha and Ščasný follow the latter. Because the Stone index depends on model parameters, the estimation of the AIDS is a nonlinear econometric problem. Specifically, the Stone index *P* satisfies:

$$\log P = \alpha_0 + \sum_i \alpha_i \log p_i + \frac{1}{2} \sum_{i,j} \gamma_{ij} \log p_i \log p_j \qquad (17.5)$$

Since prices may be potentially endogenous, Brůha and Ščasný experiment with a correction for possible price endogeneity using the general methods of moments. They specifically instrument the consumer energy prices by world energy prices and find few changes in the estimation results, probably indicating that energy prices are exogenous for a small open economy like that of the Czech Republic. Last but not least, Brůha and Ščasný (2006b) found that almost all the Engel curves for all commodities of interest and household groups are linear[7], which justifies using the linear approximate AIDS (LA-AIDS) model for their demand estimation.

Tables 17-5A and 17-5B display the estimation results from the AIDS on transportation demand. All the estimates of income elasticities of demand have

TABLE 17-5A Estimation of Income Elasticities from the AIDS.

Household group	Motor Fuels	Bus	Rail	Urban Transport
Farmer (small)	0.70	0.58	0.68	
Farmer (large)	0.63	0.66	0.65	0.64
Pensioners (small)	0.60	0.65	0.64	
Pensioners (medium)	0.60	0.65	0.64	0.58
Pensioners (large)	0.57	0.58	0.50	0.58
EA1 (small)	0.66	0.67	0.68	
EA1+ (small)	0.82	0.74	0.68	
EA2 (small)	0.64	0.55	0.84	
EA2+ (small)	0.78	0.77	0.75	
EA1 (large)	0.66	0.72	0.64	0.66
EA1+ (large)	0.82	0.75	0.69	
EA2 (large)	0.69	0.68	0.74	0.62
EA2+ (large)	0.74	0.69	0.68	0.80
Weighted average	0.707	0.681	0.665	0.685

TABLE 17-5B Uncompensated (Marshallian) Price Elasticities (with Own-Price Elasticities Shaded).

Household Group	MOTOR FUEL Demand with Respect to Price of...				BUS Demand with Respect to Price of....				RAIL Demand with Respect to Price of...				URBAN PUBLIC TRANSPORT Demand with Respect to Price of...			
	Motor fuels	Bus	Rail	Urban Public	Motor fuels	Bus	Rail	Urban Public	Motor fuels	Bus	Rail	Urban Public	Motor fuels	Bus	Rail	Urban Public
Farmer (small)	-0.51	0	0.22		0.13	-0.45	0.30		0.14	0.32	-0.47					
Farmer (large)	-0.06	-0.03	0.06	0.20	-0.03	-0.48	0.08	0.09	0.09	0.33	-0.51	-0.10	0.17	0.01	0.08	-0.43
Pensioners (small)	-0.44	0.32	0.27		0	-0.39	0.09		0.05	-0.03	-0.57					
Pensioners (medium)	-0.67	-0.04	0.11	0.01	0.18	-0.58	0.07	-0.21	0.09	0	-0.55	-0.09	0.12	0.34	0.25	-0.64
Pensioners (large)	-0.44	0.04	0.11	0.04	0.03	-0.56	0.21	-0.28	-0.10	-0.05	-0.56	0.04	-0.11	0.15	0.33	-0.51
EA1 (small)	-0.59	0.18	0.38		0.09	-0.43	0.01		0.05	-0.07	-0.47					
EA1+ (small)	-0.55	0.28	-0.07	0.20	0.01	-0.48	-0.07		-0.2	0.03	-0.47					
EA2 (small)	-0.55	0.29	0.01		-0.25	-0.48	0.26		0.08	0.16	-0.44					
EA2+ (small)	-0.52	0	-0.01		-0.05	-0.67	0.33		0.09	0.12	-0.54					
EA1 (large)	-0.60	0.28	0	0.10	-0.02	-0.19	0.19	0.06	0.17	0.2	-0.52	0.09	0.23	0.45	0.18	-0.47
EA1+ (large)	-0.62	0.10	0.24	0.11	0.12	-0.55	0.38	0.02	0.06	0.29	-0.54	-0.02	-0.07	0.17	0.1	-0.60
EA2 (large)	-0.51	0.20	-0.25	0.25	-0.02	-0.53	0.06	0.08	0.38	0.18	-0.42	0.12	0.13	0.14	0.28	-0.61
EA2+ (large)	-0.49	0.38	0.02	0.12	0.08	-0.50	-0.02	0.2	-0.25	0.25	-0.48	0.06	0.13	0.14	0.28	-0.46
Weighted average	-0.517	0.205	0.070	0.121	0.049	-0.494	0.155	0.030	0.008	0.184	-0.506	0.036	0.063	0.189	0.228	-0.526

the expected signs and range between +0.50 to +0.84 with the weighted average for all four transportation goods about +0.7. The estimates of own-price elasticities are displayed in Table 17-5A and all have expected signs as well. Interestingly enough, weighted averages of all four own-price elasticities are about –0.50. A few estimates of cross-price effects have unexpected negative signs, but they are quite small in absolute value. Overall the authors document quite large heterogeneity in behavioral responses across household groups with respect to price changes. The price and income elasticities from Table 17-5B are then plugged into the DASMOD micro-simulation model to make our tax incidence evaluations.

Inequality Measurement

Inequality of the distribution is usually measured by the Gini coefficient (Gini 1912), with its wide applications, especially in the field of income or wealth inequality measurement (see more in Brůha and Ščasný 2008b). Tax progressivity can be then measured by a specific tax progressivity index (such as the Suits index), or by computing a change in some inequality index (e.g., marginal Gini or marginal polarization index).[8]

The marginal Gini index was originally proposed by Jorgensen and Pedersen (2000), and then applied, for instance, by Wier et al. (2005) to analyze progressivity of energy taxes in Denmark. The progressivity of a marginal tax change is calculated here as the difference between the Gini index as estimated before and after a policy of interest. This index indicates in which direction concerned policy intervention might affect the original income distribution. Positive changes in the marginal Gini index then indicate a regressive burden of concerned policy, i.e., increased income disparities. As noted by Brůha and Ščasný (2008b), the reference income before the policy has to be arbitrarily chosen when changes in the Gini index for marginal changes in a certain tax are measured. Basically, the marginal index can measure a difference in after-policy income distribution compared either to gross earnings—i.e., without assuming any initial tax or transfer—or to net earnings, i.e., assuming all taxes and transfers being implemented in the system except the one of interest. In our computation of the marginal Gini index we follow the latter approach.

The Suits index compares cumulated percentages of total income (y) and cumulated percentages of total tax burden (T_x) (Suits 1977) and is computed as

$$S_x = 1 - \int_0^{100} T_x(y)dy \bigg/ K \qquad (17.6)$$

where K is normalized constant equal to 5,000 and the integral is in our approach estimated by a trapezoid rule as

$$\int_0^{100} T_x(y)dy \cong \sum_{i=1}^{n} (1/2)\left[T_x(y_i) + T_x(y_{i-1})\right] \cdot (y_i - y_{i-1}) \cdot \varpi_i \qquad (17.7)$$

with ϖ that denotes PKOEF, i.e., the weight of how each household is represented in the entire Czech population.[9]

The Suits index measures the progressivity of taxes paid and illustrates whether tax is distributed equally among households. Negative Suits index values indicate regressivity, and so reductions in the Suit index indicate increased regressivity. The diagonal, when the Suits equals zero, describes a flat-tax rate—each dollar is taxed evenly.

Brůha and Ščasný (2008a) propose an alternative index to measure tax progressivity, which they call the Jinonice index. This new index basically combines the main features of the Suits and Gini approaches. In contrast to the Suits index, which compares the cumulated percentages of total income and cumulated percentages of total tax burden, the Jinonice index compares cumulated percentages of units (households), ranked by their income, and cumulated percentage of total tax burden. Similar to the Suits index, the value of the Jinonice index is bounded by −1 and +1, but the diagonal in a graphical representation of the Jinonice index and when the index equals just zero now indicates a lump-sum tax. The Jinonice index is then computed similarly to the Suits index, but the integral is computed again by the trapezoid as[10]

$$\int_0^{100} T_j(x)dx \cong \sum_{i=1}^{n} (1/2)\left[T(\varpi_i) + T(\varpi_{i-1})\right] \cdot (\varpi_i - \varpi_{i-1}) \qquad (17.8)$$

In the distributional analysis performed in this chapter, we use the Suits and Jinonice indices for our *ex post* measurement, and the marginal Gini and Suits indices for *ex ante* policy evaluation.

Statistical Inference

The inequality or progressivity measure basically presents an estimate of the true but unknown index; as such it is a function of the underlying distribution, which is unknown. In the real world we only observe a more or less appropriate sample from that distribution. Therefore, it makes sense to derive the underlying statistical distributions for testing and inference purposes.

As mentioned by Brůha and Ščasný (2008b), the obvious approach is to use an asymptotic expansion of the underlying estimator. The problem with this approach, however, is that many measures are bounded (such as the Gini, Theil, or Suits indices), and the normal approximation is poor if an index takes a value near its boundary.

An obvious alternative is the percentile bootstrap (also known as wild bootstrap) introduced by Mills and Zandvakili (1997). The concept of percentile bootstrap is simple: repeatedly resample the original sample, compute the statistics of interest for each bit of resampled data, and then approximate its distribution with the empirical distributional function of these values. However, Davidson and Flachaire (2007) argue that the percentile bootstrap does not yield an improvement over the asymptotic expansion and that—due to the presence of influential observations—both approaches fail to give accurate results. The moon bootstrap, which resamples the block of m observations from the set of n observations, can offer a solution to this problem (2007). Brůha and Ščasný (2008b) therefore assessed the presence of influential observations when the Gini and Theil indexes for income distribution were computed for the years 1993–2005 for the Czech Republic and conclude that especially the Gini index is rather robust, i.e., a removal of an observation rarely causes a percentage change in the index greater than 1%. Benefiting from this work, then, we conclude that the percentile bootstrap can be an appropriate method for statistical inference, and we compute the confidence intervals for Gini and Suits indices to measure *ex post* progressivity by using the wild bootstrap.

Ex Post Distributional Impact Assessment

We measure tax progressivity for several taxes within the Czech tax system over the period 1993–2008. Overall we consider separately value added taxes (VAT) levied on five distinct consumer items such as motor fuels, public means of transport (urban public service, buses, and rails), food, energy (electricity, gas, solid fuels, centrally supplied heat), and remaining goods. While food items, heat supplied by central systems, and public transport services were subject to VAT at a reduced rate that amounted to 5%, motor fuels and the rest of the other goods were subject to the standard VAT rate of 23% (1993–1994), 22% (1995–2003), and 19% (2004–2008), respectively. The reduced rate was increased from 5% to 9% in 2007. Energy (electricity, gas, coal) have been subject to VAT taxation at the standard rate since January 1998.

We also account for the excise tax on fuels. Because petrol and diesel are taxed at different rates, and because we only have data on expenditures on all motor fuels in aggregate, we have to make some assumptions about the share of various types of fuel used. Following the analyses of more detailed datasets of HBS conducted in 1993, 1999, and 2005 by Brůha and Ščasný (2006b), we assume that petrol and diesel were consumed over the entire period with an 80:20 ratio. However, in our *ex post* assessment we omit the excises levied on tobacco, cigarettes, and alcohol products. Moreover, to have our time-series be consistent over the entire period, we also omit the excise taxes on electricity and solid fuels that have been in force in the Czech Republic since January 2008.[11]

In our *ex post* assessment we first compute paid taxes for all described tax items above for each household in our dataset for each year during 1993–2008. Then, we compute both the Suits and Jinonice indices as described above to measure the progressivity of each considered tax. Please note here that while zero on the Suits index indicates the flat-rate tax—even taxation of any dollar—the zero of the Jinonice index indicates the lump-sum tax, i.e., even taxation of each household unit. The negatives in both cases signal the regressivity of the tax in question.

In sum, total taxation (hereinafter labeled by TOTAL)—that consists of all considered tax items mentioned earlier except for social benefits—is quite flat, with the Suits index ranging from about –0.01 to +0.4 when the total taxation became slightly regressive (1997–2002) and then slightly progressive in the remaining years. Table 17-6 then reports the Suits indices including their 95% confidence intervals for other taxes and years that were computed by bootstrapping with 1,000 replications.

We conclude that labor taxation (LABOR) is mildly progressive (with the Suits up to +0.1), personal income tax (PIT) is more progressive than social security contributions (SSC), and social benefits are strongly regressive (the Suits is between –0.4 to –0.5). All other taxes are regressive, where VAT levied on remaining goods (other than energy, fuels, public transport services, and food) is least regressive (Suits indices are very close to zero), and VAT on energy and food are most regressive (–0.13 to –0.18) among all analyzed taxes.

Regressivity of both taxes levied on fuels (FUEL) and public transportation services (TRANSPORT SERVICE) at the time is in between the regressivity of VAT on energy and food and VAT on remaining goods. Tax on fuels is less regressive (Suits between –0.02 to –0.11) than the tax on public transportation (Suits between –0.04 to –0.16). This basically indicates that personal vehicles are used more by those with high incomes, while public transportation is used and paid for by lower income households. Similar results provide the Jinonice index, also in Table 17-6. In most cases the year-by-year differences in tax progressivity as measured by both our indices are statistically different at a 1% level if the index estimate for a given year is compared with an estimate of progressivity for later years. However, if the index estimates are properly compared for two consequent years, the tax progressivity does differ only in few cases. Both of these observations also hold for the progressivity of fuel taxation and of the public transportation tax.

Figures 17-2A and 17-2B provide another useful piece of information. These figures display a distribution of taxes paid for cumulative income (Suits), or cumulative household units (Jinonice), respectively, for two years (1995 and 2005). As described earlier, the Suits index compares the percentage of incomes held and the percentage of taxes paid; the diagonal on the Suits index equals zero, indicating a flat-tax rate. On the other hand, the Jinonice index compares the percentage of household units (ranked by net income) with the percentage of taxes paid. In this case the diagonal on the index equaling zero indicates a lump-sum tax. Here we see that, for example, the poorest households (accounting for up to 50% of total incomes) paid about 46% of total personal income tax payments, but 60% of all payments of excise tax levied on fuels and 64% of VAT levied on public transportation in 2005.

One can reasonably expect that if the income distribution is not perfectly even, more than 50% of households (if ranked according to income per family member in ascending order) usually hold 50% of total incomes. The distribution of paid taxes if measured for the ordered units (i.e., the Jinonice

TABLE 17-6 *Ex Post* Measurement of Tax Progressivity by the Suits and Jinonice Indices.

	Suits Index					Jinonice Index				
	1993	1995	2000	2005	2008	1993	1995	2000	2005	2008
Total	0.000	0.044	−0.002	0.009	−0.085	0.191	0.228	0.185	0.196	0.184
CL_05	−0.025	0.023	−0.019	−0.011	−0.110	0.173	0.210	0.169	0.178	0.165
CL_95	0.022	0.068	0.014	0.029	−0.063	0.206	0.246	0.201	0.214	0.201
Labor	0.047	0.087	0.052	0.057	0.024	0.240	0.273	0.240	0.247	0.297
CL_05	0.019	0.065	0.033	0.034	−0.004	0.222	0.254	0.222	0.227	0.277
CL_95	0.071	0.111	0.072	0.081	0.051	0.257	0.293	0.260	0.267	0.315
PIT	0.111	0.151	0.117	0.117	0.154	0.306	0.338	0.306	0.308	0.426
CL_05	0.083	0.127	0.098	0.093	0.124	0.287	0.318	0.287	0.288	0.405
CL_95	0.136	0.176	0.139	0.141	0.182	0.324	0.360	0.327	0.330	0.443
SSC	0.002	0.040	0.000	0.006	−0.062	0.195	0.225	0.187	0.195	0.212
CL_05	−0.025	0.017	−0.018	−0.016	−0.090	0.176	0.207	0.170	0.176	0.192
CL_95	0.027	0.064	0.020	0.028	−0.036	0.211	0.245	0.206	0.215	0.231
Transfers	−0.422	−0.373	−0.459	−0.444	−0.630	−0.267	−0.226	−0.321	−0.305	−0.411
CL_05	−0.452	−0.405	−0.488	−0.478	−0.671	−0.293	−0.256	−0.352	−0.341	−0.448
CL_95	−0.394	−0.339	−0.428	−0.410	−0.588	−0.243	−0.196	−0.290	−0.270	−0.372
VAT total	−0.047	−0.003	−0.052	−0.042	−0.189	0.141	0.178	0.133	0.141	0.077
CL_05	−0.072	−0.027	−0.073	−0.064	−0.216	0.122	0.158	0.115	0.121	0.053
CL_95	−0.025	0.022	−0.032	−0.019	−0.164	0.158	0.198	0.152	0.163	0.098
VAT food	−0.140	−0.126	−0.159	−0.182	−0.299	0.046	0.053	0.027	0.002	−0.032
CL_05	−0.164	−0.147	−0.175	−0.199	−0.324	0.033	0.039	0.015	−0.011	−0.048
CL_95	−0.119	−0.105	−0.144	−0.165	−0.275	0.060	0.069	0.039	0.015	−0.017

continued

TABLE 17-6 (Cont.)

	Suits Index					Jinonice Index				
	1993	1995	2000	2005	2008	1993	1995	2000	2005	2008
VAT energy + heat	-0.152	-0.132	-0.176	-0.163	-0.306	0.033	0.048	0.009	0.022	-0.040
CI_05	-0.175	-0.153	-0.194	-0.183	-0.334	0.020	0.033	-0.007	0.007	-0.057
CI_95	-0.131	-0.112	-0.158	-0.145	-0.281	0.047	0.062	0.023	0.038	-0.023
VAT rest	-0.031	0.018	-0.022	-0.014	-0.152	0.156	0.199	0.163	0.169	0.113
CI_05	-0.058	-0.007	-0.045	-0.039	-0.183	0.136	0.178	0.142	0.146	0.082
CI_95	-0.008	0.044	0.001	0.013	-0.123	0.174	0.220	0.185	0.194	0.139
Transport Service	-0.043	-0.037	-0.135	-0.141	-0.289	0.153	0.152	0.053	0.041	-0.026
CI_05	-0.073	-0.063	-0.164	-0.173	-0.326	0.130	0.125	0.025	0.010	-0.059
CI_95	-0.015	-0.008	-0.109	-0.111	-0.251	0.177	0.177	0.080	0.070	0.007
FUEL	-0.060	-0.023	-0.098	-0.084	-0.182	0.127	0.158	0.088	0.100	0.091
CI_05	-0.091	-0.056	-0.125	-0.111	-0.213	0.099	0.128	0.062	0.076	0.065
CI_95	-0.030	0.010	-0.072	-0.058	-0.151	0.153	0.186	0.112	0.124	0.116

index) rather than for the incomes they hold (the Suits) can therefore provide another useful piece of information. The measurement based on the Jinonice index actually answers the question how much tax (in relative terms) is paid by x% households with the lowest incomes per member. From Figures 17-2A and 17-2B we now draw new conclusions for the year 2005: 50% of households with the lowest incomes also paid about 28% of PIT, 42% of the excise taxes levied on fuels, and 45% of taxes levied on public transportation. These figures also indicate lower regressivity of both these taxes levied on fuels and on public transportation in 1995 compared with the later period; in 1995, 50%

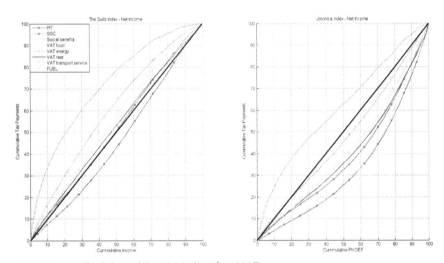

FIGURE 17-2A The Suits and Jinonice Indices for 1995 Taxes.

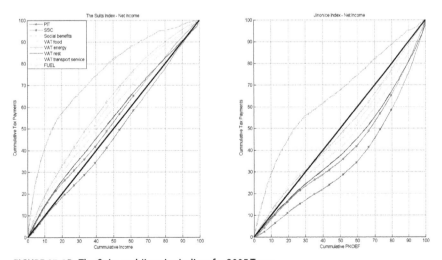

FIGURE 17-2B The Suits and Jinonice Indices for 2005 Taxes.

of households with the lowest income paid only about 26% of all PIT, and 38% of both fuel and public transportation tax. This indication is supported by the Suits and the Jinonice indices for total taxation, both of which report larger values of the indices for 2005 compared to 1995.

Ex Ante Distributional Impact Assessment

To measure tax progressivity and the effect of tax changes on income inequalities for several policy scenarios, we introduce certain changes in taxation on fuel or public transportation service (Table 17-7 describes all of our scenarios). The first policy scenario is based on higher fuel taxation resulting in a 50% increase in the final price of fuel (*Price50*). The second introduces a tax on fuels with a 50% larger rate, resulting in a 24% price increase (*Tax50*). The next three scenarios then assume the revenue-neutrality of the "*Tax50*" policy scenario by recycling all the additional revenues via the increase in tax credit (*Tax50-credit*), the lowest personal income tax cuts (*Tax50-PIT*), and a reduction in social security contributions paid by employees (*Tax50-SSC*). The last scenario increases the VAT rate levied on all three public transportation services from 5% to 10% (*VAT10*). For the sensitivity analysis we also compare our simulation results for the "*Tax50*" scenario with the model predictions based on the average values of elasticities (*Tax50-average*), or assuming no responses at all (*Tax50-noresponse*).

We use the micro-simulation model DASMOD to predict the impact of each scenario on household expenditures on fuels and on all three types of public transportation services, as well as the effects on welfare and public revenues. We compute changes in labor tax parameters (tax credit, PIT rate, or SSC rate) to keep the revenue-neutrality of the tax policy.

TABLE 17-7 Policy Scenarios.

	Before Policy	Price Increase after Policy					
		Tax50	Tax50-credit	Tax50-PIT	Tax50-SSC	Price 50	VAT10
P_fuel (CZK/l)	28.36 Kč	24.00%	24.00%	24.00%	24.00%	50.00%	
P_urban public	38.87 Kč						4.80%
P_bus	57.79 Kč						4.80%
P_rail	37.18 Kč						4.80%
PIT tax credit	7,200 Kč		8,583 Kč				
PIT (1st bracket)	12.00%			10.74%			
SSC	12.50%				11.70%		

Note: Price of fuel is expressed in CZK per liter; other prices for public transport services are expressed in CZK per normalized unit (Brůha and Ščasný 2005).

Our results of the policy scenario evaluation are the following: An increase in the excise tax on fuels by 50% (*Tax50*) would increase expenses on fuels by CZK 4 billion annually, while the expenses on public transport services would be increased only slightly (due to a weak cross-price effect). This policy results in an 11.55% reduction in fuel consumption and a corresponding positive effect on the environment. Welfare is then reduced by CZK 8.2 billion, that is, around 0.84% of total expenditures. It would involve almost CZK 6 billion of paid taxes levied on fuels and public services (mostly from fuel taxation), yet due to a reduction of revenues from VAT levied from other goods whose consumption will be reduced, total net additional revenues would be only CZK 5.22 billion. These revenues can then be used to reduce labor taxation by either increasing PIT credit by CZK 1,383 annually (*Tax50-credit*), by reducing the lowest PIT rate from 12% to 10.74% (*Tax50-PIT*), or by reducing the SCC rate from 12.5% to 11.7% (*Tax50-SSC*). All three policies are set in such a way as to ensure revenue neutrality. The recycled revenue significantly reduces the negative effect on welfare that is now –0.17% compared with –0.84% of total expenditures. The recycling would also reduce the effect on fuel consumption, however, thus increasing the expenditures on fuels. The results for all of our scenarios are summarized in Table 17-8.

TABLE 17-8 Predictions Of Impacts At Macro Level.

	Tax50	Tax50-credit	Tax50-PIT	Tax50-SSC	Price50	VAT10	Tax50-average	Tax50-noresponse
Expenditures on								
fuel (bln. CZK)	3.85	4.08	4.08	4.06	6.97	0.67	3.92	8.54
public transport (bln. CZK)	0.12	0.20	0.20	0.20	0.22	0.53	0.13	0.00
both as % of total expenditures	0.41%	0.44%	0.44%	0.44%	0.74%	0.12%	0.42%	0.88%
Welfare								
bln. CZK	–8.19	–1.63	–1.62	–1.63	–14.94	–0.65	–8.19	–8.61
% of total expenditures	–0.84%	–0.17%	–0.17%	–0.17%	–1.53%	–0.07%	–0.84%	–0.88%
Public revenues								
increase in paid taxes on fuel and public transport	5.86	6.01	6.01	6.00	10.72	0.76	5.90	8.89
total change	5.22	5.32	5.32	5.32	9.57	0.84	5.25	7.53
Consumption of fuel								
mil. L	–162	–155	–155	–156	–302	24	–160	–28
%	–11.6%	–11.1%	–11.1%	–11.1%	–21.6%	1.7%	–11.4%	–2.0%

The more strict scenario—"*Price50*"—results in an increase in expenditures on fuels by almost CZK 7 billion annually. Consumption of fuel falls more (by 21.6%), but thanks to quite a large tax rate (CZK 23.38 instead of CZK 17.19 per liter under the *Tax50* scenario), paid taxes and total revenues are CZK 10.7 and CZK 9.6 billion. Welfare is reduced by about CZK 15 billion or 1.5% of total expenditures.

Last, we run our micro-simulation model to predict impacts for two alternative *Tax50* scenarios when we plug into the model the average value of elasticities or zero elasticities, respectively, instead of the values estimated from the AIDS for several household segments. If no behavioral response is assumed (*Tax50-noresponse*), the model overestimates the impact on expenditures and public revenues. Using the averaged values of elasticities (*Tax50-average*) yields predicted values similar to when the elasticity estimates were used but would have an effect on the predictions of distributional effects.

A more detailed picture of the distributional effects is provided by our predictions as reported for several household groups. First, Table 17-9 shows a quite even distribution of impacts. The *Tax50* policy scenario increases, on average, annual expenditures on fuels and public transportation and reduces welfare in a range of CZK 800 to 980 and of CZK 1,700 to 2,060 across deciles. More inequalities are revealed, however, if we analyze the impacts across household groups defined by family status; for instance, welfare is now reduced by about CZK 1,700 CZK in non-single families living in small municipalities, while the welfare of a single family is on average about CZK 2,600. Revenue recycling via labor taxation cuts reduces welfare most in the deciles 6 and 7, in families of farmers, or in families with two economically active persons with children living in a larger municipality (EA2+_large). The welfare impact of VAT10 policy is distributed evenly, ranging between CZK 130 to 168 per year across deciles.

Table 17-10 displays the average increase in expenditures on fuels and the expenditures on all three types of public transport services along with the average reductions in welfare as expressed by a percentage of total expenditures of a given household group. We conclude here that the impact of all policies on both expenditures and welfare is quite small regardless of household group.

Finally, we measure the effect of all policies on tax progressivity by the Suits index, and then the effect of changes in fuel, transport, and labor taxation on income distribution by the marginal Gini index. We compute the Suits and marginal Gini indices for each tax first for their before-policy rates and then for their after-policy rates. While any decrease in the Suits index compared to its before-policy level indicates increased regressivity of the tax in question, any increases in the marginal Gini index signal increased income inequality being affected by the relevant marginal tax change. We highlight the former cases in Table 17-11 using bold numerals, while the latter cases are shaded. Increases in fuel taxation lead to larger regressivity of this tax, as the

TABLE 17-9 Mean Welfare Impact per Household, in Czech Crowns per Year.

	Tax50	Tax50–credit	Tax50–PIT	Tax50–SSC	Price50	VAT10
1	−2,061	−369	−396	−372	−3,759	−152
2	−2,017	−328	−341	−396	−3,679	−168
3	−2,055	−514	−516	−423	−3,751	−160
4	−2,052	−422	−405	−362	−3,746	−166
5	−1,965	−405	−418	−448	−3,585	−153
6	−1,962	−272	−259	−235	−3,580	−159
7	−1,687	−152	−145	−218	−3,080	−150
8	−2,047	−553	−533	−504	−3,737	−152
9	−1,852	−344	−338	−394	−3,380	−160
10	−1,794	−516	−513	−526	−3,280	−130
farmer_small	−1,898	−257	−290	−205	−3,463	−151
farmer_large	−2,073	−237	−143	−262	−3,781	−167
retired_small	−2,079	−771	−796	−743	−3,795	−156
retired_medium	−1,992	−739	−745	−779	−3,634	−150
retired_large	−2,163	−479	−468	−452	−3,948	−165
EA1_small	−2,613	−622	−683	−600	−4,769	−192
EA1+_small	−1,740	−403	−401	−317	−3,176	−157
EA2_small	−1,667	−345	−350	−269	−3,045	−146
EA2+_small	−1,820	−442	−391	−306	−3,321	−144
EA1_large	−1,889	−316	−345	−414	−3,451	−144
EA1+_large	−2,086	−487	−488	−522	−3,808	−155
EA2_large	−2,077	−467	−447	−509	−3,790	−153
EA2+_large	−1,773	−231	−235	−230	−3,236	−155

revenue-recycling via SSC cuts increases regressivity of this tax as well. Overall, revenue-recycling increases the progressivity of labor taxation if we consider PIT and SSC together. Income inequalities are marginally increased, especially in the case of taxation on public transport services, due to the direct effect of increased taxation of these services (*VAT10*), or due to the cross effect involved by higher fuel taxation. Income inequalities are also increased in the case of fuel taxation, but this effect is very small; in the case of recycling revenues via personal income tax cuts, the effect on inequalities disappears. Revenue-recycling also increases the income inequalities for labor taxation—if both PIT and SSC are considered together—for all three options.

TABLE 17-10 Change in Household Expenditures and Welfare as the Share of Total Expenditures.

| | Change in Welfare | | | | | | Increase in Expenditures on Fuels and Public Transport | | | | | |
	Tax50	Tax50-taxcredit	Tax50-PIT	Tax50-SSC	Price50	VAT10	Tax50	Tax50-taxcredit	Tax50-PIT	Tax50-SSC	Price50	VAT10
1	-0.88%	-0.16%	-0.17%	-0.16%	-1.61%	-0.07%	0.42%	0.45%	0.45%	0.45%	0.78%	0.12%
2	-0.84%	-0.14%	-0.14%	-0.16%	-1.53%	-0.07%	0.40%	0.43%	0.43%	0.43%	0.75%	0.12%
3	-0.86%	-0.21%	-0.22%	-0.18%	-1.56%	-0.07%	0.42%	0.44%	0.44%	0.45%	0.78%	0.12%
4	-0.85%	-0.17%	-0.17%	-0.15%	-1.55%	-0.07%	0.42%	0.45%	0.45%	0.45%	0.78%	0.13%
5	-0.86%	-0.18%	-0.18%	-0.20%	-1.57%	-0.07%	0.41%	0.44%	0.44%	0.44%	0.75%	0.12%
6	-0.84%	-0.12%	-0.11%	-0.10%	-1.53%	-0.07%	0.40%	0.44%	0.44%	0.43%	0.76%	0.13%
7	-0.78%	-0.07%	-0.07%	-0.10%	-1.42%	-0.07%	0.38%	0.41%	0.41%	0.41%	0.72%	0.12%
8	-0.89%	-0.24%	-0.23%	-0.22%	-1.63%	-0.07%	0.44%	0.47%	0.47%	0.46%	0.81%	0.13%
9	-0.79%	-0.15%	-0.14%	-0.17%	-1.44%	-0.07%	0.38%	0.41%	0.41%	0.41%	0.73%	0.12%
10	-0.80%	-0.23%	-0.23%	-0.24%	-1.47%	-0.06%	0.40%	0.43%	0.43%	0.43%	0.75%	0.11%
farmer_small	-0.84%	-0.11%	-0.13%	-0.09%	-1.54%	-0.07%	0.40%	0.44%	0.44%	0.43%	0.73%	0.12%
farmer_large	-0.82%	-0.09%	-0.06%	-0.10%	-1.50%	-0.07%	0.40%	0.43%	0.43%	0.43%	0.72%	0.12%
retired_small	-0.94%	-0.35%	-0.36%	-0.34%	-1.71%	-0.07%	0.44%	0.47%	0.47%	0.47%	0.80%	0.14%
retired_medium	-0.91%	-0.34%	-0.34%	-0.36%	-1.66%	-0.07%	0.43%	0.46%	0.46%	0.46%	0.78%	0.11%
retired_large	-0.87%	-0.19%	-0.19%	-0.18%	-1.58%	-0.07%	0.43%	0.46%	0.46%	0.46%	0.78%	0.13%
EA1_small	-1.06%	-0.25%	-0.28%	-0.24%	-1.94%	-0.08%	0.51%	0.56%	0.55%	0.55%	0.93%	0.14%
EA1+_small	-0.77%	-0.18%	-0.18%	-0.14%	-1.41%	-0.07%	0.38%	0.40%	0.40%	0.40%	0.68%	0.12%
EA2_small	-0.80%	-0.17%	-0.17%	-0.13%	-1.46%	-0.07%	0.40%	0.43%	0.43%	0.43%	0.72%	0.12%
EA2+_small	-0.76%	-0.18%	-0.16%	-0.13%	-1.38%	-0.06%	0.36%	0.39%	0.39%	0.39%	0.66%	0.11%
EA1_large	-0.81%	-0.13%	-0.15%	-0.18%	-1.47%	-0.06%	0.40%	0.43%	0.43%	0.43%	0.73%	0.12%
EA1+_large	-0.91%	-0.21%	-0.21%	-0.23%	-1.65%	-0.07%	0.44%	0.47%	0.47%	0.47%	0.79%	0.13%
EA2_large	-0.82%	-0.19%	-0.18%	-0.20%	-1.50%	-0.06%	0.40%	0.43%	0.43%	0.43%	0.72%	0.12%
EA2+_large	-0.81%	-0.11%	-0.11%	-0.11%	-1.48%	-0.07%	0.39%	0.42%	0.42%	0.42%	0.71%	0.13%

TABLE 17-11 Tax Progressivity Measurement by the Suits and the Marginal Gini Indices.

	Before policy	Price50	Tax50	Tax50– taxcredit	Tax50– SSC	Tax50– PIT	VAT10
Suits index							
PIT	0.2391	0.2391	0.2391	0.2794	0.2391	0.2816	0.2391
SSC	–0.0424	–0.0424	–0.0424	–0.0424	–0.0429	–0.0424	–0.0424
LABOR	0.0551	0.0551	0.0551	0.0616	0.0582	0.0623	0.0551
Taxes on fuels	–0.1237	–0.1244	–0.1241	–0.1243	–0.1238	–0.1245	–0.1241
VAT public transport	–0.1776	–0.1728	–0.1748	–0.1750	–0.1746	–0.1751	–0.1762
VAT food	–0.1990	–0.1990	–0.1990	–0.1989	–0.1982	–0.1990	–0.1990
VAT rest	–0.0700	–0.0693	–0.0696	–0.0700	–0.0692	–0.0702	–0.0699
Marginal Gini index							
PIT	–0.0451	–0.0451	–0.0451	–0.0451	–0.0456	–0.0451	–0.0451
SSC	0.2422	0.2422	0.2422	0.2790	0.2422	0.2813	0.2422
LABOR	0.0551	0.0551	0.0551	0.0616	0.0583	0.0625	0.0551
Taxes on fuels	–0.1269	–0.1270	–0.1269	–0.1271	–0.1266	–0.1272	–0.1271
VAT public transport	–0.2208	–0.2143	–0.2171	–0.2171	–0.2165	–0.2172	–0.1988
VAT food	–0.2007	–0.2007	–0.2007	–0.2006	–0.1999	–0.2007	–0.2007
VAT rest	–0.0761	–0.0755	–0.0758	–0.0761	–0.0754	–0.0763	–0.0760

Conclusions

The transition of the Czech economy from a centrally planned system toward a market economy brought many changes in passenger transportation. The number of cars increased, while the share of public transportation decreased. Despite the increasing burden of passenger transport on the environment, pricing of transportation has not become more strict. Lacking a nominal indexation of the excise tax on fuels, the rate of fuel tax has effectively decreased, while the final price has been constant over the entire period 1995–2008. Authorities have justified keeping the real price of fuels constant by citing potential adverse social impacts of higher taxation.

We found fuel taxation to be somewhat regressive (Suits index in the range of –0.06 to –0.11) but not as regressive as taxation on energy or food items—the most regressive of all such taxes in the Czech Republic. Analysis using the Suits index and the newly developed Jinonice index confirms this finding. Taxation on public transport services (basically by VAT only) is more regressive than taxation on fuels, but less so than energy taxes, with a Suits index in the range of –0.10 to –0.16 for most years. The weaker regressivity of transport taxation compared to energy taxation is also supported by the fact that, on average, the budget shares on fuels of the households having some expenses are more or less the same across all deciles. Larger nominal expenditures on fuels in higher deciles are due to greater ownership of passenger cars and higher incomes.

By using the micro-simulation model DASMOD, we then predict the distributional effect of several policies that would result in higher passenger transport taxation. A 50% increase in fuel taxation would increase household expenditures on passenger transport by 0.4% of total expenditures and reduce welfare by 0.8%. Both of these effects are distributed quite evenly across all income deciles. These impacts are more uneven if we predict impacts for household segments defined according to family status and size of residence. We determine that households of economically active persons with children living in smaller towns and villages are relative winners of such a policy, while families of one economically active person with children living in larger cities along with households of pensioners living in small and medium-sized municipalities would be the losers. Revenue-recycling reduces the negative effect on welfare, where the progressivity of labor tax is increased most if the revenues are recycled via cuts at the lowest rate of personal income tax. Any effect of the relevant policies on tax progressivity or income inequalities of marginal tax change is, however, very small. On the whole we predict the effect on overall income distribution and tax progressivity, but also the impacts on various household groups separately. The comprehensive database of our micro-simulation model—allowing us to predict the impacts for each household included in that database separately—lets us compute and report the effects on a wide range of distinct household segments.

Our modeling has several limitations, however. First, our static model addresses only the first-order effects, neglecting general equilibrium effects or any possible dynamic effect such as on savings. Second, our simulations do not account for the tax-interaction effect. Implicitly we thus assume a separability of preferences, i.e., that there is no effect of fuel or transport price changes on leisure/labor supply. Third, although our simulation model is embodied by several sets of elasticities as separately estimated for several household groups, the estimated demand system was always partial, consisting only of a few non-durable goods of interest. The main limitations of our model remain to be solved in future research.

Acknowledgments

This research was supported by the Ministry of Education, Youth, and Sports of the Czech Republic, Grant No. 2D06029 "Distributional and social effects of structural policies," funded within the National Research Programme II. The support is gratefully acknowledged. I wish to thank Thomas Sterner for his encouragement with this chapter, and Selma Oliveira and Hana Škopková for their assistance in formatting the manuscript. I am also grateful to the participants of the 16th Annual Conference of the European Association of Environmental and Resource Economists held in Gothenburg, June 25–28, 2008 for very useful comments. Responsibility for any errors remains with the author.

Notes

1   Statistics based on CDV (2009).
2   All values expressed in Czech crowns (CZK) are expressed in real 2005 prices using the consumer price index.
3   The households of farmers presented quite an important household segment during the communist regime and at the beginning of the 1990s. Their signifi-cance has declined, however, and since 2007 the segment is no longer included in the HBS sampling strategy. Turning to households of pensioners, they had to have a head to be retired without any earnings, i.e., the retired who were still working were not included in these three household segments.
4   Using predictions about energy consumption, the model also computes changes in environmental damage by calculating the external costs associated with differ-ences in energy use or fuel consumption.
5   Our calculation based on Paasche and Laspeyers CLIs only provides an approxi-mation of theoretically proper measures of welfare changes (see, e.g., McKenzie and Pearce 1976; Hausmann 1981; or Vartia 1983).
6   The demand system can be estimated directly from consumer preferences using the Linear Expenditure System, the AIDS system, or other flexible demand systems such as the Rotterdam translog or so-called "mixed" demand functions (Moschini and Rizzi 2006). For instance, Halvorsen (2006a) discusses the differences in estimations due to the time period covered in the analysis, micro versus macro data use, or the number of factors capturing the gross income effects. Vaage (2000) then provides a brief overview of the development of the discrete/continuous choice models as based on Hanemann (1984) or Dubin and McFadden (1984) approaches.
7   They regress the budget share for given commodity as the dependent variable on the log of expenditures as the independent variable by a simple linear regres-sion and also by using a nonparametric method to infer whether the linearity assumption is an empirical problem. They found that in most cases the linear regression line lies within the 95% confident interval of nonparametric estima-tion. The curve slope of the Engel curve differs if the estimation is corrected by socioeconomic characteristics, an observation that actually shows the importance of using such variables as explanatory variables in the regressions.
8   The change can be considered either as an absolute change or as a percentage change, which is usually normalized by the change in the total public finance reve-nues. However, the latter approach is not well-defined for revenue-neutral policies.
9   $\frac{1}{n}\sum_{i=1}^{n}\omega_i = 1$, $x_0 = L(x_0) = T(x_0) = 0$
10 $\omega_i = 100 \cdot \dfrac{\sum_{j=1}^{i}\omega_j}{\sum_{j=1}^{n}\omega_j}$, $\omega_0 = T(\omega_0) = 0$

11 These taxes were introduced in the Czech Republic to implement 2003/96/EC
    Directive on taxation on energy products and electricity. Because the rates on
    fuels were already set above the minimum rates required by the Directive, and
    because natural gas used for household heating is exempted from taxation, taxes
    on electricity and solid fuels had to be implemented into the Czech tax system.
    Due to the quite small rates of these two new taxes, their effect on households
    is expected to be very small (Ščasný and Brůha 2007). A detailed evaluation of
    the implementation of the Directive, including a summary of the impacts for the
    Czech Republic, is provided in Ščasný and Máca (2010).

## CHAPTER 18

Distributional Effect of Reducing Transport Fuel Subsidies in Iran

SANAZ ETTEHAD AND THOMAS STERNER

I ran, one of the founding members of OPEC, has the world's third-largest proven oil reserves and—by some estimates—the second cheapest gasoline in the world, after Venezuela. Iran's oil industry was nationalized in 1950. Currently, a state-owned company, NIORDC,[1] is responsible for all activities related to refining, production of different oil products, and distribution throughout Iran, as well as marketing and exports. In spite of many years' experience and good revenues, domestic gasoline production is currently not sufficient to meet domestic demand. One important reason for this is that Iran's gasoline demand has been growing at an increasing rate due to factors such as population growth, increased incomes, and low prices of fuel. Excess demand causes several economic and environmental problems. On one hand, the government has been forced to import gasoline to supply the domestic demand; in fact, it devotes a sizable share of the national budget to this purpose (2% in 2009). Also, about 4% of the total budget share is devoted to domestic gasoline subsidies (by our own estimation). Iran has one of the largest subsidy schemes relative to its GDP in the whole world. Between 2007 and 2008, oil product subsidies amounted to US$ 32 billion, or about 11% of GDP (*Ettelat* 2008).

High levels of gasoline consumption also cause environmental problems. In 2005 CO_2 emissions per capita were 6.54 metric tons. Except for some slight fluctuations, Iran's GDP has shown an upward trend during the last decade. The growth rate was 6.8% in the decade leading up to 2007 (Central Bank of Iran). We can expect continued growth in CO_2 emissions in the future as well. Increasing GDP implies more transports. Besides, an insistence on fuel price subsidies will lead to higher negative environmental externalities. The average gasoline subsidy has constituted around 80% of its import price (Houri Jafari and Baratimalayeri 2008).[2] In 2009 the gasoline subsidy constituted as much as 87% of its import price.[3] Clearly, cutting gasoline subsidies has the potential to reduce domestic consumption, but this may also have significant effects on the country's economy and global emissions. However, one must be conscious of the power of political resistance to fuel taxes (Hammar et al. 2004). Governments with a tendency for populism feel that they must keep fuel prices low to prevent domestic protests.

At the moment, it appears there is some serious disagreement between Iran's government and parliament regarding the gasoline price increase and the distributional effects of this policy. This chapter examines the distributional impact of a gasoline price increase in Iran. First, we review Iran's gasoline market and the pricing policies. We then present the results of tax incidence measurement, and the final section concludes.

Gasoline Market and Pricing Policy

Iran's CO_2 emissions from fossil fuel combustion became four times larger between 1981 and 2007, with the transport sector the largest source of emissions. Internal transportation accounts for over 25% of Iran's CO_2 emission (WRI 2005). Figure 18-1 shows carbon dioxide emissions per capita from 1980 to 2005.

Until 1979, Iran's gasoline market was balanced, with no need to import gasoline. However, in the 1980s a deficit started to appear. On the one hand, domestic supplies were reduced by about 10% as the result of refinery damages during the Iran-Iraq War from 1980–1988.[4] On the other hand, gasoline consumption soared, with an average growth rate of over 8% from 1997 to 2009—more than twice the growth rate of production (3.6%).[5]

One of the main reasons for increased gasoline demand was the rapid growth rate of population, of which almost half was age 30 or younger and looking for a good quality of life. In addition, Iran has had a rapid growth rate of domestic car production; however, this production was not focused on high

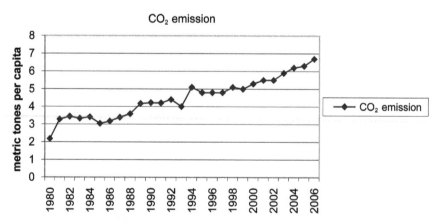

FIGURE 18-1 CO_2 Emissions in Iran, Metric Tons per Capita (1980–2005).

Source: World Bank, World Development Indicators.

fuel efficiency partly due to the low prices of gasoline. All this has resulted in rapid demand expansion (Houri Jafari and Baratimalayeri 2008). It is noteworthy that there are more than 9 million vehicle owners in Iran. According to the latest data, there are 7,000,000 private vehicles, 2,400,000 pickup trucks, 450,000 taxis, 250,000 part-time taxis, and 200,000 government vehicles that benefit from special gasoline quotas (*Economic Hamshahri* 2010).

The Iran-Iraq War of 1980–1988 undermined crude oil and petroleum product production capacity when oil fields and refineries were severely damaged. The Abadan refinery located in southern Iran was completely destroyed, and the remaining seven refineries' production capacity was diminished by 21.5% on average (NIORDC 2008).[6] Additionally, since the 1979 revolution, economic sanctions and trade restrictions imposed by the U.S. and others against Iran made it difficult to restore and develop oil fields and refineries. Only one refinery was constructed during the postwar period. According to the Ministry of Energy, in 2007 Iran's oil refineries were only designed to give a 17.5% gasoline output (MoE 2008). To sum up: Decreased production capacity thanks to damaged refineries and no investment in developing new refineries; a rapid population growth rate; and mass production of cars with old motor technology—all these factors obliged the country to increase fuel imports to meet the growing fuel demand.

The response of Iran's government to the strong increase in gasoline demand has been to allow imports to grow at more than 25% (between 1997 and 2009), until by 2009 almost 36% of gasoline demand was provided through imports.[7] Figure 18-2 shows the historical data trend of gasoline consumption and production.

However, referring to the gasoline import value, there may be some expectations of import reduction in the future. The U.S. Senate has ratified sanctions

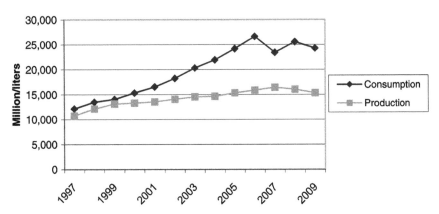

FIGURE 18-2 Gasoline Consumption (Transportation Sector) and Production in Iran, 1997–2009.

Note: Decline from 2007–2009 due to rationing.
Source: NIORDC.

against companies exporting fuel to Iran or assisting with improvements to oil refineries in Tehran (BBC 2010a); some petrol suppliers such as BP, Shell, Reliance, and Glencore have since stopped trading with Iran (BBC 2010b).

The main factor keeping gasoline consumption high is its low domestic price (until recently, about US$ 0.10/liter). Data indicate that low domestic prices of gasoline in Iran do not follow the fluctuations in international prices due to the government's intervention. Actually, the government has chosen to subsidize gasoline precisely to maintain low prices. As mentioned earlier, from 1997 to 2007, the average gasoline subsidy per liter has constituted about 78% of its import price (Houri Jafari and Baratimalayeri 2008). Specifically in 2009 the total amount of gasoline subsidy constitutes about 87% of its import price (Houri Jafari and Baratimalayeri 2008). It is estimated that between 1986 and 2006 the total paid subsidy was $47.7 billion in current values (Shahmoradi and Honarvar 2008). Between 2003 and 2006, gasoline subsidies grew sharply—from $2,516.3 million in 2003 to $10,192.9 million in 2006. Other data show that gasoline subsidies in 2005 and 2006 hit $6.6 billion and then $10.2 billion—figures that exceed oil income during those years by 10% (Houri Jafari and Baratimalayeri 2008). However, due to a rationing system launched in 2007, gasoline consumption and import has been almost constant, and it was even slightly reduced. We estimate the 2009 gasoline subsidy at $11 billion.

One might wonder why the regime does not raise the price. Presumably, one reason is the large and dangerous demonstrations that erupted when the petrol rationing in 2007 led to violent protests. Iran is clearly vulnerable, since sanctions are constantly being discussed in the United Nations (related to the nuclear program of uranium enrichment and other issues). An obvious target for sanctions would be petrol, which the country must import. For this reason the regime, not daring to cut subsidies, tested petrol rationing at the pump. Some might also wonder whether a regime such as Iran's is sensitive to public opinion, but the truth is probably that all regimes are very sensitive to public opinion (BBC Persian 2009).

In a market-based economy, gasoline shortages would lead to increased prices. However, in Iran prices remain very low. Although the nominal gasoline price increased after the war of 1980–1988, real prices are still far lower than international prices. The government says it is forced to set a low gasoline price in order to control inflation. Inflation is a controversial issue, often mentioned as a disadvantage of raising fuel prices in Iran. Expectedly, Iran's parliament voted against ratification of the gasoline subsidy cut due to their concerns about the inflation rate. They believe that this policy imposes more burden on low-income households. The current inflation rate in Iran is over 20%, and Iran's oil minister believes that a gasoline price increase would induce an inflation rate jump of 11%, taking current inflation above 30% (BBC Persian 2009).

Subsidizing the oil industry is far from an obvious way to combat inflation. This should be done through a proper monetary policy since it is a monetary issue. One can also note that several countries have high inflation rates, even among OPEC countries with low domestic fuel prices. Many EU countries have also faced sharp increases in fuel prices in recent years, but they have still managed to control the inflation rate due to proper monetary policies (Tahmassebi 2007).

According to government sources, economic damage induced by recent restrictions and sanctions from the West has threatened the social and economic situation of the country. To diminish the costs of highly subsidized gasoline prices, in 2007 the government started a plan consisting of an increase in nominal prices (from 800 Rials to 1,000 Rials)[8] and a rationing program (even though real prices started falling after 2006; see Figure 18-3). According to a recently ratified law in the Iranian Parliament Majlis (late December 2009), all subsidies for fuel, natural gas and electricity are to be cut and the acquired revenue is to be redistributed among low-income households to compensate for the post-plan jump in gasoline prices (Press TV 22 Feb 2010). In consequence, the government of Iran must gradually increase gasoline prices to the FOB prices of the Persian Gulf during the next 5 years. Therefore, the government has proposed a budget bill containing a new gasoline pricing policy, i.e. the per-liter gasoline price must reach 4,000 Rials for the next Iranian calendar year that began 21 March 2010. The Parliament opposes this, arguing that the proposed gasoline price is too high. It proposes a price of 2,500 Rials/liter gasoline that must be accompanied by a rationing system for the upcoming year (*Ettelaat* 2010b). At the moment there does not yet appear to be any accepted gasoline pricing policy to implement the above-mentioned law.

The Minister of Oil, in support of the government's proposed budget law, deems the government's suggested gasoline price to be far below the market

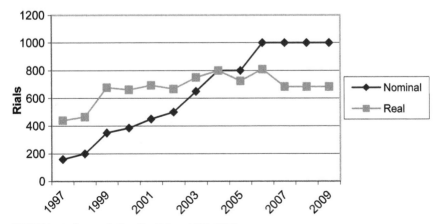

FIGURE 18-3 Domestic Gasoline Prices, 1997–2009.

Source: NIORDC.

price. He has estimated the real FOB price to be between 7500 and 8000 Rials (*Ettelaat* 2010a). In addition, some experts criticize the rationing program as a clumsy "command and control" instrument whose implementation will involve enormous bureaucracy and may still fail in reaching its goals. There is already evidence that the rationing program has failed due to manipulation and an emerging black market. To sum up, the government has chosen a method that has been tried and rejected by other countries in the past, rather than opting for a systematic fuel price increase (Tahmassebi 2007).

Research on distributional effects of fuel taxation is a relatively new research field, and we still need to know more about these effects, especially among OPEC members. As we have seen, there is a concern for the reactions from different strata of the population in Iran. Hence, there could be political resistance toward environmental taxation, particularly in the short-term when positive environmental effects are less tangible.

Income Distributional Effect of Gasoline Price Increase in Iran

Based on 2007 data from the Iran Energy Agency (MoE 2008), Figure 18-4 demonstrates that affluent households benefit from gasoline subsidies 28 times more than poor households. Therefore there should be no major issue of fairness in subsidy reduction.

In this study, the total expenditure approach is used instead of the disposable income approach (see Poterba 1991). The main data used is the household's total expenditure, fuel expenditure, and public transport expenditure by income deciles, using the National Survey of Household Income Expenditure, 2006.

To investigate the effect of gasoline taxation on income distribution, we note both direct and indirect consumption. The household's direct consumption

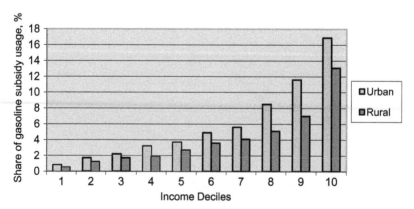

FIGURE 18-4 Share of Households in Gasoline Subsidy Consumption by Expenditure Deciles.

Source: MoE 2008.

consists of the expenditure for fuel for driving (expenditure on gasoline makes up 85% of total fuel expenditure in Iran), while indirect consumption consists of household expenditures for public transport (road transportation by bus, taxi, minibus, and so on). Indirect fuel consumption is also important when examining the tax incidence, but our data only allow us to consider the indirect effect through public transport in this chapter. Table 18-1 shows the share of income spent on gasoline and public transportation for each income decile.

Figure 18-5 shows that the first income decile's share of total expenditure on public transport is higher (1%) than the other income groups. This share falls in subsequent income deciles to only 0.39% for the richest decile. In contrast, the budget share of different income groups for gasoline is fairly constant.

In order to estimate the indirect fuel consumption we need to approximate the fuel share. Our informal surveys showed that the estimated fuel share in operating costs is about 10% for a taxi. Household budget survey data for 2006 show that taxis are the preferred carrier in all households with different income deciles—about 64% of total public transport expenditure is for taxis—so we use this approximation for total public transport. Figure 18-6 and Table 18-2 show the share of direct and indirect fuel expenditures in total expenditure in 2006. The combined gasoline expenditure shows that the tax

TABLE 18-1 Share of Income Spent on Gasoline and Public Transportation.

Income deciles	1	2	3	4	5	6	7	8	9	10
Average gasoline expenditure (US$)	4.36	7.06	10.57	12.08	15.60	19.27	24.38	30.87	40.09	73.24
Share of expenditure on gasoline (%)	0.21	0.18	0.21	0.19	0.20	0.21	0.22	0.22	0.22	0.22
Average expenditure on public transport (US$)	20.33	30.30	38.89	46.26	52.98	58.06	70.03	78.93	100.75	133.49
Share of expenditure on public transport (%)	1	0.78	0.76	0.73	0.69	0.63	0.63	0.57	0.055	0.39

Source: Authors' estimation using National Survey of Household Income Expenditure, 2006, published by the Statistical Center of Iran (SCI), 2006.

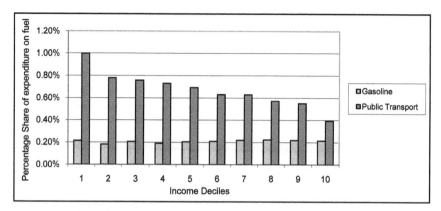

FIGURE 18-5 Share of Total Expenditure on Gasoline and Public Transport for Each Expenditure Decile, 2006.

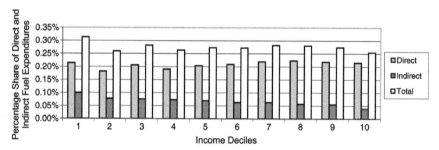

FIGURE 18-6 Direct and Indirect Gasoline Expenditure in Percentage of Total Expenditure by Income Deciles, 2006.

Source: Authors' estimates based on Iran's National Survey on Household Income and Expenditure, 2006.

TABLE 18-2 Budget Share of Direct and Indirect Gasoline Expenditure in 2006.

Income decile	Total (%)	Indirect (%)	Direct (%)
1	0.31	0.10	0.21
2	0.26	0.08	0.18
3	0.28	0.08	0.21
4	0.26	0.07	0.19
5	0.27	0.07	0.20
6	0.27	0.06	0.21
7	0.28	0.06	0.22
8	0.28	0.06	0.22
9	0.27	0.05	0.22
10	0.25	0.04	0.22

burden is fairly neutral, as it is the sum of a weak progressivity for the direct expenditure while the indirect gasoline expenditure share is still regressive.

To illustrate distributional effects, a Suits index is calculated. Figure 18-7 displays the concentration curve of the Suits index.

Figure 18-7 clearly shows that the tax is almost exactly neutral. Technically speaking it is weakly progressive regarding direct consumption only, but weakly regressive regarding both direct and indirect fuel consumption. However, the effects are minimal: the Suits coefficient is only –0.016 for total or ±0.01 for the direct effect—surely within the margin of error.

Our study of Iran has shown that the fuel tax is almost neutral, and its burden falls fairly equally on all housholds. With regard to indirect fuel consumption, there is some regressivivity, and its burden falls more heavily on low-income households than on high-income households (Suits index = –0.134). To prevent a negative distributional effect of the fuel tax on the poor, policymakers could devote some part of the revenue gained from a gasoline price increase to the improvement of public transportation.

Conclusion

During recent years, market-based climate policies have gained considerable interest, one of which is taxing fuel. However, there is resistance toward such taxation on the part of governments. The opponents of fuel taxes claim

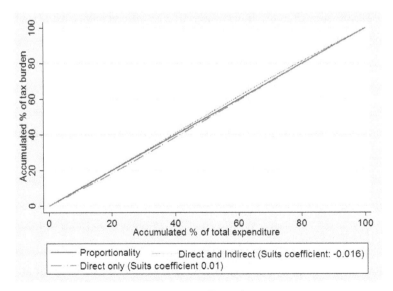

FIGURE 18-7 Concentration Curves and Suits Coefficient for Iran.

Source: Authors' calculation based on the National Survey of Household Income Expenditure, 2006 from SCI.

that raising fuel prices has a negative distributional effect and could generate public opposition and rioting.

To measure the progressivity of the fuel tax in Iran, a Suits index has been measured using data from 2006. Results show that the fuel tax is almost proportional: its burden falls fairly equally on all housholds. If only direct consumption of fuel is considered, the tax is progressive (with progressivity extremely weak). Perhaps this is because of the combination of fairly high income levels and extremely cheap fuel, which has led to fairly high levels of fuel use even among the poorest deciles. If indirect fuel use through public transport is taken into account, this tax, seen in isolation, would thus be regressive.

Considering the fact that the gasoline subsidy is so costly (in 2009 about 4% of the country's budget was spent on gasoline subsidization, and the country is faced with the embarrassing need to import fossil products), a subsidy cut is of utmost importance. Such a cut could however trigger widespread public opposition in the short term, and therefore the government is quite careful. The government's ownership of the oil industry and its intervention in the market has contributed to generating certain expectations among Iranians; when the government's revenue increases due to a rise in oil prices, Iranians expect the subsidies to become higher. It seems that the Iranian government has understood that continued subsidies are unsustainable. They lead to environmental degradation, to inappropriate and wasteful technology and consumption choices, as well as lost export earning (and worse, the need to import fuel).The government would therefore be very wise to phase out subsidies—particularly if it combined such subsidy cuts with direct income support for the poor. Such a policy would make sure that any possible negative distributional effects are taken care of.

Notes

1   National Iranian Oil Refining and Distribution Company
2   Authors' calculations for 1997–2007 using IIE data cited in Houri Jafari and Baratimalayeri (2008).
3   Authors' own calculation using data cited on p. 4 (*Ettelaat* 2010a).
4   NIORDC database (data extracted by the authors) at www.niordc.ir/index.aspx?siteid=78&pageid=463 (accessed January 15, 2010).
5   Authors' own calculation using NIORDC database.
6   Data extracted by the authors from NIORDC (National Iranian Oil and Refinery Distribution Company).
7   Authors' own calculation using NIORDC database.
8   Exchange rate at the time of writing was 1 US$ = ~10,000 IRR.

CHAPTER 19

Conclusions

THOMAS STERNER, JING CAO, EMANUEL CARLSSON, AND
ELIZABETH ROBINSON

C limate change is one of the defining issues of our time. The transport
sector is one of the biggest and fastest-growing sectors, so naturally it will
have to face very strong policy instruments to reduce emissions. It is worth
reflecting on whether this sector is easier or more difficult to deal with than
other sectors and how gasoline and diesel compare to other fuels such as coal
that also contribute to climate change. Technically it is much simpler to deal
with fuel change in large boilers than in car engines. For this reason carbon
abatement is typically a lot cheaper in industrial applications. However, policy
instruments that regulate industrial energy use (including electricity) are
politically very sensitive, since effects on competitiveness and loss of jobs are
feared. Among the fossil fuels used by households, fuels such as kerosene are
used for heating and cooking by very low income households in poor countries and are thus again very sensitive. Due to these concerns, it may still be
easier to deal with the transport sector that is largely local[1] in character and
where incidence is largely on consumers not producers. However, the distributional consequences of fuel taxation will be crucial to its political feasibility.

We have many reasons to be interested in fuel taxation: Urban planning,
public transport, local pollution, congestion, and many other aspects are
important for everyday life quality. But fuel taxes are also of particular interest
because we may well find that if we cannot deal with the political consequences
of taxing transport fuels, it will be no easier for many other sources that affect
the climate. Coal use in heating, electricity, and industry will raise issues of
competitiveness in addition to distributional concerns. Emissions from agriculture and forestry will often be so complicated and uncertain that this in
itself poses a barrier to policy.

Climate change is thus not the only reason we are interested in this issue—
but the longer our time horizon the more likely it is for climate to be a dominant factor. To deal with the problem effectively entails work on two related
sets of problems—that of creating effective international agreements on the
one hand and, on the other hand, that of instituting and implementing efficient and sustainable policy instruments at the local or national level. These

two problems are connected: One of the barriers for national action is the absence of international agreements. Some countries suspect that others will not only fail to "do their part" but in fact will benefit from any national efforts through carbon leakage. For example, if one country taxes fossil fuel use and this depresses the world fuel price, fossil fuel use abroad might increase, leaving global emission levels almost unaffected. This is the "worst scenario" for national decisionmakers. However, it is also difficult to build international treaties if there are no national experiences that serve as role models to build on.

National policies vary, and as it happens the world actually harbors quite a variety of transport fuel prices which offer us a laboratory to understand both the effect of prices on use and also the political determinants of taxation and thus price levels. Sterner (2007) shows that fuel demand is quite price-elastic—at least in the long run—and therefore that taxation is quite an effective long-run instrument to lower demand (and thus emissions). It is however an understatement to say that fuel taxation is not very popular. One of the main concerns raised in countless protests is that fuel taxation would hurt the poor. As this book is being sent to the publisher in early 2011, we have seen major recent riots in Bangladesh, India, and Bolivia in which citizens specifically asked the government to resign (in Bangladesh) and forced the government to revoke recent fuel price increases (in Bolivia). This book is dedicated to discussing this issue by looking at different countries and from a number of different methodological viewpoints. We will first summarize the findings of previous chapters and then move to a meta-analysis of the distribution of results found. This naturally leads up to some outstanding methodological issues concerning distributional effects, which we discuss. We then present a brief summary and policy conclusions.

Summary of Individual Chapters

This book consists of 17 case study chapters covering 22 countries. Chapters 2, 3, and 4, which study distributional effects in the U.S., were selected largely to illustrate different methodological approaches. Almost all the other chapters were commissioned directly for this book. This has permitted us to harmonize the structure of the chapters and methodology chosen to facilitate a comparison between countries and avoid repetition of discussions on methodology while still allowing, hopefully, for enough individual variation to keep readers fascinated by the subject.

Geographically the book has a global coverage. After the three U.S. chapters, Chapters 5 and 6 move south to Mexico and Costa Rica. Chapters 7 to 9 study the three Asian giants China, India, and Indonesia. Chapters 10 to 15 give a panorama of six African countries ranging from low-income Ethiopia to relatively middle-income South Africa; from the large but sparsely populated Mali to the small and more densely populated Ghana; and including two neighbors with different political trajectories: Tanzania and Kenya. Chapter 16 gives an overview of effects in seven European countries: Spain, the UK, Germany, France, Italy, Serbia, and Sweden. Chapter 17 complements the picture with the "formerly planned" Czech Republic. Chapter 18 analyzes Iran, and finally this Chapter 19 concludes.

In Chapter 2, Hassett, Mathur, and Metcalf measure the incidence of a carbon tax on gasoline in the U.S. and compare results using three different income measures (current income and two measures of lifetime income) to rank households. They point out that incidence calculations based upon annual income create substantial measurement problems because annual income fluctuates and varies over the lifecycle; very young or old people look poor on the basis of annual income, while in fact they spend more and may be economically quite comfortable in a lifetime context. In addition, transitory income shocks tend to bias the incidence results to be more regressive. Using current consumption as a proxy for permanent income (as is done also in many chapters in this book) does mitigate the problem, but it cannot completely correct the bias. Instead Hassett, Mathur, and Metcalf construct lifetime-corrected consumption for each household by sketching out a "typical" path of lifetime consumption based on age and education categories. They then construct the lifetime consumption for each household by looking at the ratio of current consumption of a given household to the average for their age–education group. They also examine tax incidence across regions, but the variation across regions is fairly modest.

Ideally, calculating the incidence of fuel taxes requires a great deal of information, including both the direct burden of fuel expenditure and indirect burden from consuming other commodities that use fuel as an intermediate input. Not only does the increase in fuel taxes pose a burden on fuel consumers, it also tends to lower the producer price of fuel, hurting the owners of gas stations and gasoline producers, as well as workers within the industries through lower wages. Therefore, an ideal measure of tax incidence should capture both welfare changes on the consumption side and welfare changes on the factor returns and producer side. Several chapters in this book capture various indirect effects. Hassett, Mathur, and Metcalf (Chapter 2), Datta (Chapter 8), and Kpodar (Chapter 13) use input-output approaches to capture the indirect effects of other commodity expenditures. Bento, Goulder, Jacobsen, and von Haefen (Chapter 3) examine the impacts of U.S. gasoline taxes in an econometrically based multi-market model that links the markets for new, used, and scrap vehicles. By addressing both supply and equilibrium in all three markets, they find distributional impacts depend significantly on how the tax revenues are recycled to the households:

1   Under flat recycling, lower income groups experience a welfare improvement while higher income groups suffer a welfare loss. The effects of the recycling more than fully offset any potential regressivity of the tax;

2   Under income-based recycling, the pattern of impacts is inverse U-shaped; the middle-income households experience the largest welfare loss;
3   VMT-based recycling implies a fairly flat pattern of impacts across the income distribution, although the welfare losses are greater for higher-income households.

Further, they decompose the welfare impacts into the various contributing factors: the change in gasoline price, the use of gasoline tax revenue, the net capital gain or loss associated with policy-induced changes in car prices, and changes in profit to new car producers. Interestingly, their empirical results suggest that changes in the gasoline price and transfers are by far the most important sources of the household welfare impact, while the other factors only count minimally. This suggests that, without complicated general equilibrium modeling, it is fairly safe[2] to assume that the imposition of a gasoline tax does not affect the producer price of the good; thus the entire burden of the tax falls on consumers, while the government can implement transfer regimes to achieve any desired distributional outcome. This supports many partial-equilibrium studies in this book that examine the gasoline tax incidence by focusing on the developing world, while lacking the detailed data for a complete general equilibrium analysis.

In Chapter 4, West and Williams consider the distributional effects of a gasoline tax increase using four incidence measures under three scenarios for gas tax revenue use. Based on the Consumer Expenditure Survey Data and TAXSIM model, they conclude that a gasoline tax in the U.S. is regressive if revenues are not recycled, but they also show that ignoring demand response will overstate this regressivity. The comparisons of the four welfare measures (see the section "Some Methodological Issues" later in this chapter for more details) suggest that the differences are relatively small between the most favored equivalent variation and easy-to-implement consumer surplus measures. Ignoring cross-price effects only introduces relatively small errors in incidence calculations, thus suggesting that if data is limited one can rely on simple incidence measures and ignore the cross-price effects. West and Williams emphasize strongly that using gas tax revenues to fund lump-sum transfers flips regressivity into progressivity, while the effects of funding labor tax cuts are smaller. This sheds some light on future research concerning goals related to tax incidence. We need to focus on how to adjust existing tax structures such as income tax adjustment or transfer programs to neutralize any negative distributive effects.

The three U.S. chapters together give the impression that there may indeed be some regressivity, yet several factors imply that regressivity is smaller than generally believed if proper analysis and sophisticated modeling are undertaken to include all effects. Futhermore, the regressivity results would easily be reversed if just a part of the tax proceeds were dedicated to those with lower incomes.

Institutionally most similar to the U.S. are the European countries. Sterner and Carlsson (Chapter 16) compare the fuel tax incidence in France,

Germany, Italy, Serbia, Spain, Sweden, and the UK. The overall Suits measures are slightly negative—i.e., they are regressive when using a traditional income measure, except for Serbia, a relatively low-income country. However, when expenditures are used as a proxy for lifetime income, the results are mixed—for instance in both Germany and Sweden fuel taxes turn out to be progressive instead of regressive, and adding indirect use does not overturn the results. The overall picture for Europe is of a rather neutral tax. Ščasný (Chapter 17) finds similar results (very weak regressivity) for the Czech Republic using a micro-simulation model for policy assessment. He finds that the negative distributional impacts on households are likely to be small, and again they could be reversed if mitigated through revenue recycling regimes.

Chapters 5 and 6 deal with middle-income countries in Central America. For Mexico, Sterner and Lozada find that the burden from fuel taxes is neutral or weakly regressive. The direct effect is slightly progressive, and if indirect use through public transport is included the total is weakly regressive with respect to income. Had the authors had the possibility to work with lifetime income or to look at the use of the tax proceeds, it is likely that the effects would have been fully neutral. Turning to Costa Rica, Blackman et al. find similar results: They find that the first-order distributional impacts of a fuel tax are different for gasoline and diesel: a tax on gasoline is progressive and a tax on diesel is regressive, mainly because households in low and middle socioeconomic strata devote a significant share of their expenditures to bus travel. However, even in the short run, the impact of an increase in fuel taxes on these poorer households would be modest: a 1% increase in diesel taxes would spur at most a 0.5% increase in spending on diesel, including spending via bus transportation. These results suggest that distributional concerns need not stand in the way of using fuel taxes to address vehicle-related externalities.

Next we turn to Asia: Jing Cao (Chapter 7) first estimates fuel demand elasticities based on panel data of Asian countries and finds that the Asian developing countries (China, India, Indonesia, Philippines, and Thailand) are in a similar range compared with western countries, with a short-run price elasticity of fuel demand at –0.2, income elasticity at 0.3, long-run price elasticity at –0.8, and income elasticity at 1.3. Then based on a household survey in China, Cao examines the incidence of recently launched fuel tax reform starting in China as of January 1, 2009. The direct tax burden is progressive in the case of China, though the indirect tax burden is an inverted U-shape and peaks at decile 9. Since the very poorest households cannot afford public transportation, and the very richest households rarely use public transport, the impact of a fuel tax through public transport mainly falls on the middle-class population. Using a simple spreadsheet model assuming all the households have the same elasticities, the short-run and long-run fuel tax incidences are both progressive. Of consequence for a transitional economy with a rapidly increasing number of vehicles, Cao's study also calculates the Suits index for the fuel tax in China. Irrespective of the method used, the Suits coefficient of total tax burden is strongly progressive and increases from 0.09 in 2002 to 0.24 in 2007 using the lifetime income approach, though the Suits coefficient measured using expenditure deciles is smaller than one using income deciles.

Datta (Chapter 8) also finds that in India the direct effects of a fuel tax on all fuels except kerosene and coal are progressive. Unlike most of the other chapters, Datta looks at the distributional effects of taxes on all major cooking, lighting, and transport fuels. Since solid fuels are a major source of emissions in India, and a large part of energy produced in India is thermal power, the incidence of a tax on coal has important policy implications. Second, when calculating incidence, Datta includes all indirect effects by applying an input-output model analysis, while most research (including some in this book) restricts attention to indirect effects in the public transport sector, assuming that a substantial part of the indirect effects can be captured.[3] Datta finds it is important to capture multi-sector interactions; for instance, the Suits index of a coal tax when all indirect effects are taken into account is 0.07, compared to –0.24 when only direct effects are considered. An item like food might not appear important since its fuel share is low—but the budget share for food in the lowest deciles is so high that it could still be important. In urban areas, a tax on cooking gas falls mainly on the middle deciles. Datta also decomposes the sectoral response to a coal tax: the electricity sector responds most and plays an important role in determining the incidence. Kerosene is the only major fuel that consistently remains regressive even when indirect effects are included. Transport taxes, however, are always progressive even with all indirect effects included.

Yusuf and Resosudarmo (Chapter 9) use a computable general equilibrium (CGE) model to examine the incidence of reducing vehicle fuel subsidies in Indonesia. One of the contributions of this chapter is to capture price effects on the expenditure side as well as changes on the factor returns in the income channel. The authors find similar strong evidence of progressivity, since higher income households spend more intensively on vehicle fuels and commodities closely associated with them, while they are also dependent on income from factors of production in the sectors that are heavily affected by the reform. Thus we find that transportation fuel taxes would be progressive in all three major Asian countries studied. Because these countries contain almost half the world's population, these results are significant.

The last big block of countries contains the African countries in Chapters 10 to 15. For Ethiopia (Chapter 10), Mekonnen et al. find that fuel taxes are progressive. Even for public transport they are progressive (although less so than for private transport). In all scenarios, the burden is the largest in the richest decile, so higher taxes will have more impact on households with the highest expenditure and hence may be pro-poor. Akpalu and Robinson (Chapter 11), examining the distributional impact of Ghana's fuel subsidies, find that the transport fuel subsidies that have been a frequent element of government policy are regressive, implying that taxes would be progressive. In Chapter 12, Mutua, Börjesson, and Sterner find that true for Kenya also:

fuel taxes would be strongly progressive, based on evidence from a household sample that covers only Nairobi. Including lower income and lower vehicle density areas would presumably increase the size of this effect.

Kpodar (Chapter 13) applies an input-output approach to assess the impact of oil price changes on income distribution in Mali. Both price-controlled and non-price-controlled sectors are considered. As in Datta's paper on India, it is important to keep track of how budget shares for different items vary by decile and how the pass-through from energy prices thus hits differentially either through budget shares or high pass-through. The total effects show that rich households benefit disproportionately from oil subsidies, thus removal of gas and diesel subsidies is progressive. However, removing subsidies is politically difficult. Politicians have to clearly explain the rationale to the public and gradually push forward the removal of such subsidies. Nevertheless, kerosene may need to be somewhat protected considering that the very poorest households use kerosene more extensively.

Turning to South Africa, Chapter 14, based on the South African Income and Expenditure Survey of 2000, Chitiga, Mabugu, and Ziramba conclude that fuel taxes are not as regressive in South Africa as is sometimes claimed. The analysis of indirect fuel use shows that middle-income groups spend more on fuel than other income groups, suggesting some progressivity of fuel tax as the budget share of indirect fuel increases until the seventh decile.

Chapter 15 on Tanzania brings interesting insights that are both empirical and methodological. Mkenda et al. employ two main approaches for assessing whether fuel taxation is progressive or regressive. The first approach uses both the Kakwani and Suits indices, including estimates of their standard errors to judge statistical significance of the results. Their second approach is stochastic dominance tests using the concentration curve following Yitzhaki and Slemrod (1991) and also an assessment of income elasticity of fuel based on the concentration and Lorenz curves following Kakwani (1977b). On the whole, the preliminary assessment (disregarding some differences between rural and urban areas) is that taxes on charcoal, electricity, petrol, and diesel are found to be progressive, while taxes on firewood and kerosene are regressive (though we do not know if they would have been progressive with a lifetime income approach).

Taken together, our results suggest that a fuel tax would not impose an excessive burden on the poorest households as has been argued in the literature. When all forms of fuel use (except perhaps kerosene) are taken into account, fuel taxes are in effect progressive, even if the progressivity is lower in some countries than in the low-income African countries mentioned above. This suggests that fuel tax would be an effective and desirable instrument for pollution control. Evaluation of alternative price reforms shows that increasing the price of fuel in a bid to reduce emissions is progressive. When one takes total transport costs into account, the effect is not as progressive, particularly between the middle-income and the richest households.

The European countries in Chapters 16 and 17 have already been discussed. A very important final category of countries that we would have liked to analyze is the oil producing countries in the Middle East. We have only managed to undertake an analysis of one such country, Iran. This is a particularly interesting case because Iran, like several of its neighbors, subsidizes fuel very heavily on its domestic market. In Chapter 18, Ettehad and Sterner show that fuel taxes (or abolishing subsidies) would not be progressive but neutral or weakly regressive. Iran is a middle-income country with extremely low fuel prices, and thus fuel consumption is widely spread even amongst the lowest deciles. Since the gasoline subsidy is so costly (4% of the country's budget in 2009), and since the country faces the embarrassing need to import fossil products despite being one of the world's main oil producers, a subsidy cut is nevertheless urgent. Such a cut could trigger widespread public opposition in the short term, and the government apppears to be very wary of such opposition.

Meta-analysis of Regressivity

To give this book a measure of homogeneity and comparability we imposed a certain methodological uniformity across the chapters. This was done by sending out sample analyses and by asking all authors to use the Suits (in some cases supplemented by the Kakwani) index in as many as possible of the chapters. To avoid repetition, the chapters have been edited to remove discussion of index construction and so forth, which is instead concentrated mainly in the introduction and this concluding chapter.

Introduced 1977, the Suits and Kakwani indices were inspired by the better known Gini coefficient (Gini 1912). The idea was to summarize the distributional effects of a tax or subsidy into one single coefficient. The indices used in this book are the most commonly used progressivity indices in economics and have some attractive features: zero indicates proportionality, negative values regressivity, and positive values progressivity. They can easily be calculated with average data per decile and nicely illustrated by concentration curves. Gini type indices are, however, simple geometric constructs with no clear welfare theoretic foundation, so they can be somewhat unintuitive. We have discovered during many hours of detailed work on this book one slightly unfortunate feature in the weighting and construction of Suits and Kakwani indices: because these indices are weighted by income, the relation between richer deciles becomes more important to the outcome than the relation between poorer deciles.

This can be illustrated by a simple pedagogical example, the data of which are shown in Table 19-1, where the shares of fuel tax payments in total expenditures for the 2nd and 9th decile are each 5%, while they are 1% for all other deciles. The reader will appreciate that this is a perfectly symmetrical case. Had all deciles had a 1% share we would of course have seen this as a completely

TABLE 19-1 An Example of Tax Payments with Three Different Income Distributions.

		Decile									
		1	*2*	*3*	*4*	*5*	*6*	*7*	*8*	*9*	*10*
	Fuel tax rate	1%	5%	1%	1%	1%	1%	1%	1%	5%	1%
Income distribution	Low inequality	7.5	8	8.5	9	9.5	10	10.5	11	12	14
	Mid inequality	3	4	5	6	7	8	11	14	17	25
	High inequality	0.1	0.3	0.6	1	2	6	7	13	20	50

neutral tax. An increase in a tax on fuel would have affected all deciles equally. In our case, two deciles are affected more—one the second richest and one the second poorest. So one may ask: What value would we now want from our index? Disregarding possible differences in the marginal utility of money, one could point to the fact that there is symmetry, and one would therefore classify this as neither progressive nor regressive. If one were somehow to think that effects on decile 2 were more important than effects on decile 9—for instance due to a pro-poor welfare bias—then one might want our summary index to qualify a tax in this case as regressive.

Now let us consider what actually happens to the Suits and Kakwani indices: First of all we need to understand that the Suits index compares a given tax distribution to the underlying income distribution. If the tax makes the given distribution less uneven, it is a progressive tax. For illustrative purposes we can imagine three very different income distributions in Table 19-1.

The first distribution has a low level of inequality—the income share of the richest decile is less than twice that of the poorest, while this ratio goes to 8 for middle inequality and 500 for the high levels of inequality assumed. Given this distribution of tax burden, we get a whole continuum of different results depending on the distribution of income. For a very low level of inequality, the outcome is somewhat progressive (0.022). The middle level of income inequality yields a proportional outcome (rounded to three decimals). In the most uneven income distribution, the result is clearly regressive (−0.107), and as we see in Figure 19-1, the dashed grey curve for high inequality has a very large portion above the 45° line, which reflects the fact that the 5% tax on the rich decile 9 has a much larger geometric influence than the 5% on decile 2.

For an international comparison of fuel taxes in countries with very different income distributions, this property needs to be understood. In reality, the percentage distribution of tax burden is never as extreme as in our example, but these effects will nonetheless be real in countries where middle-income earners have a higher fuel share than both the richest and the poorest. The progressive relation between low and middle deciles will have a lower

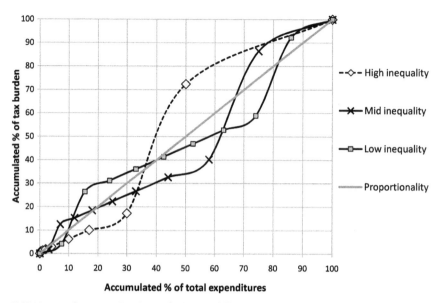

FIGURE 19-1 Concentration Curves for Low, Middle, and High Income Inequality.

impact on the coefficient than the regressive relation between mid-deciles and the highest deciles. The Kakwani index is less biased towards the richer deciles but still somewhat. For our purposes, such an effect is not desirable. Rather, we would like each decile to have an equal importance, i.e., weighted after its share in the total population.

To achieve this, we can supplement the standard indices with a new index, one which is just based on the taxes paid by each decile. If for instance we are interested in comparing a fuel tax with a proportional value added tax, we will judge any tax where budget shares increase as progressive and any tax where budget shares decrease as regressive. Instead of the income weight used in the Suits index, the Equal Weight index uses equal population weights instead so that each decile, or tenth of the population, also gets one tenth of the weight in the index summation. Note that this is very similar to the Jinonice index introduced by Ščasný in Chapter 17.

$$EW = 1 - \frac{\sum_{i=1}^{10}(1/2)\left[t_{x_i} + t_{x_{i-1}}\right](p_i - p_{i-1})}{5000}$$

Here, t_{x_i} is the accumulated percent of the fuel tax's share in total expenditures, for each decile i. p_i is the accumulated percentage share of population for decile i. By definition, each decile is 10 percent of the population, so the accumulated share for the first decile is 10 percent, for the second decile it is 20 percent and so on. A comparison with the formula for the Suits index in Chapter 1 shows clearly that it is the unequal weighting that makes the difference.

TABLE 19-2 Suits, Kakwani, and Equal Weight Indices for Different Income Distributions.

Inequality	Suits	Kakwani	Equal Weight
Low	0 .023	0 .016	0
Med	0 .044	0 .038	0
High	−0.095	0 .004	0

Table 19-2 shows the Suits, Kakwani and our Equal Weight indices for the three different illustrative income distributions. It shows that the indices calculated depend on the underlying income distribution and that particularly the Suits index is quite sensitive since it even changes sign in our high inequality case. The relationship between the index and the degree of inequality is however quite complex because it depends on a geometric construction without welfare foundations. This provides some argument for looking at our Equal Weight index.

Note however that the assumptions both concerning distribution and expenditure shares in table 19-2 were quite extreme and normally we would not expect to get completely different interpretations. Table 19-3 shows Suits, Kakwani and Equal-Weight coefficients for the countries studied in the book. For this cross-country comparison, we have chosen to calculate the coefficients from the expenditure approach, defining the tax burden as the share of paid fuel tax in total expenditures. Exceptions are the US data from years 2001 and 2003 and Mexico for 2004–2005, where the income approach is used since expenditure data were unavailable.

We see in Table 19-3 that the difference between the indices is seldom large. While it is true that the Kakwani gives slightly more progressive results than Suits, and the Equal Weight index again gives even more progressive results than the Kakwani, the differences are small—the simple averages in Table 19-3 (for those countries that have all three measures) are 0.01, 0.02, and 0.06 for Suits, Kakwani, and Equal Weight respectively. In only three cases is there a sign change, but there—in Spain, the UK, and the U.S. 2001—all the index figures are so close to zero anyway that any reasonable characterization finds them proportional or neutral, neither progressive nor regressive.

If we are to summarize the country results, we find regressivity in Italy and the U.S. (at least in two out of three studies; the third is slightly less conclusive), with weak regressivity in Mexico (where income, and not lifetime measures were used). If we consider as neutral countries where the absolute value of the indices is smaller than 0.02, then we have quite a large group of countries where a fuel tax would be neutral: The Czech Republic, France, the UK, Iran, and Spain. In the remaining 14 countries—a definite majority both by number and even more so by population—we find that a fuel tax would be progressive. Table 19-3 also includes the ordinary Gini measure of income inequality, but there is no very obvious relation. Instead there is quite a clear relationship with income per capita, as illustrated in Figure 19-2: The lower the income, the more fuel taxes are progressive.

TABLE 19-3 Progressivity/Regressivity Indices for Selected Countries[a]

Country	Time	Suits	Kakwani	Equal Weight	Gini
Brazil**	1996–1997	0.067	0.090	0.234	55.00
Chile	2007	0.132	0.127	0.188	52.00
China	2002–2007	0.205	0.203	0.180	41.50
Costa Rica	2005	0.023	0.051	0.189	47.20
Czech Republic	2005	−0.011	−0.012	−0.011	25.80
Ethiopia	2004–2005	0.245	0.220	0.259	29.80
France	2006	−0.024	−0.019	−0.007	32.70
Germany	2006	0.008	0.017	0.047	28.30
Ghana	1999	0.136	0.119	0.145	42.80
Great Britain	2006	−0.004	0.005	0.028	36.00
India	2003–2004	0.067	0.067	0.070	36.80
Iran	2006	−0.016	−0.012	−0.014	38.30
Italy	2006	−0.110	−0.098	−0.074	36.00
Kenya	2004–2005	0.225	0.206	0.228	47.70
Mexico	2004–2005	−0.100	−0.068	−0.028	48.10
Peru	2006	–	–	0.503	49.60
Serbia	2007	0.061	0.056	0.097	–
South Africa	2000	0.164	0.141	0.311	57.80
Spain	2006	−0.002	0.001	0.007	34.70
Sweden	2004–2006	0.064	0.068	0.078	25.00
Tanzania	2007	–	0.520	–	34.60
U.S.	1996–1998	−0.194	−0.172	−0.137	40.80
U.S.	2001	−0.14	−0.031	0.007	40.80
U.S.	2003	−0.253	−0.220	−0.226	40.80

Notes: Most indices are from the data in this book's chapters. For the U.S., data from Chapters 2–4 are used (1996–1998 data from West and Williams; 2001 data from Bento et al.; 2003 data from Hassett et al.). *Gini-coefficients included for comparison calculated from UN development report data, 1992–2007. (http://hdrstats.undp.org/en/indicators/161.html accessed September 29, 2010). **From Living standards measurements survey, Brazilian Institute of Geography and Statistics (IBGE). (http://go.worldbank.org/2OEIODS4U0 accessed October 7, 2010). [a] For Czech Republic, coefficients are based on data from households only with fuel expenditures larger than zero (which probably gives more regressivity). For Peru, we have no income or total expenditure data, making it impossible to calculate Suits or Kakwani. The data for almost every country include both direct and indirect fuel tax components. Some studies' more advanced models include all indirect effects; others are simpler and include only public transport expenditures besides direct fuel use (e.g. Brazil, Chile, China and the European countries). See respective chapters for details.

The distribution effects studied are a combination of direct distribution effects based on private consumption of fuels and indirect use such as in public transport. The former is almost always a progressive effect. The indirect effects on public transport, however, vary significantly across countries. Public transport is often used more intensively by the poor or by middle classes, and so the distributional effect of fuel taxes through this route is more likely to be regressive, though in some very low-income countries even public transport is too expensive for the very poor. Our chapters show that in South Africa and Mexico, such indirect fuel taxes are regressive; in China they have an inverted-U shape; while in India they are progressive. Overall the aggregate results tend to be less progressive. In the case of Kenya the authors find that when including public transport expenditures, the Suits index is 0.225 using a lifetime income approach, which is less than half of the Suits coefficients (0.483) for the direct tax burden. Similarly for the Ethiopia case (Chapter 10), the Suits coefficient for total transport fuel tax paid (including both private and public transport burden) is 0.245, roughly a third of the value for private transport fuel tax paid (0.668). In addition, if the poorest households spend a high fraction of total expenditures on other types of fuel—diesel, kerosene, or coal for heating or other purposes—the results tend to be less progressive if these fuels are included.

Figure 19-2 plots the Equal Weight index against income and shows that in most countries a tax on fuel would be progressive. There is a clear relationship with income per capita. In poor countries the progressivity is much more pronounced. The higher the average income in a country, the less progressivity there is, and in the richest of countries there is some regressivity. The reader is reminded that these measures do not generally include all indirect effects, nor

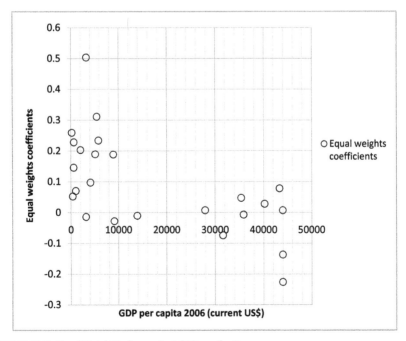

FIGURE 19-2 Equal Weight Index against GDP per Capita.

Note: For Tanzania the Kakwani index is used since it was the only value available estimate.

do they include adaptation mechanisms such as individual decile elasticities, nor do they include the distributional effects of environmental damage. Both the latter factors would tend to give more progressivity. Finally, they do not of course consider the use of the funds collected.

Another salient point is that the Equal Weights index gives equal weight to all deciles. Use of Kakwani or Suits would give less weight to the very poorest deciles and provide an estimate that is more regressive. The reason for this is that many of the expenditure share curves are hump-shaped with respect to income so that the very lowest and highest deciles have lower fuel expenditures. With Suits (and Kakwani) indices the deviation in the richest deciles is given greater weight than the one at the bottom of the distribution. With our modified index they are given equal weight. (One could of course also construct indices that gave greater weight to the lowest deciles—but there is no single way of doing this that is obviously superior to others, and we thus leave this for future research).

Some Methodological Issues

As noted in the survey of chapters, there are a number of methodological issues each of which can be decisive for the overall conclusion of whether or not a fuel tax is regressive. This is particularly so for those countries where the degree of regressivity or progressivity is fairly close to neutral so that small changes can push the conclusion either way. In this section we will discuss these issues one by one.

Selecting Appropriate Welfare Measures

An ideal measure of fuel tax incidence would reflect all general equilibrium changes in welfare due to fuel tax changes. This would include the direct effects on fuel consumers due to the rising consumer price of fuel and the indirect effects of rising prices of sectors using fuel as intermediate inputs, thus imposing an indirect burden on consumers as well. In addition, there would be capitalized effects that affect factor owners of resources that are complements to or substitutes for fuel. Therefore with changing fuel prices, we get a whole series of other price changes. For example, the value of used cars will change depending on how relatively fuel efficient they are. If fuel becomes very expensive, then gas guzzlers will depreciate faster. Patterns of transportation will change with concomitant changes in property values. There will also be labor market effects due to shifts in the demand for public transport and so forth. Further examples are effects on shareholders (as well as business partners or employees) of gas stations and refineries subsequently affecting workers in these sectors through lower wages. However, to calculate such economy-wide effects would require a finely calibrated general equilibrium model to assess the direct and indirect effects. Most CGE models are based on the representative household assumption, making such a task very difficult, since we would of course have to differentiate different household types. In addition, as West and Williams mention in Chapter 4, both demand and supply elasticities for all the relevant industries and the distribution of ownership of firms in such industries would be required. Such an investigation is often outside the scope of many research projects in developing countries with poor data availability. Therefore, for simplicity, studies commonly assume that the supply of consumer goods is perfectly elastic, implying that consumers bear the entire tax burden. A similar assumption applies to workers as well. These assumptions allow us to focus on the incidence for households alone. In practice, of course, these assumptions do not hold; thus the results of incidence on households that consume fuel will be biased upwards, while households that own firms or work in the refinery industries that supply fuel will be prone to bias downwards. A possible extension to adjust these biases might rely on a micro-simulation CGE model based on information combining a social accounting matrix and household expenditure survey data.

In principle, one would also like to account for the distributional effects of (avoided) pollution. A fuel tax will typically reduce and restructure traffic, in turn changing the extent and spatial distribution of (urban) pollution, finally having distributional effects depending on residential distribution. Typically, people with fewer economic resources spend much more time in heavily polluted areas. They might thus stand to gain more by a reduction in pollution. Understandably, the extent of this reduction is quite hard to evaluate, and few studies attempt to do so. For climate change the effects are in any case not directly local—still, distributional effects occur at the global level because some people and some countries will be much more affected than others. But there is no relationship to the actual distribution of driving patterns—only to total emissions.

In practice, the incidence of a fuel tax is typically assessed by examining the variation across economic strata of some measure of changes in households' welfare due to the tax normalized by pre-tax absolute levels of welfare. In Chapter 4, West and Williams discuss three methodological issues regarding welfare measures:

1 **Equivalent Variation (EV) approach:** The most rigorous measure of welfare change due to the imposition of tax changes, which implicitly accounts for all cross-price effects through the cross-price derivatives of the indirect utility function. The applications of EV for studying the incidence of tax policies have been conducted by Cornwell and Creedy (1997), and Brännlund and Nordström (2004), but we have found no study calculating EV measures on fuel tax incidence (a challenge for future work). However, the difficulty with this approach is estimating the indirect utility function, which is often unavailable.

2   **Consumer Surplus (CS) approach:** The consumer surplus approach requires less information than the EV approach, and the researcher only needs to know expenditure on the taxed good before policy change, percentage change in price, change in income induced by the tax, and own-price elasticity for any goods whose price changes. This method does impose more assumptions, thereby bringing some bias to the estimates. For instance, due to the income effects, the welfare change measured under the CS approach is the area under the uncompensated demand curve, while the welfare change measured under the EV approach is the area under the compensated demand curve; the two are often different, unless the income elasticity of demand is zero, which is generally not realistic at all in the case of fuel, where income elasticity is very high in some developing countries, in particular long-run elasticities.

3   **"No adaptation" approach:** As West and Williams mention in Chapter 4, many incidence articles ignore demand responses altogether, simply summing incidence over all goods of the price changes plus the income changes after imposing the policy (Metcalf 1999; Poterba 1991; West 2004; and Walls and Hanson 1999). However unless all demand elasticities (including both cross-price and own-price) are equal to zero, this measure will differ from both the EV and CS approach.

An additional and closely related methodological issue concerns whether one should measure *variation across deciles in the price elasticity of fuel demand or assume that demand elasticities do not vary across households*. Most studies in this book use the average demand elasticity for all households to calculate tax incidence. Šťasný (Chapter 17) estimated different elasticities for different income groups, and West and Williams found in Chapter 4 that demand is more elastic in lower income groups—a feature that reduces fuel tax regressivity, suggesting that the average approach usually used might overstate the incidence on those lower income groups with relatively elastic demand, and understate the incidence on those higher income groups with relatively inelastic demand. However, with micro-level data it is questionable how reliable the own-price elasticities at the decile level may be. Even aggregate elasticity estimates based on household data are notoriously sensitive (Dahl and Sterner 1991a), while by-decile estimates are likely to be even more sensitive to household characteristics, education, age, and so on, and omitted variables may become important sources for bias. It would require considerably more work in the future to compare how these by-decile demand elasticity estimates differ from the aggregate demand elasticities in both developed and developing countries.

Lifetime Income versus Annual Income Approach

Studies that use annual income are somewhat more prone to support the common assertion that fuel taxes are regressive in western countries (KPMG

Peat Marwick 1990). However, this approach has been criticized on the grounds that households make consumption choices on the basis of their "lifetime" or "permanent" income rather than annual income (Friedman 1957). As Poterba pointed out, "...data on income dynamics suggest a surprising degree of instability in the annual income distribution" (1991). Therefore, using transition data based on an annual income approach seems to overstate the regressivity compared with a "lifetime" income approach (Poterba 1991; Walls and Hanson 1999; Metcalf 1999; Hassett et al. 2009). In this book, we report and compare both measures where available but feel somewhat more confident using expenditure data as a proxy for lifetime income to sort households into economic strata and normalize changes in welfare (see Chapters 5, 7, 11, and 14).[4]

Fuel Tax Incidence, Revenue Recycling, and Induced Technological Change

An ideal study of fuel tax incidence would also consider effects from recycling of fuel tax revenue. Metcalf (1999) and Wiese et al. (1995) suggest the incidence of fuel taxes will depend on how the tax revenue is recycled. Different rebates to households according to various criteria such as income would be an effective way to mitigate any regressivity. Alternatively one may imagine that other taxes are lowered in a budget neutral tax reform, or else a budget deficit might be reduced or public spending increased. Researchers might analyze the distributional incidence of a reduced budget deficit or increased provision of public goods. In addition, tax incidence studies should study whether a fuel tax policy would induce fuel-saving technological change in the auto industry, thereby improving its competitiveness. Cleaner cars would also bring benefits to households through indirect mechanisms, with different effects across deciles. For example, rich people, often the first buyers of new cars, might be affected more—or at least earlier—by the availability of cleaner cars. Chapter 3 particularly demonstrates that revenue recycling effects can easily offset regressivity.

System Borders

Another issue touched on above is where system boundaries for a study should be drawn. When governments impose a new fuel tax or increase pre-existing tax rates, that tends to affect the value of certain capital goods. These capitalized effects, mentioned above, may be quite important. They will hurt primarily whatever groups in society are the owners of property that is most connected with fuel intensity. We tend to think of low-income owners of poorly insulated houses or old fuel-guzzling cars, but one may equally imagine reasons why fuel taxes might primarily hit the rich. As we have seen, the balance is an empirical question that may well vary from country to country and is not necessarily easy to capture since it requires CGE modeling sensitive to the consumption and ownership patterns of different economic strata. Another important methodological issue that pushes the boundaries of what most researchers are able to do is the inclusion of the distributional aspects of *external benefits of reduced fuel consumption.* These include improved local environmental air quality or congestion effects, the benefits of which are unevenly distributed across income groups. Also, as Datta notes in Chapter 8, if diesel and gasoline prices are raised, a potentially unforeseen effect is abuse: kerosene and other similar products can be used to adulterate diesel, smuggling could increase, and so on. Naturally, the degree of progressivity will differ depending on whether only fuel taxes alone are raised or if the increase applies to all fossil products.

Finally, we note a number of technical measurement issues of concern when we use household survey data for tax incidence studies. For example, many researchers prefer to adjust data for differences in family size and composition by using equivalence scales. Household needs grow with additional members, but there are "economies of scale" in consumption and therefore needs do not grow proportionally. The need for a car, for instance, does typically not increase very much as the household grows from one to four or five persons since that many people easily fit in the same car. Equivalence scales assign to each household a value that is supposed to be proportional to its needs. The measures usually take into account the size and composition of the household (e.g., number of children and their age). A range of equivalence scales exist (Atkinson et al. 1995).[5]

Some of the studies in this book use equivalence scale to adjust total expenditure on fuel for different family sizes. Since the deciles are intended to reflect individuals' or households' standard of living, with a given level of total income or expenditure, household size matters in incidence calculation. Though both West and William (Chapter 4) and Cao (Chapter 7) mention that results are similar when using different equivalence scale methods, results might change in some circumstances. As an example of the impact when using different adjustment methods for the household's size and composition, Table 19-4 shows progressivity coefficients for two countries (Serbia and the U.S.) and three types of adjustment.

First, the no adjustment approach means that the economic well-being of each household is assessed only from its total expenditure, regardless of the household size and the age of its members. This can be considered as the least desirable alternative, since it clearly overstates the economic well-being of a large family. Second, the per capita adjustment is calculated by dividing

TABLE 19-4 Progressivity Coefficients for Different Types of Household Adjustments.

Adjustment type	Serbia*			USA**		
	Suits	*Equal Weights index*	*Kakwani*	*Suits*	*Equal Weights index*	*Kakwani*
No adjustment	0.081	0.173	0.083	−0.084	−0.039	−0.066
Per capita adjustment	0.046	0.070	0.050	–	–	–
Equivalence scale	0.061	0.097	0.065	−0.143	−0.114	−0.129

*Data from Chapter 16. **Data from Chapter 3.

household expenditures by the number of members, then arranging the households according to total per capita expenditures (only used for Serbia). This approach is an improvement, but it assumes that the economic needs for children and adults are equal. Last, the equivalence scale approach adjusts for both size and composition in the household, then arranging the households according to equivalence scale adjusted total expenditures. In our example in Table 19-4, the OECD equivalence scale was used for Serbia, while the OECD-modified equivalence scale was used for U.S. (for more information about this see endnote 5 and OECD 2010). Table 19-4 shows clearly that the type of adjustment does not have a large impact on the result, though the differences are somewhat larger for the Equal Weights index. In the cases studied, the type of adjustment seems to affect the magnitude, not the sign, of the coefficient.

Finally, another important technical issue is sample selection bias: the household survey sometimes leaves out the poorest and the richest households (simply because they do not make themselves available for the gathering of data). Sample selection bias may arise and lead to some bias in incidence estimations. And one should not forget that we have other sources of sampling error than selection bias. For example, the household budget surveys have sometimes fairly uncertain estimates for expenditures on fuel and public transport in a specific decile (large confidence intervals due to small number of observations).

There are thus numerous possibilities for future research, but the numerically most important are likely to be the use of the proceeds from taxation—or alternatively the baseline point of comparison—which ought not to be no tax but rather an alternative tax generating the same amount of revenue.

Conclusions and Further Reflections for Environmental Policy

Mitigating carbon emissions to avoid excessively destructive climate change while we experience economic growth is going to require carbon intensities to be reduced sharply. This will require a whole series of measures; no sector will be excepted. Fuel use in the transport sector tends to grow rapidly, with income elasticities of at least one, and in fast growing developing countries such as China or India often much more. Since price elasticities are lower than income elasticities, prices will have to rise very substantially in order to both reduce demand by stimulating technical innovations and lifestyle changes at the same time as providing incentives for the development of new non-fossil fuels (or carbon capture and storage and similar technologies). By implication, considerably higher fuel taxes will be needed, but such taxes are often resisted. Although the resistance may in reality be driven by the vested interests of various special groups, anti-fuel tax intellectual argument tends to center on the claim that fuel taxes would be regressive and hurt the poor.

Our book set out to judge this argument. We find no foundation for such a claim not only in poor but even average income countries. In the major

Asian economies as well as in the African countries studied, fuel taxation in itself would be a progressive tax. In some of these countries where fuels are subsidized, abolishing these subsidies would be a progressive change in itself. In intermediate and richer countries the picture is more mixed. In most of these countries we still find that a fuel tax would be fairly neutral, but in some, like the U.S., it might be regressive. We should reiterate that to some extent our results have a bias towards regressivity, so if we were to include more adaptation mechanisms and capitalized effects, they would tend to lean even more toward the progressive side. Also, the use of expenditures rather than stated income tends to give somewhat more progressive numbers. On the other hand there could also be mechanisms pointing toward regressivity—particularly the inclusion of more indirect effects.

The effects discussed in the previous two paragraphs are the effects of the tax itself. But the tax generates revenues, and one must therefore discuss their use. If new revenues were used to reduce a regressive tax or simply refunded on a per capita basis, it would be easy to make an overall fuel tax reform progressive, even if the direct effects of the tax, in isolation, were regressive.[6] But the question remains as to whether there is demand for, and indeed political will for, progressive tax reforms. Although it is possible to make the whole tax system more progressive, few countries in fact have strongly progressive tax systems. Both the U.S. and Sweden have, for instance, during the last decades cut taxes much more for high-income earners—implying more regressive tax reform. That trend raises further questions as to why the regressivity argument comes up so forcefully in the context of fuel taxes, which after all are relatively unimportant compared to the sum of all other taxes. Perhaps the political logic goes the other way. Broad popular groups in the middle and lower middle of the income distribution range have found themselves powerless to stop a general regressive movement, with increased wage differentials and with tax reductions for the richest. Perhaps that creates such mistrust against all tax reform that when the opportunity comes to criticize one tax it is easy to mobilize an opinion. Moreover, people can certainly see and understand the immediate impact of a proposed tax that increases the price of a specific good or service, and yet they do not necessarily recognize that *any* given tax must be offset either by lower taxes elsewhere or lower budget deficits.

To mobilize public opinion, the most important ingredients are a well-organized lobby with strong interests and a reasonably receptive audience. The well-organized lobby may well be provided by oil companies, trucking and transport companies, car companies, and maybe other entities who are served well by fuel-intensive activities. The middle classes (very broadly defined) may provide the receptive audience, particularly in light of the findings in this book and elsewhere that expenditure shares for fuel are in many cases hump-shaped, with the middle classes paying somewhat more per share than the very poorest and the very richest. If these middle classes compare themselves more with the richest than with the poorest, perceived regressivity will be magnified.

This tendency is often reinforced by anecdotal evidence concerning some *individual motorists* who are adversely affected by fuel taxes. Clearly people who live in low density rural areas and have long commutes will be negatively affected. One must however keep in mind that not all people in these areas have long commutes, so overall effects can be better judged by the type of methods used in this book rather than by reliance on individual anecdotal evidence. The population density argument is quite often heard. Basically it says that countries with low population density (like the U.S., Canada, or Sweden) would need more fuel per person given larger driving distances. However, low density rural areas actually do not play such a prominent role in the aggregate—exactly because so few people live there. It is much more important to consider the local population density and commuting distances in heavily populated areas, and it is no coincidence that the U.S., with its cheap fuel, has so much more sprawl than Europe. Developing fuel-efficient urban infrastructure is of course one of the slowest processes; nonetheless, it is vital to understand that the design of our communities also depends on fuel prices in the long run. It is also important for the policymaker to avoid the temptation to compensate all parties affected. Inevitably, people with long commutes will be hurt by fuel taxation (in fact the taxation is intended to provide an incentive for people to commute less). If politicians introduce mechanisms to compensate all those who are hurt, it tends to make policies very expensive and ineffective, as well as opening up avenues for rent seeking.

Interestingly, though the popular belief in all countries appears to be focused on how regressive fuel taxes are, our studies do not typically find this. They find neutral or very weakly regressive results in richer countries (which could easily be neutralized through appropriate use of revenue) and quite strongly progressive evidence in the developing countries—China, India, Ethiopia, Indonesia, Ghana, Mali, Kenya (Nairobi at least), and several more. That our intuition can be here demonstrated empirically is unsurprising; in most developing countries, the very poorest households cannot afford to own a car at all. Fuels have more of a "luxury" character, and hence fuel taxes are more progressive.

In really low income countries the progressivity is, as we have seen, quite clear; yet we have often met considerable resistance in these countries—even from our own colleagues. Colleagues from Africa for instance quite often start by believing that fuel taxes are regressive. Our own impression—which has not been tested or verified in any way—is that middle-class academics may implicitly compare themselves with the rich rather than with the very poor; this leads them to think in terms of regressivity. One might also propose that although many governments purport to speak for the poor, the very poorest often lack effective representation and voice. Instead, the urban middle classes are the most powerful and worrying factor for many governments. When people do take to the streets, they may well use radical slogans like "fuel tax is not good for the poor"—but one should not accept every protest slogan verbatim.

We also often hear the argument that when fuel prices rise, all prices rise. The dynamic described by quite a few colleagues is one of repressed inflation.

Shopkeepers would like to raise prices of all kinds of goods but are unable to do so because of price controls. Once the state does allow some prices to rise, this creates an opportunity to raise others. Often the first price to rise has in fact been petroleum products, since they are in many countries imported, and so politicians find themselves at the mercy of the vagaries of a rather erratic market. Historically, when oil prices did jump suddenly (for instance in 1973, 1979, 1990, or 2008) oil-importing countries experienced real losses quite distinct from a change in domestic taxation (which only implies local transfers of wealth). Yet for the non-economist this point may not be so obvious. These dramatic, historic cases of fuel price escalation were often followed by inflation and harsh times, and by loss in real income. In environments of controlled prices, governments could veto many price increases but were generally forced to accept the large increases in the price of imported petroleum products. In an environment of repressed inflation, the acceptance of one price increase often had the effect of opening the sluice gates to more price increases. Any industry that could cite increased transport or fuel costs as a valid reason would also raise prices—and anecdotal evidence has it that the prices would typically rise more than would be motivated simply by the increased input cost.

Periods of recession, inflation, or general malaise may well hit the poorest harder, and the mechanisms described may have helped make fuel price increases particularly unpopular. We have here focused on fuel taxes, but it might perhaps be more natural to write about carbon taxes in general.[7] An efficient climate policy should of course in principle tax carbon emissions at the margin equally, irrespective of source. But because much coal is used in industry or to make electricity, taxation raises a number of issues of (perceived) competitiveness, the analysis of which is not within the realm of this book. On the whole, our impression is that few energy carriers have a use profile that is easier to deal with than gasoline. With fuels such as kerosene, coal, and electricity we have both industrial users (and issues of employment) and also incidence on very poor households' demand for heating and lighting. Thus we believe that the case for carbon pricing will not necessarily be easier than the case for automobile fuel taxation.

Another more sophisticated argument that sometimes comes up is that even if the poorest lose 1% of their income to an increased fuel tax and the rich lose 5%, the loss to the poor might be bigger. This is basically an allusion to the marginal utility of money and the curvature of the utility function. As such it may be true, but once again everything comes down to the choice of reference point. We would argue that the realistic choice in many African economies is not between a fuel tax and no tax. The states need money to provide public goods like infrastructure, health, education, law, and order. They already have very high rates of dependency on foreign aid and can hardly increase those. A realistic choice might be between an increase in value added tax (VAT), which is essentially a proportional tax, and an excise tax on fuel that is somewhat progressive in the sense we have defined it. It may then be true that the poor have a higher marginal utility of income, but they should still prefer a fuel tax to a general increase in VAT.

In the months when this book is being finished, there is much talk of green growth, lifestyle changes, and the "Sputnik moment"—or the challenge of rising up to find the technologies of the future. Few politicians will speak in favor of a fuel tax, but it is exactly that kind of instrument (and more broadly taxes on all climate gases) that is needed to guide technologies and lifestyles. It would have been unfortunate if such policies would have hurt the poor, but as our work shows it seems we need not worry too much on that account. We do however need to think very hard about how policies should be designed and presented publicly. It is not a good idea to be complacent as long as prices are high. High fuel or energy prices may in some sense be "good for the environment," but if they are high due to monopolies, inefficiency, corruption or similar reasons then the high prices will be severely resented. We see in many cases how unpopular monopoly companies are, whether when charging high electricity rates (and offering poor service), or when operating large airlines. Their high monopoly prices tailored to businessmen may be good for the environment in one sense, but it is companies like Ryanair who win popular support.

For the policymaker, similar logic applies. Important lessons can be drawn from the case of Bolivia, where higher fuel prices were announced in 2010. When the government ended a six-year price freeze, gasoline jumped from 3.74 to 7.51 Bolivianos/liter (with similar increases applied to other fuels) according to Supreme Decree 748, December 26, 2010. The announcement led to such violent protests that President Evo Morales had to revoke all the price changes just a few days later on December 31. We do not have a study for Bolivia, but in Peru, which shares many economic features, we found that a fuel tax would be strongly progressive, see Table 19-4. The current government under Morales definitely rules the country on a mandate from the low-income Aymara and other indigenous peoples. Most of Morales' supporters do not have cars and probably would benefit from a fuel tax that is in their context close to a luxury tax—so long as the proceeds are honestly used to provide public goods or to replace more regressive taxes such as value-added taxation. If not even Morales can raise a fuel tax, then one suspects that this is partly due to having marketed the tax reform poorly.

It is important to make energy taxes and fuel prices transparent to make sure there is no suspicion of corruption.[8] It is important to depoliticize the process of tariff and tax implementation—in Bolivia the president and vice-president have recently appeared to be personally involved in a day-to-day management of fuel prices. It is usually preferable if the politicians formulate a long-term policy that is then implemented by civil servants according to clear and transparent rules and formulae. Likewise, gradual and pre-announced policies are usually easier to accept than sudden and unannounced policies. Trying to shelter consumers from imported price increases is prohibitively expensive for a low-income country and economically irrational. For this reason it is important to liberalize and introduce market prices and limit political ambitions to the setting of a fuel tax component. This tax, in turn, needs to be set according to explicit and clear formulae and should not rise too fast over time. Policymakers also need to explain in a very transparent and pedagogical manner how fuel tax proceeds will be used in order to increase acceptability and in fact build a positive interest and feeling of ownership in such taxes as an integral part of a popular climate policy.

Acknowledgments

Useful comments are gratefully acknowledged from Ashokankur Datta, Don Fullerton, Mark Jacobsen, Adolf Mkenda, Daniel Slunge, and Sarah West.

Notes

1   Another important category of transport fuel is the bunkers used by shipping and aviation. Aviation is growing particularly fast. Both are currently untaxed—largely due to the fact that they are outside national jurisdiction.
2   Actually, in equilibrium they find that a substantial fraction of the loss in welfare might be borne by producers (24% in the central case, but this of course depends on the revenue recycling considered).
3   Another point is that many studies that do not have full input-output data do not know the exact input coefficient for fuel in public transport, and they often estimate this through interviews or literature review, thus possibly introducing some element of bias.
4   An ideal way to further improve this approach is to construct lifetime income for all households as done by Metcalf (1999).
5   Several alternative scales are available. For instance, the "Oxford or OECD equivalence scale," which has a value of 1 for the first adult, 0.7 for each additional adult, and 0.5 for each child. Or, the "OECD-modified scale," adopted by EUROSTAT in the late 1990s, assigns a value of 1 to the household head, 0.5 to each additional adult member, and 0.3 to each child. Finally, a third alternative is the "square root scale" based on the square root of household size.
6   Fullerton et al. (2011) discuss how indexing of transfer programs such as Social Security in the U.S. (to account for changes in energy prices on cost of living) offsets regressivity across the lowest income groups when they are classified by annual expenditures as a proxy for permanent income.
7   See for instance Fullerton (2009) for a book that focuses on case studies of carbon rather than fuel taxation.
8   Various authors can provide more thoughts on the reduction of fuel (and energy) subsidies in general and the case of Bolivia in particular. See for instance work by Armin Wagner at GTZ (www.gtz.de/fuelprices), or Bacon and Levy at the World Bank, or Coady, Gillingham and others at the IMF.

References

Aasness J., and E. R. Larsen. 2002. *Distributional and Environmental Effects of Taxes on Transportation.* Discussion Paper No. 321. Statistics Norway, Research Department.

Abouleinein, S., H. El-Laithy, and H. Kheir-El-Din. 2009. The Impact of Phasing Out Subsidies of Petroleum Energy Products in Egypt. Working Paper No. 145 (April).

Addison, T., and R. Osei. 2001. Taxation and Fiscal Reform in Ghana. UNU-WIDER Discussion Paper 2001/97, Helsinki.

Agras, Jean, and Duane Chapman. 1999. The Kyoto Protocol, CAFE Standards, and Gasoline Taxes. *Contemporary Economic Policy* 17: 296–308.

Ahola, H., E. Carlsson, and T. Sterner. 2009. Är bensinskatten regressiv. *Ekonomisk Debatt* 37(2): 71–77.

Alfaro, M. del R., and G. Ferrer. 2001. Concentración de monóxido de carbono en San José. *Informe Técnico. Universidad Nacional.*

Aligula, E. M. 2006. Policy Options for Reducing the Impact of Energy Tariffs in Kenya. National Economic and Social Council. Final Draft.

Aligula, E. M., Z. Abiero-Gariy, J. M. Mutua, F. Owegi, C. Osengo, and R. Olela. 2005. Urban Public Transport Patterns in Kenya: A Case Study of Nairobi City. Survey Report, Special Report Number 7, Nairobi Kenya.

Allenby, Greg, and Peter Lenk. 1994. Modeling Household Purchase Behavior with Logistic Normal Regression. *Journal of the American Statistical Association* 89: 669–679.

Alpizar, F., and F. Carlsson. 2003. Policy Implications and Analysis of the Determinants of Travel Mode Choice: An Application of Choice experiments to Metropolitan Costa Rica. *Environment and Development Economics* 8: 603–619.

Alves, Denisard C.O. and Rodrigo De Losso da Silveira Bueno, 2003. Short-Run, Long-Run and Cross Elasticities of Gasoline Demand in Brazil, *Energy Economics,* Vol. 25 (March), pp. 191–99.

Amegashie, J. A. 2006. The Economics of Subsidies. *Crossroads* 6: 7–15

Ananthapadmanabhan, G., K. Srinivas, and V. Gopal. 2007. Hiding Behind the Poor, GREENPEACE INDIA.

ARESEP (Autoridad Reguladora de Servicios Públicos). 2008. Informe preliminar de estudio de oficio, modalidad autobuses. Anexo No. 1: Cuadro de ajustes tarifarios según resoluciones y fechas de corte, durante el período de inclusión.

Armington, Paul S. 1969. A Theory of Demand for Products Distinguished by Place of Production. International Monetary Fund, Staff Paper # 16:159–176.

Arndt, Channing. 1996. An Introduction to Systematic Sensitivity Analysis via Gaussian Quadrature. Technical Report 305 Center for Global Trade Analysis, Department of Agricultural Economics, Purdue University,

Asensio, J., A. Matas, and J. Raymond. 2003. Petrol Expenditure and Redistributive Effects of its Taxation in Spain. *Transportation Research A* (37)1: 49–69.

Atkinson, A. B., L. Rainwater, and T. M. Smeeding. 1995. Income Distribution in OECD Countries. *OECD Social Policy Studies* 18. Paris.

Austin, David, and Terry Dinan. 2005. Clearing the Air: The Costs and Consequences of Higher CAFE Standards and Increased Gasoline Taxes. *Journal of Environmental Economics and Management* 50: 562–582.

Automobile Association. 2009. *Fuel Price Report – June* 2006, Available (Online): www.theaa.com/onlinenews/allaboutcars/fuel/2006/June2006.doc (accessed March 02, 2009).

Avalos, A. 2007. "Accidentes son causa de 20% de las pensiones por invalidez." La Nación, November 15.

Azar, C. 2005. Post-Kyoto climate policy targets: costs and competitiveness implications, *Climate Policy* 5 pp 309–328.

Azis, I. J. 2006. A Drastic Reduction of Fuel Subsidies Confuses Ends and Means. *ASEAN Economic Bulletin* 23, no. 1: 114–36.

Bacon, R., and M. Kojima. 2006a. Coping with Higher Oil Prices?; Energy Sector Management Assistance Program, Report 323/06, c/o Energy and Water Department, The World Bank Group, Washington

———. 2006b. Phasing Out Subsidies: Recent Experiences with Fuel in Developing Countries. Viewpoint No. 37199, World Bank.

Bacon, Robert. 2005. The Impact of Higher Oil Prices on Low-Income Countries and on the Poor. UNDP/ESMAP Report. Washington: The World Bank.

Baig, T., A. Mati, D. Coady, and J. Ntamatungiro. 2007. Domestic Petroleum Product Prices and Subsidies: Recent Developments and Reform Strategies. IMF Working Paper, Fiscal Affairs Department, WP/07/71.

Baland, J. M., P. Bardhan, S. Das, D. Mookherjee, and R. Sarkar. 2007. Managing the Environmental Consequences of Growth: Forest Degradation in the Indian Mid-Himalayas. In *India Policy Forum*, ed. by S. Bery, B. Bossworth, and A. Panagariya. New Delhi: Sage Publications, 215–77.

Baldasano, J. M., E. Valera, and P. Jiménez. 2003. Air Quality Data from Large Cities. *The Science of the Total Environment* 307: 141–165.

Baranzini A., J. Goldemberg, and S. Speck. 2000. A Future for Carbon Taxes. *Ecological Economics* 32(3): 395–412.

Barnes, D. F., R. V. D. Plas, and W. Floor. 1997. Tackling the Rural Energy Problem in Developing Countries. *Finance and Development*.34(2)(June): 11–15.

Basri, M. Chatib and Arianto Patunru. 2006. Bulletin of Indonesian Economic Studies Survey Of Recent Developments. 42, no. 3 (2006): 295–319.

Baumol W. J., and W. E. Oates. 1988. The Theory of Environmental Policy. Cambridge University Press, Cambridge, UK.

BBC News. 2007. Iran fuel rations spark violence. http://news.bbc.co.uk/2/hi/middle_east/6243644.stm (accessed June 27, 2007).

———. 2010a. U.S. Senate Backs New Sanctions against Iran. January 29. http://news.bbc.co.uk/2/hi/americas/8486441.stm (accessed January 29, 2010).

———. 2010b. U.S. Puts Economic Squeeze on Iran. April 16. http://news.bbc.co.uk/2/hi/americas/8625621.stm (accessed August 2010).

BBC Persian. 2009. Oil Minister Mirkazemi Says Gasoline Price Increase Will Induce 11% Jump in Inflation Rate. October 9. www.bbc.co.uk/persian/business/2009/10/091009_ka_petrol_mirkazemi.shtml (accessed January 2, 2010).

Berkovec, James. 1985. New Car Sales and Used Car Stocks: A Model of the Automobile Market. *RAND Journal of Economics* 16: 195–214.

Berry, S. 2006. Whither the Public Distribution System? Rediff News, January 10, 2006.

Berry, Steven, James Levinsohn, and Ariel Pakes. 1995. Automobile Prices in Market Equilibrium. *Econometrica* 63: 841–890.

———. 2004. Differentiated Products Demand Systems from a Combination of Micro and Macro Data: The New Car Market. *Journal of Political Economy* 112: 68–105.

Bolnick, B., and J. Haughton. 1998. Tax Policy in Sub-Saharan Africa: Examining the Role of Excise Taxation. African Economic Policy Discussion Paper No. 2 (July).

Bovenberg, A. L., and L. Goulder. 2001. Neutralizing the Adverse Industry Impacts of CO_2 Abatement Policies: What Does It Cost? In *Distributional and Behavioral Effects of Environmental Policy*, eds. C. Carraro and G. E. Metcalf. Chicago: University of Chicago Press, pp. 45–85.

Boyd, R. and N. D. Uri. 1991. The Impact of a Broad-based Energy Tax on the U.S. Economy. *Energy Economics*, 13(4):. 258–273

BPS. 2006. Tingkat Kemiskinan di Indonesia, 2005–2006. Technical report Biro Pusat Statistik. Berita Resmi Statistik, Jakarta.

Brännlund, R., and J. Nordström. 2004. Carbon Tax Simulations Using a Household Demand Model. *European Economic Review* 48(1): 211–33.

Brooks N., and R. Sethi. 1997. The Distribution of Pollution: Community Characteristics and Exposure to Air Toxics. *Journal of Environmental Economic Management* 32: 233–50.

Brůha J. and M. Ščasný. 2005. Analysis of Distributional Effects due to Regulation in Energy and Transport Consumption Area. Working Paper prepared for the Final Report of R&D Project 1C/4/73/04 (Environmental and Economic Effects of Economic Instruments in Environmental Protection). Prague, mimeo (in Czech): IREAS and Charles University Environment Center,

———. 2006a. Environmental Tax Reform Options and Designs for the Czech Republic: Policy and Economic Analysis. p. 750. In *Critical Issues in Environmental Taxation: International and Comparative Perspectives*, eds K. Deketelaere, L. Kreiser, J. Milne, H. Ashiabor, and A. Cavaliere. Volume III. March 2006. Richmond, UK: Richmond Law and Tax Publisher Ltd.

———. 2006b. Distributional Effects of Environmental Regulation in the Czech Republic. Paper prepared for The 3rd Annual Congress of Association of Environmental and Resource Economics (AERE). Kyoto, July 4–7, 2006.

———. 2008a. Distributional Effects of Environmentally-Related Taxes: Empirical Applications for the Czech Republic. Paper presented at the 16th Annual Conference of the European Association of Environmental and Resource Economists, Gothenburg, June 25–28.

———. 2008b. Tax Progressivity Measurement: Empirical Applications for the Czech Republic. In: *Modelling of Consumer Behaviour and Wealth Distribution*, eds M. Ščasný, M. Braun Kohlová et al. Matfyzpress, Praha, 157–79.

Bull, Nicholas, Kevin A. Hassett, and Gilbert E. Metcalf. 1994. Who Pays Broad-based Energy Taxes? Computing Lifetime and Regional Incidence. *The Energy Journal* 15(3): 145–164.

Bunch, David, David Brownstone, and Thomas F. Golub. 1996. A Dynamic Forecasting System for Vehicle Markets with Clean-Fuel Vehicles, in David Hensher, Jenny King and Tae Hoon Oum, eds, World Transport Research: Proceedings of the 7th World Conference on Transport Research (Oxford: Permagon), Vol. 1, pp. 189–203.

Bureau, B. 2011. Distributional Effects of a Carbon Tax on Car Fuels in France. *Energy Economics* 33(1):121–50.

Burtless, Gary, and Jerry Hausman. 1978. The Effects of Taxation on Labor Supply: Evaluating the Gary Income Maintenance Experiment. *Journal of Political Economy* 86: 1103–30.

Burtraw, Dallas, Alan Krupnick, Karen Palmer, Anthony Paul, Michael Toman, and Cary Bloyd. 2003. Ancillary Benefits of Reduced Air Pollution in the U.S. from Moderate Greenhouse Gas Mitigation Policies in the Electricity Sector. *Journal of Environmental Economics and Management* 45(3): 650–673.

Buse A., and W. H. Chan. 2000. Invariance, Price Indices, and Estimation in Almost Ideal Demand Systems. *Empirical Economics* 25: 519–39.

The Business Analyst. 2011. "NPA: Oct-Dec 2010 Fuel Subsidy Was \$70million." http://thebusinessanalystgh.com/index.php?option=com_content&view=article&id=35:npa-oct-dec-2010-fuel-subsidy-was-70million&catid=1:latest-news&Itemid=70 (accessed June 26, 2011).

Cantero, M. 2008. "Autobuseros amenazan con paralizar servicio a usuarios." La Nacion. July 5.

Cao, J., and J. Zeng. 2010. Estimates from a Consumer Demand System: Implications for the Incidence of Fuel Tax Reform in China. Working Paper. Tsinghua University.

Casler, S. and A. Rafiqui. 1993. Evaluating fuel tax equity: Direct and indirect distributional effects. *National Tax Journal* 46: 197–205.

Caspersen, E., and Gilbert Metcalf. 1993. Is a Value Added Tax Progressive? Annual Versus Lifetime Measures. *National Tax Journal* 47(4): 731–746.

CBO (Congressional Budget Office). 2007. Trade-offs in Allocating Allowances for CO_2 Emissions. www.cbo.gov/ftpdocs/80xx/doc8027/04-25-Cap_Trade.pdf (accessed May 17, 2011).

CDV (Centrum Dopravního Výzkumu). 2009. Study on Transport Trends From Environmental Viewpoints in the Czech Republic 2008. Transport Research Centre, Brno. August.

Cedi Post. 2010. "Ghana Government Spends Gh¢12 Million on Fuel Subsidy." March 30.

Celis, R., J. Echeverría, and A. Conejo. 1996. Informe final economia ambiental: Sistema de ordenamiento de la Gran Area Metropolitana (San José, SOAGAM).

Central Bank of Iran. 2009. http://cbi.ir/simplelist/4454.aspx (accessed August 12, 2009).

CESifo Group. 2009a. Tax Rates on Unleaded Petrol, 2006: http://www.cesifo-group.de/portal/page/portal/ifoHome/a-winfo/d3iiv/_DICE_details?_id=6745571&_thid=12084169&_cat=c

———. 2009b. Tax Rates on Unleaded Diesel, 2006: http://www.cesifo-group.de/portal/page/portal/ifoHome/a-winfo/d3iiv/_DICE_details?_id=6745571&_thid=12084021&_cat=c

Chabé-Ferret, Sylvain. 2005. L'impact distributif des politiques agricoles des pays développés au Brésil: une analyse non paramétrique (unpublished; Clermont-Ferrand: Centre d'Études et de Recherches sur le Développement International).

Chen, D., J. Matovu and R. Reinikka. 2001. A Quest for Revenue and Tax Incidence. In *Uganda's Recovery: The Role of Farms, Firms and Government*, eds R. Reinikka and P. Collier. World Bank. Washington D.C.

Chernick, H., and A. Reschovsky. 1992. Is the Gasoline Tax Regressive? Discussion Paper. Institute for Research on Poverty, 980–92.

———. 1997. Who pays the gasoline tax? *National Tax Journal*, 50, 2: 233–259.

Christensen L. R., D. W. Jorgenson and L. J. Lau. 1975. Transcendental Logarithmic Utility Functions. *American Economic Review* 65(3): 367–383.

CIA (Central Intelligence Agency). 2009. *The World Factbook.* www.cia.gov/library/publications/the-world-factbook/fields/2172.html (accessed June 30, 2009).

Clements, Benedict J., Sanjeev Gupta, and Hong-Sang Jung. 2003. Real and Distributive Effects of Petroleum Price Liberalization: The Case of Indonesia. IMF Working Paper 03204. International Monetary Fund, Washington, DC.

Coady, D., and D. Newhouse. 2005. Ghana: Evaluation of the Distributional Impacts of Petroleum Price Reforms. Technical Assistance Report (unpublished), International Monetary Fund, Washington, DC.

———. 2006. Ghana: Evaluating the Fiscal and Social Costs of Increases in Fuel Prices. In *Poverty and Social Impact Analysis of Reforms: Lessons and Examples from Implementation*, eds A. Coudouel, A. A. Dani, and S. Paternostro. Part 879, World Bank Publications, pp. 520.

Coady, D., M. El-Said, R. Gillingham, K. Kpodar, P. Medas, and D. Newhouse. 2006. The Magnitude and Distribution of Fuel Subsidies: Evidence from Bolivia, Ghana, Jordan, Mali, and Sri Lanka. IMF Working Paper No. 06/247. IMF, Washington, D.C.

Coady, D., R. Gillingham, R. Ossowski, J. Piotrowski, S. Tareq, and J. Tyson. 2010. Petroleum Product Subsidies: Costly, Inequitable, and Rising. IMF Staff Position Note 10/05 (Washington: International Monetary Fund).

Colatei, D. and J. I. Round. 2000. Poverty and Policy: Experiments with a SAMbased CGE Model for Ghana. Paper presented at the XIII International Conference on Input-Output Techniques, 21–25 August. Macerato, Italy.

Cooper, J. 2003. Price Elasticity of Demand for Crude Oil: Estimates for 23 Countries, OPEC Review, Vol. 27 (March), pp. 1–6.

Corden W. M., and J. P. Neary. 1982. Booming Sector and De-industrialisation in a Small Open Economy. *The Economic Journal* 92(December): 825–48.

Cornwell A., and J. Creedy. 1997. Measuring the Welfare Effects of Tax Changes Using the LES: An Application to a Carbon Tax. *Empirical Economics* 22: 589–613.

Dahl, C. 1986. Gasoline Demand Survey. *Energy Journal* 7: 67–82.

Dahl, C., and T. Sterner. 1991a. Analyzing Gasoline Demand Elasticities: A Survey. *Energy Economics* 13: 203–10.

———. 1991b. A Survey of Econometric Gasoline Demand Elasticities. *International Journal of Energy Systems* 11, 53–76.

Dahl, C. 1995. Demand for transportation fuels: a survey of demand elasticities and their components. *The Journal of Energy Literature* 1, 3–27.

Davidson, O., Tyani, L., Afrane-Okesse, Y., 2002. Climate Change, Sustainable Development and Energy: Future Perspectives for South Africa. OECD.

Davidson, R., and E. Flachaire. 2007. Asymptotic and Bootstrap Inference for Inequality and Poverty Measures. *Journal of Econometrics* 141: 141–66.

DEAT (Department of Environmental Affairs and Tourism). 2000. Initial National Communication under the United Nations Framework Convention on Climate Change. Pretoria.

Deaton, A., and J. Muellbauer. 1980. An Almost Ideal Demand System. *The American Economic Review* 70: 312–26.

Deaton, Angus. 1997. *The Analysis of Household Surveys: A Microeconometric Approach to Development Policy* (Baltimore: Johns Hopkins University Press for the World Bank).

Drollas, L. 1984. The demand for gasoline: further evidence. *Energy Economics* 6, 71–82.

Dubé, Jean-Pierre. 2004. Multiple Discreteness and Product Differentiation: Demand for Carbonated Soft Drinks. *Marketing Science* 23: 66–81.

Dubin J., and D. McFadden. 1984. An Econometric Analysis of Residential Appliance Holdings and Consumption. *Econometrica* 52: 345–362.

Economic Hamshahri. 2010. "9 Million Cars Waiting for a Decision." No. 5062, year 18. February 21, p. 4.

EIA (Energy Information Administration). 2006. Energy Market Impacts of Alternative Greenhouse Gas Intensity Reduction Goals. Washington DC, EIA.

———. 2008. Emissions of Greenhouse Gases in the United States 2007. Washington DC, EIA.

EMTCTA (Ethiopian Ministry of Transport and Communications Transport Authority). 2009. Ethiopian Ministry of Transport and Communications Transport Authority. Unpublished report.

Espey, M. 1996. Explaining the Variation in Elasticity Estimates of Gasoline Demand in the United States: A Meta-analysis. *Energy Journal* 17: 49–60.

Ethiopian Petroleum Enterprise. 2009. *Addis Ababa Retail Price Build-up (August, 2004–1981); Country-wide Consumption of Petroleum Products during the Years 1985/86–1996/97; Importation of Petroleum Products (1998/99–2007/08),* Unpublished.

Ettelaat. 2008. "$100 Billion Subsidy in Iran." December 2, No. 24353, p. 19.

———. 2010a. "Oil Minister's Reasons for Importing Extra Gasoline and Gasoil." January 11, No. 24661, p. 4

———. 2010b. "Gasoline Price Will Reach 4000 Rials by the Next Year." January 17, No. 24666, pp. 1and 4.

Eurostat. 2009. Dataset for Inequality For Income Distribution: Gini Coefficients. http://epp.eurostat.ec.europa.eu/tgm/table.do?tab=table&plugin=0&language=en&pcode=tessi190 (accessed June 30, 2009).

FAO (Food and Agriculture Organization of the United Nations). 2005. *The State of Food Insecurity in the World, Eradicating World Hunger: Key to Achieving the Millennium Development Goals.* Food and Agriculture Organization of the United Nations, Rome, Italy

Feenberg D., and E. Coutts. 1993. An Introduction to the TAXSIM Model. *Journal of Policy Analysis and Management* 12(1): 189–94.

Feldstein, M. 2001. "Achieving Oil Security: A Practical Proposal." *The National Interest*. November 1.

Feng, Y., D. Fullerton, and L. Gan. 2005. Vehicle Choices, Miles Driven, and Pollution Policies. NBER Working Paper No. 11553. Cambridge, MA.

Fofana, I., M. Chitiga, and R. Mabugu. 2009. Oil Prices and the South African Economy: A Macro–meso–micro Analysis. *Energy Policy* 37(12): 5509–18.

Formby, J. P., T. G. Seaks, and W. J. Smith. 1981. A Comparison of Two New Measures of Tax Progressivity. *The Economic Journal* 91(364): 1015–1019.

French Ministry of Economy, Industry and Employment. 2009. *Le pétrole – textes de reference, analyses,* Available (Online): www.industrie.gouv.fr/energie/petrole/f1e_petr.htm (accessed February 27, 2009).

Friedman, M. A. 1957. *A Theory of the Consumption Function*. Princeton: Princeton University Press.

Fullerton, D. 1995. Why Have Separate Environmental Taxes? NBER Working Paper No. 5380.

———. 2009. (ed.) *The Distributional Effects of Environmental and Energy Policy*, Aldershot, UK: Ashgate Publishers.

———. 2011. Six Distributional Effects of Environmental Policy. NBER Working Paper No. 16703.

Fullerton, D. and G. E. Metcalf. 1998. Environmental Taxes and the Double Dividend Hypothesis: Did You Really Expect Something for Nothing? *Chicago-Kent Law Review*. 73(1): 221–56.

———. 2002. Tax Incidence. NBER Working Paper No. 8829.

Fullerton, D., A. Leicester, and S. Smith. 2008. Environmental Taxes. NBER Working Paper No. 14197.

Fullerton, D., G. Heutel, and G. Metcalf. 2011. Does the Indexing of Government Transfers Make Carbon Pricing Progressive? Forthcoming in the proceedings issue of *American Journal of Agricultural Economics*.

Gangopadhyay, S., B. Ramaswami and W. Wadhwa. 2005. Reducing Subsidies on Household Fuels in India: How Will It Affect the Poor? *Energy Policy*, 33, 2326–2336.

Gately, D., and H. G. Huntington. 2001. The Asymmetric Effects of Changes in Price and Income on Energy and Oil Demand. Economic Research Report. Department of Economics, New York University, New York.

Gately, D. and S. Streifel. 1997. The Demand for Oil Products in Developing Countries, World Bank Discussion Paper No. 359 (Washington).

Gemmell, N., and O. Morrissey. 2005. Distribution and Poverty Impacts of Tax Structure Reform in Developing Countries: How Little We Know. *Development Policy Review* 23(2): 131–44.

Gini, C. 1912. Variabilità e mutabilità. Reprinted in *Memorie di metodologica statistica*, ed. Pizetti E. Salvemini, T. Rome: Libreria Eredi Virgilio Veschi (1955).

Goldberg, P. K. 1995. Product Differentiation and Oligopoly in International Markets: The Case of the U.S. Automobile Industry. *Econometrica* 63: 891–951.

———. 1998. The Effects of the Corporate Average Fuel Efficiency Standards in the US. *Journal of Industrial Economics* 46: 1–33.

Goodwin, P. 1992. A Review of New Demand Elasticities with Special Rreference to Short and Long Run Effects of Price Changes. *Journal of Transport Economics and Policy* 26, 155–63.

Goodwin, P., J. Dargay, and M. Hanly. 2004. Elasticities of Road Traffic and Fuel Consumption with Respect to Price and Income: A Review. *Transport Reviews* 24(3): 275–92.

Goulder L. H. 1992. Carbon Tax Design and U.S. Industry Performance. In *Tax Policy and the Economy*, vol. 6, ed. J. Poterba, 59–104.

———. 2002. Introduction. In *Environmental Policy Making in Economies with Prior Tax Distortions*, Edward Elgar Publishing Ltd., Cheltenham, UK and Northampton, MA.

Goulder, L., and R. C. Williams III. 2003. The Substantial Bias from Ignoring General Equilibrium Effects in Estimating Excess Burden, and a Practical Solution. *Journal of Political Economy* 111: 898–927.

Goulder, L., I. W. H. Parry, R. C. Williams III and D. R. Burtraw. 1999. The Cost-Effectiveness of Alternative Instruments for Environmental Protection in a Second Best Setting. *Journal of Public Economics* 72: 329–60.

Gourieroux, C., and A. Monfort. 1996. *Simulation-Based Econometric Methods*. Oxford University Press. New York, NY.

Government of Kenya. 2004. Sessional Paper No. 4 on Energy. Ministry of Energy. Nairobi: Government Printer.

Graczyk, D. 2006. Petroleum Product Pricing in India: Where Have All the Subsidies Gone? International Energy Agency Working Paper.

Graham, D., and S. Glaister. 2002. The Demand for Automobile Fuel: A Survey of Elasticities, *Journal of Transport Economics and Policy* 36, 1–26.

———. 2004. Road Traffic Demand: A Review. *Transport Review* 24, 261–274.

Greene, W. H. 2003. *Econometric Analysis*, 5th edition. Prentice Hall: Upper Saddle River, NJ.

GTZ (Die Deutsche Gesellschaft für Technische Zusammenarbeit). 2007. *International Fuel Prices 2007*. www.gtz.de/de/dokumente/en-international-fuelprices-final2007.pdf (accessed March 2, 2009).

Gundimeda, H., and G. Köhlin. 2008. Fuel Demand Elasticities for Energy and Environmental Policies: Indian Sample Survey Evidence. *Energy Economics* 30(2): 517–46.

Gupta, G., and G. Köhlin. 2006. Preferences for Domestic Fuel: Analysis with Socio-Economic Factors and Rankings in Kolkata, India. *Ecological Economics* 57: 107–21.

Halvorsen, B. 2006a. Review of the "Empirics of Residential Energy Demand" by Bengt Kriström. Discussion paper prepared for the OECD Workshop on Household Behavior and Environmental Policy: Empirical Evidence and Policy Issues. OECD, June 15–16.

———. 2006b. When Can Micro Properties Be Used to Predict Aggregate Demand? Discussion Papers 452. Statistics Norway.

Hamermesh, D. 1993. *Labor Demand*. Princeton University Press, Princeton, NJ.

Hammar, H., Å. Löfgren, and T. Sterner. 2004. The Political Economy Obstacles to Fuel Taxation. *The Energy Journal* 25(3): 1–17.

Hanemann, W. M. 1984. Discrete/continuous Models of Consumer Demand. *Econometrica* 52(3): 541–61.

Hanly, M., J. Dargay, and P. Goodwin. 2002. Review of Income and Price Elasticities in the Demand for Road Traffic. Department of Transport, London.

Hannon, B. 1982. Analysis of the Energy Cost of Economic Activities: 1963 to 2000. *Energy Systems Policy* 6(3):249–278.

Hannon, B., Richard G. Stein, B. Z. Segal, and Diane Serber. 1978. Energy and Labor in the Construction Sector. Science 202 (204):37–47

Harrison, David. 1995. *Climate Change: Economic Instruments and Income Distribution. OECD, Paris.*

Hassett, Kevin A., A. Mathur, and G. E. Metcalf. 2009. The Incidence of a U.S. Carbon Tax: A Lifetime and Regional Analysis. *The Energy Journal* (30)2: 155–78.

Hausman J. A. 1981. Exact Consumer's Surplus and Dead-weight Loss. *American Economic Review* 71 (1981): 662–676.

Hausman, Jerry, and Whitney Newey. 1995. Nonparametric Estimation of Exact Consumers Surplus and Deadweight Loss. *Econometrica* 63: 1445–1476.

Heckman J. 1979. Sample Selection Bias as Specification Error. *Econometrica* 47: 153–161.

Heien D., and C. Wessels. 1990. Demand System Estimation with Microdata: A Censored Regression Approach. *Journal of Business and Economic Statistics* 8: 365–1771.

Hendel, Igal. 1999. Estimating Multiple-Discrete Choice Models: An Application to Computerization Returns. *Review of Economic Studies* 66: 423–446.

Herendeen, R.A., C. Ford and B. Hannon. 1981. Energy Cost of Living, 1972–1973. *Energy* 6(12):1433–1450.

Hernández, R., and A. B. Hernández. 1999. Estimacion de la demanda de combustibles para Costa Rica (1975–1997). CINPE. Working Paper.

Herrera, J., and S. Rodríguez. 2005. Segundo informe de calidad del aire de la ciudad de San José año 2004–2005. Escuela de Ciencias Ambientales, Universidad Nacional, Costa Rica.

———. 2008. Informe de calidad del aire de la ciudad de San José, Año 2007. Escuela de Ciencias Ambientales, Universidad Nacional, Costa Rica.

HM Revenue & Customs. 2008. *Hydrocarbon Oils: Historical Duty Rates*, Available (Online): http://customs.hmrc.gov.uk/channelsPortalWebApp/downloadFile?content ID=HMCE_PROD1_023552 (accessed March 02, 2009).

Hoerner J., and B. Bosquet. 2001. *Environmental Tax Reform: The European Experience.* Center for a Sustainable Economy, Washington DC.

Hoerner J., and G. Erickson. 2000. Environmental Tax Reform in the States: A Framework for Assessment. *State Tax Notes* (July 31): 311–19.

Hope, E., and B. Singh. 1995. Energy Price Increases in Developing Countries: Case Studies of Colombia, Ghana, Indonesia, Malaysia, Turkey, and Zimbabwe. World Bank Policy Research Working Paper No. 1442 (March).

Horridge, Mark. 2000. ORANI-G: A General Equilibrium Model of the Australian Economy. Centre of Policy Studies/IMPACT Centre Working Papers op-93. Monash University, Centre of Policy Studies/IMPACT Centre. http://ideas.repec.org/p/cop/wpaper/op-93.html (accessed May 25, 2011).

Houri Jafari H. and A. Baratimalayeri. 2008. The Crisis of Gasoline Consumption in Iran's Transportation Sector. *Energy Policy* 36(2008): 2536–2543.

Hsu, Shi-Ling, Joshua Walters, and Anthony Purgas. 2008. Pollution Tax Heuristics: An Empirical Study of Willingness to Pay Higher Gasoline Taxes. *Energy Policy* 36(9): 3612–3619.

Hudson, E.A. and D.W. Jorgenson. 1974. U.S. Energy Policy and Economic Growth, 1975–2000. *Bell Journal of Economics* 5(0): 461–514.

IAM (Institute of Advanced Motorists). 2007. *Fuel Price Report – October 2006*

IEA (International Energy Agency). 2006. World Energy Outlook 2006. International Energy Agency/OECD, Paris.

———. 2009. IEA Energy Prices and Statistics, Quarterly statistics, 2009.

———. 2010. *Oil Market Report.* International Energy Agency/OECD, Paris.

IEA/OPEC/OECD/World Bank (International Energy Agency, Organization of the Petroleum Exporting Countries, Organization for Economic Co-Operation and Development, and The World Bank). 2010. *Analysis of the Scope of Energy Subsidies and Suggestions for the G-20 Initiative.* IEA, OPEC, OECD, World Bank Joint Report. Prepared for submission to the G-20 Summit Meeting, Toronto (Canada), June 26–27.

Ikhsan, M., T. Dartanto, Usman, and H. Sulistyo. 2005. Kajian Dampak Kenaikan Harga BBM 2005 terhadap Kemiskinan. LPEM Working Paper, 2005. www.lpem.org

Ikiara, M., A. Mkenda, D. Slunge, and T. Sterner. 2007. Potentials for Environmental Fiscal Reforms in Tanzania and Kenya: Emerging Practice and Some Suggestions for EFR-Reviews in Low Income countries. The Eighth Global Conference on Environmental Taxation, October 18–20, 2007, Munich, Germany.

ILRI (International Livestock Research Institute). 2007. Vulnerability, Climate Change and Livestock—Research Opportunities and Challenges for Poverty Alleviation. International Crops Research Institute for the Semi-Arid Tropics (ICRISAT), Nairobi, Kenya.

IMF (International Monetary Fund). 2005. *Regional Economic Outlook: Sub-Saharan Africa* (Washington).

———. 2008. *International Financial Statistics Database.* Washington, DC: IMF.

———. 2009. *World Economic and Financial Surveys.* World Economic Outlook Database, September 2006 Edition. www.imf.org/external/pubs/ft/weo/2006/02/data/weorept.aspx?sy=2006&ey=2006&scsm=1&ssd=1&sort= (accessed June 30, 2009).

IPCC (Intergovernmental Panel on Climate Change). 2007. Contribution of Working Group I to the Fourth Assessment Report. Geneva Switzerland: IPCC.

Jacobsen, M. R. 2007. Impacts of Changes in CAFE Standards in an Econometrically Based Simulation Model. Unpublished manuscript, University of California, San Diego.

Jobson, E. 2007. Volvo Hybrid Technology. April 26, 2007 Presentation at Chalmers University of Technology, Gothenburg, Sweden.

Johansson, Olof, and Lee Schipper. 1997. Measuring the Long Run Fuel Demand of Cars: Separate Estimations of Vehicle Stock, Mean Fuel Intensity, and Mean Annual Driving Distance. *Journal of Transport Economics and Policy* 31: 277–292.

Johnstone, N., J. Echeverría, I. Porras, and R. Mejias. 2001. The Environmental Consequences of Tax Differentiation by Vehicle Age in Costa Rica. *Journal of Environmental Planning and Management* 44(6): 803–814.

Jorgensen, S., and T. B. Pedersen. 2000. Danish Income Distribution and Net Contribution to Public Finances. The 6th Nordic Seminar on Micro Simulation Models.

Jorgenson, D.W. and P. J. Wilcoxen. 1992. Reducing U.S. Carbon Dioxide Emissions: An Assessment of Different Instruments. Harvard Institute of Economic Research, Discussion paper # 1590.

Kakwani, N. C. 1977a. Measurement of Tax Progressivity: An International Comparison. *The Economic Journal* 87(345): 71–80.

———. 1977b. Applications of Lorenz Curves in Economic Analysis. *Econometrica*, 45, pp.719–727.

Kasten, R., and F. Sammartino. 1988. The Distribution of Possible Federal Excise Tax Increases. Congressional Budget Office,

Keeling, C. D., and T. P. Whorf. 2004. *Atmospheric CO_2 from Continuous Air Samples.* Carbon Dioxide Research Group, Scripps Institution of Oceanography, University of California, La Jolla, California.

Keynes, John Maynard. 1923. A Tract on Monetary Reform, ch. 3. Macmillan and Co. London.

King, Mervyn. 1980. An Econometric Model of Tenure Choice and Demand for Housing as a Joint Decision. *Journal of Public Economics* 14(2): 137–159.

Kleit, Andrew. 2004. Impacts of Long-range Increases in the Corporate Average Fuel Economy (CAFE) Standard. *Economic Inquiry* 42(2): 279–94.

KNBS (Kenya National Bureau of Statistics). 2010. Economic Survey, 2010. Government Printer, Nairobi.

Kosmo, M. 1987. Money to Burn? The High Costs of Energy Subsidies. World Resources Institute, Washington, DC.

KPMG Peat Marwick. 1990. Changes in the Progressivity of the Federal Tax System, 1980–1990. Prepared for the Coalition Against Regressive Taxation. Washington, DC.

Kpodar, K. 2006. Distributional Effects of Oil Price Changes on Household Expenditures: Evidence from Mali. IMF Working Paper 06/91.

KRA (Kenya Revenue Authority). 2007. Customs and Domestic Taxes Department. www.kra.go.ke/ (accessed April 24, 2009).

Krueger, Dirk, and Fabrizio Perri. 2002. Does Income Inequality Lead to Consumption Inequality? Evidence and Theory. NBER. Working Paper No. 9202.

La Gaceta. 2009. No. 91, May 13.

Lambert, P. J. 2001. *The Distribution and Redistribution of Income*. Manchester University Press.

Larsen, B. and A. Shah. 1992. World fossil fuel subsidies and global carbon emissions, World Bank Policy Research Working Paper 1002.

Leigh, D., and M. El Said. 2006. Fuel Price Subsidies in Gabon: Fiscal Cost and Distributional Impact. IMF Working Paper No. 06/243 (October).

Leontief, W. 1986. *Input-Output Economics*, 2nd Edition. New York: Oxford University Press.

Longhurst, R., S. Kamara, and J. Mensurah. 1988. Structural Adjustment and Vulnerable Groups in Sierra Leone. Institute of Development Studies (IDS) Bulletin 19(1): 25–30.

Lyon, A., and R. Schwab. 1995. Consumption Taxes in a Life-Cycle Framework: Are Sin Taxes Regressive? *Review of Economics and Statistics* 77(3): 389–406.

Löfgren, Å. and K. Nordblom. 2008. Puzzling tax attitudes and labels, *Applied Economics Letters*, 99999:1, April 2008.

Mabugu, R., M. Chitiga, and H. Amusa. 2009. The Economic Consequences of a Fuel Levy Reform in South Africa. *South African Journal of Economic and Management Sciences* 12(3): 280–96.

Mannering, F. 1986. A Note on Endogenous Variables in Household Vehicle Utilization Equations. *Transportation Research* B 20(1): 1–6.

Mannering, F., and Clifford Winston. 1985. A Dynamic Empirical Analysis of Household Vehicle Ownership and Utilization. *RAND Journal of Economics* 16: 215–236.

Manning, Chris and Kurmya Roesad. 2006. Survey of Recent Developments. *Bulletin of Indonesian Economic Studies* 42, no. 2 (2006): 143–70.

Marland, Gregg, Tom Boden, and Robert J. Andres. 2008. *National CO_2 Emissions from Fossil Fuel Burning, Cement Manufacture, and Gas Flaring: 1751–2005*. Carbon Dioxide Information Analysis Center, Oak Ridge National Laboratory, Oak Ridge, Tennessee. http://cdiac.ornl.gov/ftp/trends/emissions/ind.dat (accessed May 24, 2011).

McDonald, Scott, and Melt van Schoor. 2005. A Computable General Equilibrium (CGE) Analysis of the Impact of an Oil Price Increase in South Africa, PROVIDE Working Paper No. 1 (Elsenburg: The Western Cape Department of Agriculture).

McFadden, Daniel, and Kenneth E. Train. 2000. Mixed MNL for Discrete Response. *Journal of Applied Econometrics* 15: 447–470.

McKenzie GW, Pearce IF. 1976. Exact Measures of Welfare and the Cost of Living, *Review of Economic Studies*, October 1976, 43, 465–68.

Metcalf, G. E. 1999. A Distributional Analysis of Green Tax Reforms. *National Tax Journal* 52(4): 655–81.

———. 2005. Tax Reform and Environmental Taxation. NBER Working paper No. 11665.

———. 2007. A Proposal for a U.S. Carbon Tax Swap: An Equitable Tax Reform to Address Global Climate Change. The Hamilton Project, Brookings Institution. October.

———. 2008. An Empirical Analysis of Energy Intensity and Its Determinants at the State Level. *The Energy Journal* 29(3): 1–26.

Metcalf, G. E., S. Paltsev, J. M. Reilly, H. D. Jacoby, and J. Holak. 2008. Analysis of U.S. Greenhouse Gas Tax Proposals. MIT Joint Program on the Science and Policy of Global Change Report # 160.

Mills, J. A., and S. Zandvakili. 1997. Statistical Inference via Bootstrapping for Measures of Inequality. *Journal of Applied Econometrics* 12: 133–150.

Ministry of Finance. 2010. *Nota Keuangan dan RAPBN 2010* (Financial Note and National Revenue and Expenditure Plan 2010). Indonesian Ministry of Finance, Jakarta.

Ministry of Petroleum and Natural Gas, Government of India. 2006. Report of the Committee on Pricing and Taxation of Petroleum Products Ministry of Petroleum and Natural Gas, Government of India.

Mkenda, A. F., J. K. Mduma, V. Leyaro and W. Ngasamiaku. forthcomming. Elasticities of Fuel Prices in Tanzania: Estimates from Cross-sectional Data. EfDT Project, Dar es Salaam.

MoE (Iranian Ministry of Energy, Office of Energy and Power Affairs). 2008. Iran's Energy Balance 2007, http://pep.moe.org.ir/_pep/Documents/taraz-%20Iran%20 1386_20090427_114349.pdf (accessed September 10, 2009).

MoFED (Ministry of Finance and Economic Development of the Federal Democratic Republic of Ethiopia). 2007. *Plan for Accelerated and Sustained Development to End Poverty*. Addis Ababa.

MoME (Ministry of Mines and Energy of the Federal Democratic Republic of Ethiopia). 2007. *Biofuel Development and Utilisation Strategy of Ethiopia*. Addis Ababa, Ethiopia.

Moschini G. 1995. Units of Measurement and the Stone Index in Demand System Estimation. *American Journal of Agricultural Economics* 77: 63–68.

Moschini G., and P. L. Rizzi. 2006. Coherent Specification of a Mixed Demand System: The Stone-Geary Model. Unpublished manuscript. Department of Economics, Iowa State University.

Msafiri, M. J. 2005. Roadside Concentration of Gaseous and Partculate Matter Pollutants and Risk Assessment in Dar es Salaam, Tanzania. Environmental Monitoring and Assessment 104: 385–407.

Muellbauer J. 1975. Aggregation, Income Distribution and Consumer Demand. *Review of Economic Studies* 62: 525–43.

———. 1976. Community Preferences and the Representative Consumer. *Econometrica* 44: 979–99.

Musgrave, R., K. Case, and H. Leonard. 1974. The Distribution of Fiscal Burdens and Benefits. *Public Finance Quarterly* 2. 259–311.

Mutua, J. M., M. Borjesson, and T. Sterner. 2009. Transport Choice, Elasticity, and Distributional Effects of Fuel Taxes in Kenya. In *Critical Issues in Environmental Taxation* Volume VII, eds. Lin Heng Lye, Janet E. Milne, Hope Ashiabor, Larry Kreiser and Kurt Deketelaere. Oxford University Press, Oxford, pp. 167–86.

MWR (Ministry of Water Resources, National Meteorological Services Agency). 2001. *Initial National Communication of Ethiopia to the United Nations Framework Convention on Climate Change (UNFCCC)*, Addis Ababa, Ethiopia

National Research Council. (NRC). 2002. Effectiveness and Impact of Corporate Average Fuel Economy Standards. National Academy Press.

NCAER (National Council for Applied Economic Research). 2005. *Comprehensive Study to Assess the Genuine Demand and Requirement of SKO (Special Kerosene Oil)*. Report. National Council for Applied Economic Research, New Delhi.

————. 2006. Study of Macroeconomic Impact of High Oil Prices, Petrofed Report. National Council for Applied Economic Research.

New Economics Foundation. 2004. NEF. Price of Powers. New Economics Foundation, London, 2004. Available at: www.neweconomics.org

Nicholson, Kit, Bridget O›Laughlin, Antonio Fransisco, and Virgulino Nhate. 2003. Fuel Tax in Mozambique Poverty and Social Impact Analysis (Washington: World Bank).

NIORDC (National Iranian Oil Refining & Distribution Co.). 2008. Statistics, Chapter 5. Gasoline Production and Consumption 1989–2007. www.niordc.ir/uploads/78_23_fasl5,6.pdf (accessed June 2009).

————. 2009. مصرف بنزين آورده صادرات اصلی طی ساهای 1353-87, Gasoline consumption 1974–2008 www.niordc.ir/index.aspx?siteid=78&pageid=464 (accessed January 2010).

OECD (Organisation for Economic Co-operation and Development). 1994. *The Distributive Effects of Economic Instruments for Environmental Policy*. OECD, Paris.

————. 2010. What Are Equivalence Scales? www.oecd.org/dataoecd/61/52/35411111.pdf (accessed October 10, 2010).

OKQ8. 2008. Prisstatistik. Available (Online): www.okq8.se/privat/pastationen/drivmedel/priser/prisstatistik (accesed March 02, 2009).

Oktaviani, R., D. Hakim, S. Sahara and H. Siregar. 2005. The Impact of Fiscal Policy on Indonesian Macroeconomic Performance, Agricultural Sector and Poverty Incidences (A Dynamic Computable General Equilibrium Analysis). Report to the Poverty and Economic Policy (PEP) Network 2005. www.pep-net.org/

Osei, R. D., and P. Quartey. 2005. Tax Reforms in Ghana. UNU-WIDER Research Paper No. 2005/66 (December).

Oum, T. 1989. Alternative demand models and their elasticity estimates. *Journal of Transport Economics and Policy* 23, 163–187.

Paltsev, Sergey, John M. Reilly, Henry D. Jacoby, Angelo C. Gurgel, Gilbert E. Metcalf, Andrei P. Sokolov and Jennifer F. Holak. 2007. Assessment of U.S. Cap-and-Trade Proposals, MIT Joint Program on the Science and Policy of Global Change, Report # 146.

Parry, I. W. H., and Wallace E. Oates. 2000. Policy Analysis in the Presence of Distorting Taxes. *Journal of Policy Analysis and Management* 19(4): 603–13.

Parry, I. W. H., and Kenneth Small. 2002. Does Britain or America Have the Right Gasoline Tax? Working Paper. *Resources for the Future and University of California-Irvine.*

————. 2005. Does Britain or the United States Have the Right Gasoline Tax? *American Economic Review* (95)4: 1276–89.

Parry, I. W. H., and G. Timilsina. 2008. How Should Passenger Travel in Mexico City be Priced? RFF Discussion Paper 08-17.

Parry I. W. H., and R. C. Williams III. 1999. A Second-best Evaluation of Eight Policy Instruments to Reduce Carbon Emissions. *Resource and Energy Economics* 21: 347–73.

Parry, I. W. H., M. Walls and W. Harrington. 2007. Automobile Externalities and Policies. *Journal of Economic Literature* XLV: 373–99.

Pearce, David. 1991. The Role of Carbon Taxes in Adjusting to Global Warming. *The Economic Journal* 101(ISSUE): 938–948.

Pearson, Ken and Charming Arndt. 2000. Implementing Systematic Sensitivity Analysis Using GEMPACK. GTAP Technical Papers 474 Center for Global Trade Analysis, Department of Agricultural Economics, Purdue University, 2000.

Pechman, J. A. 1985. Who Paid the Taxes, 1966–1985?, Washington, DC.: *The Brookings Institution.*

Pertamina. 2008. Annual Report 2008: Road to Excellence. Pertamina: Jakarta.

Petrin, Amil. 2002. Quantifying the Benefits of New Products: The Case of the Minivan. *Journal of Political Economy* 110: 705–29.

Portney, P. R., and R. N. Stavins. 2000. *Public Policy for Environmental Protection.* Resources for the Future.

Poterba J. M. 1989. Lifetime Incidence and the Distributional Burden of Excise Taxes. *American Economic Review: Papers and Proceedings* 79(2): 325–30.

———. 1991. Is the Gasoline Tax Regressive? In *Tax Policy and the Economy,* vol. 5, edited by D. Bradford. Cambridge: MIT Press, 145–165.

Presstv. 2010. Iran Parliament Ratifies $279B Budget. www.presstv.ir/detail.aspx?id=8 8423&sectionid=351020102 (accessed February 22, 2010).

Rafey, W. 2010. How to Pass a Gas Tax, *Harvard Political Review,* Fall issue.

Rajemison, H., S. Haggblade, and S. D. Younger. 2003. Indirect Tax Incidence in Madagascar: Updated Estimates Using the Input-Output Table, Cornell Food and Nutrition Policy Program

Ramanathan, R. 1999. Short- and Long-Run Elasticities of Gasoline Demand in India: An Empirical Analysis Using Cointegration Techniques, *Energy Economics,* 21, 321–330.

Ramanathan, R., and S. Geetha. 1998. Gasoline Consumption in India: An Econometric Analysis. First Asia Pacific Conference on Transportation and the Environment, Singapore.

Randem B. G., B. Ulvestad, I. Burstyn, and J. Kongerud. 2004. *Respiratory Symptoms and Airflow Limitation in Asphalt Workers.* Occupational and Environmental Medicine 61(4): 367–9.

Raupach, M. R., G. Marland, P. Ciais, C. Le Quere, J.G. Canadell, G. Klepper and C. B. Field. 2007. Global and regional drivers of accelerating CO_2 emissions, *Proceedings of National Academy of Sciences the USA* 104, 10288–10293.

Reyes Tépach, M. 2007. Investigador Parlamentario El impacto en los hogares del país por la aplicación del impuesto local a las ventas finales a la gasolina y el diesel, Cámara de Diputados; Servicios de Investigación y Análisis; Centro de Documentación; Subdirección de Economía. Mexico. www.diputados.gob.mx/cedia/sia/se/ SE-ISS-17-07.pdf (Accessed March 1st 2011).

Robinsson, E., T. Sterner and W. Akpalu. 2009. "Debate on Fuel Subsidies", *Business and Financial Times of Ghana,* 06 July 2009.

Rogat, J., and T. Sterner. 1998. The Determinants of Gasoline Demand in Some Latin American Countries. *International Journal of Global Energy Issues* 11(1–4): 162–70.

Royanian, M. 2010. Head of Iran's fuel transportation organization, cited in Economic Hamshahri, P.4, no 5062, year 18, 21/02/2010, translated by the author

Sahn, D. E., and S. D. Younger. 2003. Estimating the Incidence of Indirect Taxes in Developing Countries. In *The Impact of Economic Policies on Poverty and Income Distribution: Evaluation Techniques and Tools*, eds. F. Bourguignon and L.A. Pereira da Silva L.A. World Bank and Oxford University Press.

————. 2009. Fiscal Incidence in Africa: Microeconomic Evidence. In *Poverty in Africa: Analytical and Policy Perspectives*, eds. A. Fosu, G. Mwabu, and E. Thorbecke. Nairobi University Press and Africa Economic Research Consortium.

Santos, G., and T. Catchesides. 2005. Distributional Consequences of Gasoline Taxation in the United Kingdom. *Transportation Research Record* 1924: 103–11.

Ščasný, M. 2006. Distributional Aspects of Environmental Regulation: Theory and Empirical Evidence in the Czech Republic. Ph.D. thesis. Prague: Institute of Economic Studies, Faculty of Social Sciences, Charles University.

Ščasný, M. and J. Brůha. 2003. Public Finance Aspects of a Green Tax Reform in the Czech Republic. Working Paper presented at the Seventh Joint Meeting of Tax and Environmental Experts. November, Paris (COM/ENV/EPOC/DAFFE/CFA(2003)126, OECD).

————. 2004. Assessing Proposals for the Environmental Tax Reform in the Czech Republic. Paper presented at the 13th Annual EAERE Conference Budapest, 25–28 June 2004, Hungary.

————. 2007. *Predikce sociálních a ekonomických dopadů návrhu první fáze ekologické daňové reformy České republiky* (Prediction of Social And Economic Impacts of the First Phase of Environmental Tax Reform in the Czech Republic). Study prepared for the Czech Ministry of the Environment. Charles University Environment Center, Prague, April.

————. 2008. Ex ante Modelling of Tax Incidence on Income Distribution And Tax Payments in the Czech Families. Paper presented at the Conference Income Distribution and the Family, University of Kiel, September 1–3, 2008.

Ščasný, M., and Máca, V. 2010. Market-Based Instruments in CEE Countries: Much Ado about Nothing. *Rivista di Politica Economica*, Issue VII-IX, Year XCIX, pp. 59–82.

Schipper, L., M. Figueroa, M. Espey, and L. Price. 1993. Mind the Gap: The Vicious Circle of Measuring Automobile Fuel Use. *Energy Policy* 21, 1173–1190.

Schmalansee, Richard, and Thomas Stoker. 1999. Household Gasoline Demand in the United States. *Econometrica*, 67:645–62.

Scholes Dr R J, van der Merwe M, CSIR Division of Water, Environment & Forestry Technology and John J, Oosthuizen R, CSIR. 2000. National State of the Environment Report – South Africa: climatic and atmospheric change: overview. 2000. www.ngo.grida.no/soesa/nsoer/issues/climate/index.htm

SCI (Statistical Center of Iran). 2006. National Survey of Household Income Expenditure 2006.

————. 2007. Average annual food and no-food costs of households by expenditure deciles 2006 www.amar.org.ir/default-2128.aspx?tabid=2128&&Id=204 (accessed May 2009).

Sevigny, M. 1998. Taxing Automobile Emissions for Pollution Control. Edward Elgar Publishing Ltd., Cheltenham, UK and Northhampton, MA.

Shahmoradi, A., and A. Honarvar. 2008. Gasoline Subsidy and Consumer Surplus in the Islamic Republic of Iran. OPEC *Energy Review* 32(9)(September): 232–49.

Shonkwiler J. S., and S. Yen. 1999. Two-step Estimation of a Censored System of Equations. *American Journal of Agricultural Economics* 81: 972–982.

Simone, Alejandro. 2004. Mali: Assessing the Poverty Impact of Macroeconomic Shocks (unpublished; Washington: International Monetary Fund).

Skatteverket (Swedish Tax Agency). 2008. *Historiska Skattesatser,* Available (online): www.skatteverket.se/download/18.19b9f599116a9e8ef36800022940/skattesatser+med+historik_2008.pdf (accessed March 02, 2009).

Small, K., and Kurt Van Dender. 2007. Fuel Efficiency and Motor Vehicle Travel: The Declining Rebound Effect. *The Energy Journal* 28(1): 25–50.

Sowa, Nii K. 2006. The Role of Subsidies as a Means to Increase Welfare. http://assets.panda.org/downloads/role_subsidies_increase_welfare_2.pdf (accessed May 27, 2011).

SPI, Swedish Petroleum Institute. 2008. Bensinpriser 1981–2006. Available (online): www.spi.se/statistik.asp?omr=1&kat=1 (accessed March 02, 2009).

State of the Environment: South Africa. 2000. State of the Environment, South Africa—Climatic and Atmospheric Change: Overview. www.ngo.grida.no/soesa/nsoer/issues/climate/index.htm (accessed June 16, 2011).

Steininger, K. W., B. Friedl, and B. Gebetsroither. 2007. Sustainability Impacts of Car Road Pricing: A Computable Equilibrium Analysis for Austria. *Ecological Economics* 63: 59–69.

Stern, N. H. 2006. The Economics of Climate Change. www.hm-treasury.gov.uk/independent_reviews/stern_review_economics_climate_change/stern_review_report.cfm (accessed May 17, 2011).

Sterner, T. 1985. Structural Change and Technology Choice. *Energy Economics* 7(2): 77–86.

———. 1989a. Factor Demand and Substitution in a Developing Country: Energy Use in Mexican Manufacturing. *Scandinavian Journal of Economics* 91(4): 723–39.

———. 1989b. Oil Products in Latin America: The Politics of Energy Pricing. *Energy Journal* 10(2): 25–45.

———. 1989c. Les Prix de l'Energie en Afrique. *Revue de l'Energie* 415: 1–11.

———. 1999. The Market and the Environment: The Effectiveness of Market Based Policy Instruments for Environmental Protection, Edward Elgar.

———. 2002. *Policy Instruments for Environmental and Natural Resource Management.* Resources for the Future Press: Washington, DC.

———. 2006. Survey of Transport Fuel Demand Elasticities. Swedish Environmental Protection Agency report no. 5588., Stockholm.

———. 2007. Fuel Taxes: An Important Instrument for Climate Policy. *Energy Policy* 35, 3194–3202.

Sterner, T., and C. Dahl. 1992. Modeling Transport Fuel Demand. In *International Energy Economics*, ed. T. Sterner. London: Chapman and Hall, 65–79.

Stiglitz, J. 2006. A New Agenda for Global Warming. *Economists' Voice* 3(7), Article 3.

Stolberg, Sheryl Gay. 2007. Bush Proposes Goal to Reduce Greenhouse Gas Emissions. New York Times, June 1, 2007.

Stone, J. R. N. 1954a. Linear Expenditure Systems and Demand Analysis: An Application to the Pattern of British Demand. *Economic Journal* 54: 511–27.

———. 1954b. The Measurement of Consumers Expenditure and Behaviour in the United Kingdom: 1920–1938 Vol. 1, *Cambridge University Press*, Cambridge, UK (1954).

Sugema, I., M. Hasan, R. Oktaviani, Aviliani, and H. Ritonga. 2005. Dampak Kenaikan Harga BBM dan Efektivitas Program Kompensasi. INDEF Working Paper, 2005. www.indef.or.id/xplod/upload/pubs/BBM.PDF (accessed May 29, 2011).

Suits, D. B. 1977. Measurement of Tax Progressivity. *American Economic Review* 67(4): 747–752.

Sykes, D., W. J. Smith, and J. P. Formby. 1987. On the Measurement of Tax Progressivity: An Implication of Atkinson Theorem. *Southern Economic Journal* 53: 768–776.

Symons, E., J. Proops, and P. Gay. 1994. Carbon Taxes, Consumer Demand, and Carbon Dioxide Emissions: A Simulation Analysis for the UK. *Fiscal Studies* 15: 19–43.

Tahmassebi, C. H. 2007. Gasoline Markets in Iran: An Economic Perspective on Issues and Solutions. *Middle East Economic Survey* Vol. L(50). www.mees.com/postedarticles/oped/v50n50-5OD01.htm (accessed September 25, 2009).

Tanzania, Government of the United Republic of. 1997. National Environmental Policy. Office of the Vice-President. Dar es Salaam.

The Chronicle. 2009. Fuel Price Change and Comcomitant Smuggling. Editorial. March 26, 2009. http://allafrica.com/stories/200903260977.html (accessed October 13, 2009).

The *Guardian.* 2008. "Oil Prices: Europe Threatened with Summer of Discontent over Rising Cost of Fuel." Tuesday, June 10. www.guardian.co.uk/business/2008/jun/10/oil.france (accessed June 1, 2011).

Theil, H. 1975. The Theory and Measurement of Consumer Demand (Vols I and II). Amsterdam: North-Holland.

Timilsina, G., and H. Dulal. 2008. Fiscal Policy Instruments for Reducing Congestion and Atmospheric Emissions in the Transport Sector: A Review. World Bank Policy Research Paper 4652. Washington, DC.

Townsend, Joy. 1998. The Role of Taxation Policy in Tobacco Control, in The Economics of Tobacco Control: Towards an Optimal Policy Mix, ed. by I. Abedian and others (Cape Town, South Africa: Applied Fiscal Research Center), pp. 85–101.

Train, Kenneth E. 1986. *Qualitative Choice Analysis: Theory, Econometrics, and an Application to Automobile Demand.* MIT Press. Boston, MA.

———. 2003. *Discrete Choice Analysis with Simulation.* Cambridge University Press. Cambridge, UK.

Train, Kenneth E., and Clifford Winston. 2007. Vehicle Choice Behavior and the Declining Market Share of U.S. Automakers. *International Economic Review* 48(4): 1469–96.

Tullock, Gordon. 1967. Excess Benefit. Water Resources Research 3(2):643–644.

UNDP/ESMAP. 2003. Access of the Poor to Clean Household Fuels, UNDP/ESMAP.

UNECE (United Nations Economic Commission for Europe, Committee on Environmental Policy. 2007a. Environmental Performance Reviews—Republic of Serbia. Second Review. New York and Geneva. www.unece.org/env/epr/epr_studies/serbiaII.pdf (accessed June 29, 2009).

———. 2007b. Trends in Europe and North America 2005—Serbia and Montenegro. www.unece.org/stats/trends2005/profiles/SerbiaAndMontenegro.pdf (accessed June 30, 2009).

UNEP (United Nations Environment Programme). 2006. Integrated Assessment of the Energy Policy: With Focus on Transport and Household Energy Sectors. In *Sustainable Trade and Poverty Reduction: New Approaches to Integrated Policy Making at National Level, Country Report for Kenya.* UNEP, ETB, Geneva.

———. 2010. Sub-Saharan Africa Lead Matrix. Partnership for Clean Fuels and Vehicles. http://www.unep.org/transport/pcfv/regions/Africa.asp (accessed May 27, 2011).

UNEP/GRID-Arendal, Temperature and CO2 concentration in the atmosphere over the past 400 000 years, *UNEP/GRID-Arendal Maps and Graphics Library*, http://maps.grida.no/go/graphic/temperature-and-co2-concentration-in-the-atmosphere-over-the-past-400-000-years (Accessed 3 March 2011)

URT (United Republic of Tanzania). 2004. Public Expenditure Review of Environment. Financial Year 2004. The Vice President's Office, Dar es Salaam.

Vaage K. 2000. Heating Technology and Energy Use: A Discrete/continuous Choice Aapproach to Norwegian Household Energy Ddemand. *Energy Economics* 22(6): 649–66.

Valadkhani, Abbas, and William F. Mitchell. 2002. Assessing the Impact of Changes in Petroleum Prices on Inflation and Household Expenditures in Australia, *Australian Economic Review*, Vol. 35 (June), pp. 122–32.

Van Heerden, J., R. Gerlagh, J. Blignaut, M. Horridge, S. Hess, R. Mabugu, M. Mabugu. 2006. Searching for Triple Dividends in South Africa: Fighting CO_2 Pollution and Poverty while Promoting Growth. *The Energy Journal* 27: 113–41

Vartia Y. O. 1983. Efficient Methods of Measuring Welfare Change and Compensating Income in Terms of Ordinary Demand Functions. *Econometrica* 51(1983): 79–98.

Vega, E., R. Mejías, C. Camacho, G. Barrantes. 2004. Informe: Cuantificación de las externalidades en el transporte con tecnologías limpias. Instituto de Políticas para la Sostenibilidad (IPS). Heredia, Costa Rica.

Verhoef, E. 1994. External Effects and Social Costs of Road Transport. *Transportation Research A* 28A(4): 273–87.

von Haefen, Roger H., D. Matthew Massey, and Wiktor L. Adamowicz. 2005. Serial Nonparticipation in Repeated Discrete Choice Models. *American Journal of Agricultural Economics* 87: 1061–76.

Von Moltke, Anja, McKee Colin, and Trevor Morgan. 2004. eds., *Energy Subsidies: Lessons Learned in Assessing Their Impact and Designing Policy Reforms* (Sheffield, England: Greenleaf Publishing for United Nations Environment and Programme).

Wallart, N. 1999. *The Political Economy of Environmental Taxes*. Edward Elgar.

Walls M., and J. Hanson. 1999. Distributional Aspects of an Environmental Tax Shift: The Case of Motor Vehicle Emissions Taxes. *National Tax Journal* 52(1): 53–65.

Warr, P. 2006. The Gregory Thesis Visits the Tropics. *Economic Record* 82(257): 177–94.

West, S. E. 2004. Distributional Effects of Alternative Vehicle Pollution Control Policies. *Journal of Public Economics* 88: 735–575.

West, S., and R. C. Williams III. 2004. Empirical Estimates for Environmental Policymaking in a Second-best Setting. Working paper. Macalester College and University of Texas at Austin.

———. 2005. The Cost of Reducing Gasoline Consumption. *American Economic Review* 95: 294–299.

———. 2007. Optimal Taxation and Cross-Price Effects on Labor Supply: Estimates of the Optimal Gas Tax. *Journal of Public Economics* 91(April): 593–617.

WHO (World Health Organization). 2005. WHO Air Quality Guidelines for Particulate Matter, Ozone, Nitrogen Dioxide and Sulphur Dioxide. Global Update 2004: Summary of Risk Assessment. www.euro.who.int/__data/assets/pdf_file/0005/78638/E90038.pdf (accessed on June 16, 2011).

Wier, M., K. Birr-Pedersen, H. K. Jacobsen, and J. Klok. 2005. Are CO_2 Taxes Regressive? Evidence from the Danish Experience. *Ecological Economics* 52: 239–51.

Wiese, A., A. Rose, and G. Schluter. 1995. Motor-fuel Taxes and Household Welfare: An Applied General Equilibrium Analysis. *Land Economics* 71: 229–34.

Williams III, R. C. 2005. An Estimate of the Second-Best Optimal Gasoline Tax Considering Both Efficiency and Equity. Working Paper, University of Texas. Presented at the EAERE conference in Gothenburg, June 2008. http://*agecon.tamu.edu/research/seminars/pdf/Williamspaper.pdf* (accessed May 18, 2009).

Williams, R., D. Weiner, and F. Sammartino. 1998. Equivalence Scales, the Income Distribution, and Federal Taxes. Congressional Budget Office Technical Paper 1999-2.

Willig, R. 1976. Consumer's Surplus Without Apology. *American Economic Review* 66: 589–97.

World Bank. 2001. Pakistan: Clean Fuels. ESMAP Report No. 246/01. Washington: World Bank, Energy Sector Management Assistance Programme.

———. 2003. World Bank, 2003, Iran: Medium Term Framework for Transition, Converting Oil Wealth to Development, *Economic Report* No. 25848-IRN (Washington).

———. 2006. *Making the New Indonesia Work for the Poor.* Washington, DC.

———. 2009. Country Classifications. http://siteresources.worldbank.org/DATASTA-TISTICS/Resources/CLASS.XLS (accessed May 21, 2011).

———. 2010a. CO_2 Emissions (Metric Tons per Capita). http://datafinder.worldbank.org/co2-emissions (accessed February 15, 2010).

———. 2010b. Irans CO_2 Emissions per Capita. www.google.com/publicdata?ds=wb-wdi&met=en_atm_co2e_pc&idim=country:IRN&dl=en&hl=en&q=Iran+co2+emission (accessed February 15, 2010)

———. 2010c. GDP Growth (Annual %). http://datafinder.worldbank.org/gdp-growth-annual viewed (accessed February 15, 2010).

WRI (World Resource Institute). 2005. Carbon Dioxide Emissions by Economic Sector, 2005 http://earthtrends.wri.org/pdf_library/data_tables/cli2_2005.pdf (accessed May 15, 2009).

———. 2009a. Population, Health and Human Well-being—Population: Population Density. http://earthtrends.wri.org/searchable_db/results.php?years=2005-2005&variable_ID=431&theme=4&cID=63,70,91,238,166,173,189&ccID= (accessed June 2, 2011).

———. 2009b. Energy and Resources—Transportation: Passenger cars per 1000 people. http://earthtrends.wri.org/searchable_db/results.php?years=2003-2003&variable_ID=290&theme=6&cID=63,70,91,166,173,189&ccID=0 (accessed June 2, 2011).

———. 2009c. Population, Health and Human Well-being—Urban and Rural Areas: Urban Population as a Percent of Total Population. http://earthtrends.wri.org/searchable_db/results.php?years=2005-2005&variable_ID=448&theme=4&cID=63,70,91,202,166,173,189&ccID= (accessed June 2, 2011).

Yitzhaki, S., and J. Slemrod. 1991. Welfare Dominance: An Application to Commodity Taxation. *American Economic Review* 81: 480–96.

Yitzhaki, S., and W. Thirsk. 1990. Welfare Dominance and the Design of Excise Taxation in the Côte d'Ivoire, *Journal of development economics*, 33: 1–18.

Younger, S. D. 1993. Estimating Tax Incidence in Ghana: An Exercise Using Household Data. Cornell Food and Nutrition Policy Program Working Paper 48. CFNPP Publications Department, Ithaca, NY.

Younger, S. D., D. E. Sahn, S. Haggblade, and P.A. Dorosh. 1999. Tax Incidence in Madagascar: An Analysis Using Household Data. *World Bank Economic Review* 13: 303–30.

Yusuf, A. 2006. Constructing an Indonesian Social Accounting Matrix for Distributional Analysis in the CGE Modelling Framework. Working Paper in Economics and Development Studies No 200604. Department of Economics, Padjadjaran University. http://econpapers.repec.org/RePEc:unp:wpaper:200604 (accessed May 29, 2011).

———. 2008. INDONESIA-E3: An Indonesian Applied General Equilibrium Model for Analyzing the Economy, Equity, and the Environment. Working Papers in Economics and Development Studies No. 200804. Department of Economics, Padjadjaran University.

Yusuf, A., and B. Resosudarmo. 2008. Mitigating Distributional Impact of Fuel Pricing Reform: The Indonesian Experience. *ASEAN Economic Bulletin* 25(1): 32–47.

Index